Screening, Identification, and Quantification of Nutritional Components and Phytochemicals in Foodstuffs

Screening, Identification, and Quantification of Nutritional Components and Phytochemicals in Foodstuffs

Editor

Dario Donno

MDPI • Basel • Beijing • Wuhan • Barcelona • Belgrade • Manchester • Tokyo • Cluj • Tianjin

Editor
Dario Donno
Dipartimento di Scienze Agrarie,
Forestali e Alimentari
Università degli Studi di Torino
Grugliasco
Italy

Editorial Office
MDPI
St. Alban-Anlage 66
4052 Basel, Switzerland

This is a reprint of articles from the Special Issue published online in the open access journal *Foods* (ISSN 2304-8158) (available at: www.mdpi.com/journal/foods/special_issues/foodstuffs_components).

For citation purposes, cite each article independently as indicated on the article page online and as indicated below:

LastName, A.A.; LastName, B.B.; LastName, C.C. Article Title. *Journal Name* **Year**, *Volume Number*, Page Range.

ISBN 978-3-0365-2022-3 (Hbk)
ISBN 978-3-0365-2021-6 (PDF)

Contents

About the Editor

Dario Donno

Dario Donno, Master graduated in chemistry and with a PhD in agricultural, forest and food sciences. He is a post-doctoral research assistant at the Department of Agricultural, Forest and Food Sciences, University of Turin. His studies mainly focus on qualitative analysis and analytical strategies (e.g., chromatographic and spectrophotometric analyses coupled with multivariate statistical analysis) for the identification and quantification of the bioactive compounds in different plant materials (buds, leaves, fruits, bark) from temperate, sub-tropical and tropical areas and agri-food industry derived-products for their quality and traceability evaluation. He is the author of more than 150 scientific and technical publications in national and international journals and books, including around 80 in Web of Science and Scopus peer-reviewed journals. He has also actively participated in national and EU research projects.

Preface to "Screening, Identification, and Quantification of Nutritional Components and Phytochemicals in Foodstuffs"

The interest in nutritional components, bioactive compounds, and phytochemicals in potential functional foods continues to grow, powered by the identification of health-promoting properties and potential applications of nutraceutical substances. These products may range from isolated nutrients, agri-foods, dietary supplements, and diets to genetically engineered foods, herbal products, and processed foods.

In recent years, nutraceutical foods, food supplements, herbal preparations, plant foods, and derived products have become very attractive to the food industry, prompting their use as replacements for synthetic chemicals and nutraceuticals, but neglected and underutilized plant resources are suffering from less attention and research, and their nutritional, economic, and socio-cultural potentials are not fully exploited. The identification and quantification of nutritional substances and bioactive compounds in foods and the evaluation of their biological activities are important to gauge their efficacy as dietary interventions and in healthy applications.

This book provides readers with a good overview of the status and exciting developments in this field. It includes papers focused on modern analytical instrumentation and new methods and biological tests applied to the evaluation of plant foods, derived products, herbal products, and food supplements and the phytochemical characterization of innovative natural sources of bioactive compounds and relative health-promoting properties.

The guest editor would like to thank all the colleagues and contributors that published their works in this Special Issue, as well as the reviewers that evaluated the submissions, assuring the high quality of the published studies. The guest editor would also like to thank the publisher, MDPI, and the editorial staff of *Foods* for their high-quality, constant, and professional support as well as for their invitation to edit this Special Issue.

Dario Donno
Editor

foods MDPI

Editorial

Screening, Identification, and Quantification of Nutritional Components and Phytochemicals in Foodstuffs

Dario Donno

Dipartimento di Scienze Agrarie, Forestali e Alimentari, Università degli Studi di Torino, 10095 Grugliasco (TO), Italy; dario.donno@unito.it; Tel.: +39-011-670-8751

Foods confer many health-promoting benefits to humans for the treatment/prevention of different diseases. The interest in bioactive compounds, nutritional components, and phytochemicals in potential health-promoting foods continues to grow, powered by the identification of health-promoting properties and potential applications of nutraceutical substances. These products may range from isolated nutrients, agri-foods, dietary supplements, and diets to genetically engineered "designer" foods, herbal products, and processed foods. Today, nutraceutical products may be both traditional foods (e.g., meat products, fish, vegetables, fruits, chocolate, grains, and tea) and non-traditional foods (e.g., added ingredients or products derived from agricultural breeding). These products have become very attractive to the food industry, prompting their use as replacements for synthetic chemicals and nutraceuticals, but natural resources are suffering from less attention and research, and their nutritional, economic, and socio-cultural potentials are not fully exploited.

Studies on natural resources and foodstuffs are very important to discover new sources for natural antioxidants, health-promoting materials, nutraceuticals, and potential functional foods. The identification and quantification of nutritional substances and bioactive compounds in foods and the evaluation of their biological activities are important to gauge their efficacy as dietary interventions and in healthy applications. In the last few years, studies in the field of natural biomolecules and health-promoting foods have been focusing on naturally health-positive molecules available in food, herbal preparations, plants, and derived products. Their utilization in foodstuffs could increase quality, safety, and added value; for this reason, traditional and innovative techniques for extraction/purification and identification/quantification of bioactive compounds and nutritional substances using eco-friendly analytical strategies need to be stimulated and developed to improve production yields.

This Special Issue provides readers with a good overview of the status and important insights and developments in this field. It includes review and research papers focused on modern analytical approaches, traditional and innovative techniques, and biological tests applied to the screening, identification, and quantification of nutritional components and phytochemicals in foodstuffs together with the evaluation and valorization of innovative natural sources of bioactive molecules and relative health-promoting properties.

Turrini et al. [1] investigated, for the first time, an alternative method to produce *Ribes nigrum* bud derivatives, a category of botanicals marketed as plant food supplements in the European community. Pulsed ultrasound-assisted extraction (PUAE), using a food-grade solvent according to green chemistry principles, was employed and compared to the conventional extraction method. Untargeted polyphenolic fingerprints (UV-Visible and fluorescence) coupled with chemometrics were employed to quickly screen the best extraction conditions, evaluated by the design of experiment (DoE) method. The polyphenolic fraction of the optimized PUAE extract was quantified by targeted HPLC fingerprint and its antiradical activity was determined. PUAE on a lab pilot reactor was proven to be the most practical approach for a rapid (20 min vs. 21 days maceration) and efficient

Citation: Donno, D. Screening, Identification, and Quantification of Nutritional Components and Phytochemicals in Foodstuffs. *Foods* **2021**, *10*, 125. https://doi.org/10.3390/foods10010125

Received: 22 December 2020
Accepted: 6 January 2021
Published: 8 January 2021

extraction of bioactive polyphenolics from *Ribes nigrum* buds, encouraging the scale up to an industrial scenario.

Jang et al. [2] studied the optimization and validation of an analytical method for separating flavonoid isomers in common buckwheat sprout extract (CSE). Factors such as range, linearity, precision, accuracy, the limit of detection, and limit of quantification were evaluated for each standard using high-performance liquid chromatography (HPLC). Based on resolution and symmetry, a column temperature of 40 °C with 0.1% (*v*/*v*) acidic water and acetonitrile as mobile phases, at a flow rate of 1 mL/min were determined to be the optimal analytical conditions. The developed method was used to analyze flavonoids in CSE, with isomers satisfactorily separated and simultaneously quantified.

Lavefve et al. [3] determined the stability of polyphenolics in five products (ice pop, oatmeal bar, graham cracker cookie, juice, and gummy product) prepared with wild blueberry (WBB) powder that can be added to the formulation of foods to encourage consumption of health-promoting polyphenolics. Samples stored at 21 °C, 4.4 °C, or −20 °C (ice pops only) were analyzed at 0, 2, 4, 6, and 8 weeks for polyphenolic content and percent polymeric color. Total anthocyanins decreased over storage and storage temperatures in all products. However, the ice pop and the refrigerated juice both retained over 90% of their initial total anthocyanin content. The refrigerated oatmeal bar also showed good retention of anthocyanins (86%), but the gummy product retained only 43% and 51% when stored at 4.4 °C or 21 °C, respectively. The lower amount of polyphenolic compounds recovered in the gummies stored at 4.4 °C compared to 21 °C may be attributed to reduced extraction efficiency as a result of gel hardening at refrigerated temperature. Chlorogenic acid and flavonols were generally more stable than anthocyanins throughout storage.

Dini et al. [4] validated, for the first time, a highly specific automated enzymatic method to quantify the acetic acid in wine vinegar, in terms of linearity, precision, repeatability, and uncertainty measurement. The results were compared to the Community method of analysis. Regression coefficient ≅ 1 and the normal distribution of residuals in the ANOVA test confirmed the method's linearity. Lower limit of detection—LLOD (0.946 ppm) and lower limit of quantification—LLOQ (2.00 ppm) defined the method's sensitivity. The results of the tested and the Community methods, linearly distributed in the Shapiro–Wilk test, confirmed the method's repeatability. The few anomalous data in the Huber test were due to random errors. The high selectivity of the enzymatic method, which exclusively measures acetic acid concentration, determined the significant differences between the two tests, examined in the accuracy determination. The enzymatic method can be considered applicable since its precision and uncertainty were lower than the Community method values (relative percentage deviations = 10%). The enzymatic method compared to the Community method reduces the analysis time and the risk of errors due to operators (avoiding pipetting errors and wrong calculations), minimizes solvent and the sample consumption, and guarantees assay quality through method standardization.

Bonasia et al. [5] assessed the chemical composition (minerals, organic acids, free sugars, volatile, and phenolic compounds) of six garlic landraces collected from Puglia Region (Foggia Province) along with their main morpho-biometrical traits. A commercial genotype was also considered as a reference standard. The landraces showed large variability, but in general high morphological standards, high levels of cations and phenols, and low levels of volatile-(S)-compounds in comparison with the commercial genotype and the literature values. "Aglio di Peschici" and "Aglio Rosso di Monteleone di Puglia" were very rich in minerals and phenols (mainly ferulic acid and iso-rhamnetin). The increase in knowledge of the chemical properties of these garlic landraces could represent a tool for encouraging the consumption of a food product. At the same time, the consumption of these landraces would stimulate their cultivation and could highly contribute to protection against the risk of erosion of agro-biodiversity through *in situ*/on-farm conservation.

Beccaro et al. [6] investigated chestnut cultivars grown in the same pedoclimatic conditions and on the same clonal rootstock. Chestnuts were characterized by sensory, spectrophotometric, and chromatographic analysis to determine the phytochemical composition and nutraceutical properties. A multivariate approach, including principal component analysis and conditional inference tree models, was also performed. The multivariate approach showed that phenolic acids and tannins were the bioactive classes with the highest discriminating power among different genotypes, and that genotype is a significant variable ($p < 0.05$). Furthermore, most of the analyzed chestnut cultivars showed a content of bioactive compounds similar to or higher than the main hazelnut, walnut, and almond varieties. Chestnut agrobiodiversity could be intended as strictly associated to the genotype effect and underlines the large variability within the genus *Castanea*, and therefore, the importance of in farm and ex situ conservation of local germplasm is part of a global strategy aimed at increasing the levels of agrobiodiversity.

Suleria et al. [7] determined the polyphenol content and their antioxidant potential in twenty different fruit peel samples in an ethanolic extraction, including their comprehensive characterization and quantification using the LC-MS/MS and HPLC. The application of LC-ESI-QTOF-MS/MS tentatively identified and characterized a total of 176 phenolics, including phenolic acids (49), flavonoids (86), lignans (11), stilbene (5), and other polyphenols (25) in all twenty peel samples. From HPLC-PDA quantification, the mango peel sample showed significantly higher phenolic content, particularly for phenolic acids and flavonoids, as compared to other fruit peel samples. These results highlighted the importance of fruit peels as a potential source of polyphenols. This study provided supportive information for the utilization of different phenolic rich fruit peels as ingredients in food, feed, and nutraceutical products.

Krawęcka et al. [8] assessed the effect of adding 0%, 5%, 10%, 15%, and 20% oat (1,3)(1,4)-β-D-glucans to physicochemical properties, as well as the cooking and sensory qualities of durum wheat pasta. Additionally, 5% of xanthan gum and vital gluten was added to improve the cooking and sensory qualities of pasta. The study showed that the addition of β-glucans led to an increase of the water absorption index (WAI), water solubility index (WSI), and viscosity of products. At the same time, an increase in the content of fat, ash, and dietary fiber was observed. The addition of (1,3)(1,4)-β-D-glucans influenced the cooking quality of the pasta, extending the minimum cooking time and increasing the loss of dry matter. At the same time, the color of the product changed. In the case of cooked pasta, the addition of β-glucans decreased the brightness and increased the yellowness and redness. It was found that the products enriched with 10–15% of β-glucans, as well as 5% of xanthan gum and vital gluten would yield functional pasta that may offer health benefits beyond its nutritional value. Furthermore, this could influence high cooking and sensory quality.

Müller et al. [9] evaluated the micro- and macronutrient composition of representative, randomly mixed samples of the 15 different hazelnut cultivars. Protein, fat, and fiber contents were determined using established methods. Fatty acids, tocopherols, minerals, trace elements, and ultra-trace elements were analyzed using gas chromatography, high-performance liquid chromatography, and inductively coupled plasma triple quadrupole mass-spectrometry, respectively. The different hazelnut varieties contained valuable amounts of fat, protein, dietary fiber, minerals, trace elements, and α-tocopherol, however, in different quantities. The variations in nutrient composition were independent of growth conditions, which were identical for all hazelnut varieties. Therefore, each hazelnut cultivar presented a specific nutrient profile.

Distefano et al. [10] addressed the effects of two storage temperatures, namely 10 °C (T10) and 20 °C (T20), on main quality and functional traits of three cherry tomato cultivars ("Eletta", "Sugarland", and "Ottymo"), after 0 (S0), 7 (S7), and 14 (S14) days of storage. At T10 both fruit weight and firmness were better retained during storage. At S14, T10 promoted fruit Chroma and overall fruit color deviation (ΔE^*ab). Total polyphenols content (TPC) of fruits peaked at S7 then declined at S14 (by 16%), with the highest values recorded at T10. Lycopene showed a similar trend, but with a higher average concentration recorded at T20. β-carotene content peaked at S14, irrespective of the storage temperature. At S14, the concentrations of phytoene and phytofluene were higher at T20, but the opposite was found at S7. "Sugarland" and "Ottymo" showed the highest ΔE*ab during storage, with the former cultivar proving to have the highest TPC and lycopene content, whereas "Eletta" did so for phytoene and phytofluene. The results suggested that unravelling the possible functional interactions among these three carotenoids would allow for a better orientation of breeding programs, targeting the phytochemical evolution of tomatoes during refrigerated storage.

Donno et al. [11] investigated the main analytical strategies commonly utilized in the agri-food industry, often using complementary technologies with different purposes. This overview presented the most recent MS-based techniques applied to food analysis. An entire section was dedicated to the recent applications of high-resolution MS. Covered topics included liquid (LC)– and gas chromatography (GC)–MS analysis of natural bioactive substances, including carbohydrates, flavonoids and related compounds, lipids, phenolic compounds, vitamins, and other different molecules in foodstuffs from the perspectives of food composition, food authenticity, and food adulteration. The results represented an important contribution to the utilization of GC–MS and LC–MS in the field of natural bioactive compound identification and quantification.

In recent years, the agri-food industry has applied new analytical strategies for a full characterization of foods since consumers and regulatory agencies required a molecular characterization. Validated analytical approaches using high-performance systems were improved to ensure traceability and quality of food. Chromatographic techniques coupled to suitable detection strategies are one of the most effective tools to separate the single molecules and develop a specific profile ("fingerprint"). If chromatography, coupled to detection systems as mass spectrometry, is further combined with chemometrics, clearer patterns might be developed for analytical fingerprints.

Funding: This research received no external funding.

Institutional Review Board Statement: Not applicable.

Informed Consent Statement: Not applicable.

Acknowledgments: Guest Editor would like to thank all the colleagues and contributors that published their works in this Special Issue as well as the reviewers that evaluated the submissions assuring a high quality for the published studies. Guest Editor would also like to thank the publisher, MDPI, and the editorial staff of Foods for their high, constant, and professional support as well as for their invitation to edit this Special Issue.

Conflicts of Interest: The authors declare no conflict of interest.

References

1. Turrini, F.; Donno, D.; Beccaro, G.L.; Zunin, P.; Pittaluga, A.; Boggia, R. Pulsed ultrasound-assisted extraction as an alternative method to conventional maceration for the extraction of the polyphenolic fraction of *Ribes nigrum* buds: A new category of food supplements proposed by the Finnover project. *Foods* **2019**, *8*, 466. [CrossRef] [PubMed]
2. Jang, D.; Jung, Y.S.; Kim, M.-S.; Oh, S.E.; Nam, T.G.; Kim, D.-O. Developing and validating a method for separating flavonoid isomers in common buckwheat sprouts using HPLC-PDA. *Foods* **2019**, *8*, 549. [CrossRef] [PubMed]
3. Lavefve, L.; Brownmiller, C.; Howard, L.; Reeves, D.; Adams, S.H.; Chen, J.-R.; Diaz, E.C.; Mauromoustakos, A. Changes in polyphenolics during storage of products prepared with freeze-dried wild blueberry powder. *Foods* **2020**, *9*, 466. [CrossRef] [PubMed]
4. Dini, I.; Di Lorenzo, R.; Senatore, A.; Coppola, D.; Laneri, S. Validation of rapid enzymatic quantification of acetic acid in vinegar on automated spectrophotometric system. *Foods* **2020**, *9*, 761.
5. Bonasia, A.; Conversa, G.; Lazzizera, C.; Loizzo, P.; Gambacorta, G.; Elia, A. Evaluation of garlic landraces from Foggia province (Puglia region; Italy). *Foods* **2020**, *9*, 850. [CrossRef] [PubMed]
6. Beccaro, G.L.; Donno, D.; Lione, G.G.; De Biaggi, M.; Gamba, G.; Rapalino, S.; Riondato, I.; Gonthier, P.; Mellano, M.G. *Castanea* spp. agrobiodiversity conservation: genotype influence on chemical and sensorial traits of cultivars grown on the same clonal rootstock. *Foods* **2020**, *9*, 1062. [CrossRef] [PubMed]
7. Suleria, H.A.R.; Barrow, C.J.; Dunshea, F.R. Screening and characterization of phenolic compounds and their antioxidant capacity in different fruit peels. *Foods* **2020**, *9*, 1206. [CrossRef] [PubMed]
8. Krawęcka, A.; Sobota, A.; Sykut-Domańska, E. Physicochemical, sensory, and cooking qualities of pasta enriched with oat β-glucans, xanthan gum, and vital gluten. *Foods* **2020**, *9*, 1412. [CrossRef] [PubMed]

9. Müller, A.K.; Helms, U.; Rohrer, C.; Möhler, M.; Hellwig, F.; Glei, M.; Schwerdtle, T.; Lorkowski, S.; Dawczynski, C. Nutrient composition of different hazelnut cultivars grown in Germany. *Foods* **2020**, *9*, 1596. [CrossRef] [PubMed]
10. Distefano, M.; Arena, E.; Mauro, R.P.; Brighina, S.; Leonardi, C.; Fallico, B.; Giuffrida, F. Effects of genotype, storage temperature and time on quality and compositional traits of cherry tomato. *Foods* **2020**, *9*, 1729. [CrossRef] [PubMed]
11. Donno, D.; Mellano, M.G.; Gamba, G.; Riondato, I.; Beccaro, G.L. Analytical strategies for fingerprinting of antioxidants, nutritional substances, and bioactive compounds in foodstuffs based on high performance liquid chromatography–mass spectrometry: An overview. *Foods* **2020**, *9*, 1734. [CrossRef] [PubMed]

foods

MDPI

Article

Pulsed Ultrasound-Assisted Extraction as an Alternative Method to Conventional Maceration for the Extraction of the Polyphenolic Fraction of *Ribes nigrum* Buds: A New Category of Food Supplements Proposed by The FINNOVER Project

Federica Turrini [1], Dario Donno [2,*], Gabriele Loris Beccaro [2], Paola Zunin [1], Anna Pittaluga [1] and Raffaella Boggia [1]

[1] Department of Pharmacy, University of Genoa, Viale Cembrano 4, 16148 Genoa, Italy; turrini@difar.unige.it (F.T.); zunin@difar.unige.it (P.Z.); pittalug@difar.unige.it (A.P.); boggia@difar.unige.it (R.B.)

[2] Department of Agriculture, Forestry and Food Science, University of Torino, Largo Braccini 2, 10095 Grugliasco (TO), Italy; gbeccaro@unito.it

* Correspondence: dario.donno@unito.it

Received: 6 September 2019; Accepted: 6 October 2019; Published: 10 October 2019

check for updates

Abstract: In this research, for the first time, an alternative method to produce *Ribes nigrum* bud derivatives is presented. Pulsed ultrasound-assisted extraction (PUAE), using a food-grade solvent according to green chemistry principles, has been employed and compared to the conventional extraction method. Traditionally, bud derivatives, a category of botanicals marketed as plant food supplements in the European Community, are produced by macerating meristematic tissues of trees and plants mainly spontaneously collected. Buds are a challenging raw material for the UAE, since meristematic tissues are much softer and fragile than their corresponding adult phenological stage. It is therefore important to assess whether the polyphenolic fraction, very susceptible to degradation, is conserved after UAE. Untargeted polyphenolic fingerprints (UV-Visible and fluorescence) coupled with chemometrics are employed to quickly screen the best extraction conditions, evaluated by the design of experiment (DoE) method. The polyphenolic fraction of the optimized PUAE extract was quantified by targeted HPLC fingerprint and its antiradical activity was determined. PUAE on a lab pilot reactor was proven to be the most practical approach for a rapid (20 min vs. 21 days maceration) and efficient extraction of bioactive polyphenolics from *Ribes nigrum* buds, encouraging the scale up to an industrial plan.

Keywords: phenolic compounds; bud derivatives; *Ribes nigrum* glyceric macerate; green chemistry; pulsed ultrasound-assisted extraction; untargeted spectroscopic fingerprint; targeted chromatographic fingerprint

---

1. Introduction

FINNOVER "Innovative strategies for the development of crossborder green supply chains" is an Interreg ALCOTRA Italy/France transfrontier cooperation project (2017–2020) whose main target is the "green" implementation of some agro-industrial chains [1]. Particularly, the project supports the creation of both innovative and eco-sustainable production chains of botanicals in order to valorize the biodiversity of the ALCOTRA territory. One of the main natural products studied in FINNOVER are bud derivatives, which represent a relatively new category of botanicals marketed, in the European Community, as plant food supplements according to the Directive 2002/46/EC of the European

Parliament [2,3]. Bud derivatives are obtained by cold maceration in solvents (i.e., ethanol and glycerol) of fresh meristematic tissues of trees and plants (i.e., buds and young sprouts) as reported in the European Pharmacopoeia VIII edition (2014) [2,4,5]. These botanicals are still poorly studied, although they are widely used in gemmotherapy, a branch of phytotherapy that exploits the properties of these plant extracts for medicinal purposes [6]. The peculiarity of meristematic tissues in this particular phenological stage concerns their fragile texture and the high content of compounds which constitute the bud phytocomplex. In fact, these substances, including mainly flavonoids, enzymes, vitamins, aminoacids, nucleic acids, and plant hormones, are often present only in trace in the corresponding adult tissues [5].

In this research, a sonochemical application with a green chemistry approach was presented. Particularly, pulsed ultrasound-assisted extraction (PUAE) was employed as alternative method to quickly produce new bud derivatives in comparison to the long traditional maceration in solvent (21 days) taking under control their total phenolic fraction and antiradical activity to monitor possible PUAE-induced degradations. *Ribes nigrum* buds (RNB) were used as case study due to their common use in herbal medicine for their potential health properties. The most important industrial products of *R. nigrum* are its berries, which contain very high amounts of bioactive compounds, particularly flavonoids, phenolic compounds, and anthocyanins [7–9]. However, *R. nigrum* bud derivatives also contain high amounts of polyphenols, representing more than 60% of the bud phytocomplex [7], and they are widely used for inflammatory, circulatory, respiratory, and cutaneous disorders [10].

Ultrasounds (UAE—ultrasound-assisted extraction) together with microwaves (MAE—microwave-assisted extraction), supercritical fluids (SCF), and pulsed electric fields (PEF), are emerging "green" extraction technologies [11]. According to the six principles of the green extraction introduced by Chemat and colleagues [12] and to the twelve principles of green chemistry set by the Environmental Protection Agency of USA [13], these eco-compatible extraction techniques, with respect to conventional methods, aim to reduce the environmental impact in terms of time and energy. Moreover, they reduce both the quantities of solvents employed, preferring alternative solvents (water or food-grade solvents), and the generation of waste, hazardous substances, and consequently pollution [11,14]. In particular, UAE is a relatively simple, cheap, and efficient alternative to conventional extraction techniques whose main benefits are faster kinetics and increased extraction efficiency [15,16]. In fact, UAE allows one to quickly extract, with high reproducibility both on small and large scale, a wide variety of bioactive compounds (i.e., aromas, pigments, antioxidants, and organic and mineral compounds) from several animal tissues, plants, or food matrices [15,17]. However, the effects of ultrasound on the extraction yield may be linked to the nature of the matrix. Therefore, the experimental conditions of UAE must be optimized for each matrix [18]. In a solid/liquid media, the ultrasound waves originate the cavitation phenomena, a succession of different phases of compression and rarefaction which generates cavitation bubbles in the liquid. The implosion of the cavitation bubbles on the surface of the solid material generate microjets, at very high temperature and pressure, which destroy the wall cells of the matrix, with the consequent recovery of the intracellular content in the extraction solvent. There are several mechanisms involved (i.e., fragmentation, erosion, sonoporation, detexturation, capillarity) which independently or in combination influence the final ultrasound extraction yield [17]. When UAE is used in a pulsed mode (PUAE), the ultrasound processor works intermittently during the entire extraction process (active time vs. inactive time). This extraction mode reduces the operating temperature, allowing the extraction of thermolabile compounds and decreasing the possibility to produce alterations (i.e., oxidation products) in the final extract [19].

Untargeted phytochemical fingerprints coupled with chemometrics [2,20], particularly the UV-Visible and fluorescence spectra of each extract as multivariate response variables, have been employed to quickly screen the best experimental conditions of PUAE investigated by the design of experiment (DoE) method [21]. Finally, the extract obtained by PUAE in the optimized conditions has been characterized using targeted phytochemical fingerprinting by HPLC [2,22,23] in order to identify and quantify the main polyphenolic compounds. The polyphenolic fraction has been selected as the marker of activity and degradation susceptibility in order to make a comparison with the corresponding *R. nigrum* glyceric macerate (RNGM), representing the commercial product.

2. Materials and Methods

2.1. Bud Collection

RNB were collected from plants cultivated in the Bronda valley (Cuneo, Italy), particularly in the municipality of Pagno (44.597,7.424–44.598,7.424) and Brondello (44.603,7.422–44.603,7.418) in March 2018. Buds, after being certified by a botanical expert, were employed by an Italian Company of food supplements (Geal Pharma, Bricherasio, Turin, Italy) for the production of the corresponding RNGM. Particularly, the fresh merystematic tissues were immediately used after their collection in order to preserve their bioactive compounds. Both the traditional procedure for preparing glyceric macerates, according to the European Pharmacopoeia VIII edition, and an alternative method that exploits the action of ultrasounds have been used and compared for the production of the corresponding extracts (RNGM and RN8).

2.2. Chemicals

Ethanol and glycerol were supplied by VWR International S.r.l (Milan, Italy) and GealPharma (Bricherasio, Turin, Italy), respectively. All the standards employed for the HPLC analysis were purchased by Sigma-Aldrich (St. Louis, MO, USA). The purity of all the standards employed was ≥95%. Ultrapure water (18 MΩ) was produced by a Millipore Milli-Q system (Bedford, MA, USA) and used throughout.

2.3. Traditional Preparation of R. nigrum Glyceric Macerates

RNGM was produced according to the indications of the European Pharmacopoeia VIII edition (2014), referring to the procedure reported in the French Pharmacopoeia [4]. Particularly, a mixture of glycerol/ethanol 96% (1:1 *w/w*) as extraction solvent and a solid–solvent ratio 1:20 between buds and solvent (considering the dry weight) were employed. The phytocomplex extraction from RNB involved several steps: a cold maceration for 21 days, followed by a preliminary filtration, a manual pressing, and, at the end, a second filtration with filter paper (Whatman n. 1) after 2 days of decanting [2,5]. The obtained extracts, which represent the commercial product marketed by GealPharma, were stored at 4 °C in the dark until further analysis.

2.4. Alternative Method to Produce R. nigrum Bud Derivatives: Pulsed Ultrasound-Assisted Extraction (PUAE)

PUAE was carried out directly by an Hielscher UP200St (Teltow, Germany) in pulsed mode, with an ultrasonic titanium probe (7 mm diameter) able to transfer, with high efficiency, the acoustic energy into the treated media [2,24,25]. Fresh RNB were finely ground by a Grindomix 200 M (Retsch, Haan, Germany), for 20 s at 5000 rpm, and then sieved by a 150 µm sieve. Twenty grams of a glycerol/ethanol 96% mixture 1:1 *w/w* were added to 1 g (dry weight) of ground RNB in a polypropylene 50 mL centrifuge tube. The samples were processed in a 200 W ultrasonic processor at a constant frequency of 26 kHz, with an amplitude level of 30%, optimized in a previous paper from the authors [26], keeping temperature under control always below 70 °C. The pulse duration and pulse interval refer to "on" and "off" times of the sonicator. The same mixture of glycerol/ethanol 96% (1:1 *w/w*) and the same solid–solvent ratio 1:20 between buds and solvent (considering the dry weight), as described in the Section 2.3, were employed. The duty cycle (pulse) and the extraction time (65% and 20 min, respectively) were optimized by applying DoE (Table 1). The obtained suspension was then filtered by Buchner (Whatman n. 44 paper) and the filtrate was centrifuged at 3000 rpm for 10 min. The obtained solutions were stored at 4 °C until the analysis time.

Table 1. Experimental matrix of the Faced Central Composite Design, the experimental plan (in brackets), and the obtained response variable (Y).

Experiment	Experimental Conditions		Response Variable
	X_1 Duty Cycle (%)	X_2 Extraction Time (min)	Y PC_1 Scores
RN1	−1 (50)	−1 (10)	3.222515159
RN2	+1 (80)	−1 (10)	2.256792019
RN3	−1 (50)	+1 (20)	0.834250201
RN4	+1 (80)	+1 (20)	−0.535350397
RN5	−1 (50)	0 (15)	0.587361402
RN6	+1 (80)	0 (15)	−1.165491045
RN7	0 (65)	−1 (10)	−0.327200585
RN8	0 (65)	+1 (20)	−2.638815847
RN9	0 (65)	0 (15)	−2.234060907

2.5. *Untargeted Fingerprints of The R. nigrum Phytocomplex*

2.5.1. UV-Visible Spectroscopy

An Agilent UV-Vis spectrophotometer Cary 100 (Varian Co., Santa Clara, CA, USA) with 0.5 nm resolution, was employed to record all the UV-Vis spectra of the extracts and the corresponding glyceric macerates of *R. nigrum*. Before being analyzed, all the samples were properly diluted 1:20 in the same extraction mixture (glycerol/ethanol 96% 1:1 *w*/*w*). The total spectrum of each analyzed sample was collected in duplicate at room temperature (25 ± 1 °C), against a blank solution (i.e., the extraction mixture), using rectangular quartz cuvettes with 1 cm path length. For each sample, the resulting spectra were averaged and used as vector of variables to build the data matrix.

2.5.2. Fluorescence Spectroscopy

The excitation–emission fluorescence spectra were recorded in duplicate at room temperature (25 ± 1 °C) by a Perkin-Elmer LS55B luminescence spectrometer (Waltham, MA, USA) using the traditional right angle fluorescence spectroscopic technique [27]. A standard cell holder and a 10 mm quartz SUPRASIL® cell with volume of 3.5 mL by PerkinElmer were used. The emission spectra were recorded in the range of 450–800 nm, exciting samples at a fixed wavelength (λ ex = 430 nm) [20]. Both the excitation and the emission monochromator slits were set to 10 nm, with high gain and 600 nm/min of speed. The same dilution of all the samples, at the ratio of 1:20 with the solvent, was evaluated. For each sample, the resulting emission spectra were averaged and used as vector of variables to build the data matrix combing them together with the previous described UV-Vis spectra.

2.6. *Experimental Design and Multivariate Data Analysis*

DoE was used to optimize the experimental conditions of PUAE from RNB. A Faced Central Composite Design ($2k + 2^k + 1$) was applied with the aim to estimate the constant, the linear terms, the interactions between variables, and the quadratic terms, according to the following model [21]:

$$Y = b_0 + b_1X_1 + b_2X_2 + b_{12}X_1X_2 + b_{11}X_1^2 + b_{22}X_2^2$$

The experimental plan, illustrated in Table 1, summarizes the conditions of the nine experiments performed (namely from RN1 to RN9). The minimum, intermediate, and maximum value of each variables are labeled as −1, 0, and +1, respectively. A data matrix $M_{9,1402}$ consisting of nine rows (the nine samples/extracts obtained by DoE) and 1402 columns (the vector of 701 absorbance values at different wavelengths in the range of 230–500 nm of the UV-Vis spectra plus 701 fluorescence emissions in the range of 450–800 nm of the fluorescence spectra), was prepared and further analyzed by the PCA, a multivariate statistical technique of unsupervised pattern recognition. The scores on PC1 have been used as a response variable of the experimental design (Y). In detail, the standard normal variate (SNV) transform, or row autoscaling, was previously performed on the spectral data in order to revise both the baseline shifts and the global intensity variations [28]. Subsequently, PCA was performed using the nonlinear iterative partial least squares (NIPALS) algorithm on the column-centered data [29]. After the PCA, the scores on PC1, explaining the 90.3% of the total variance, were extracted and used as response to elaborate DoE. DoE and multivariate data analysis were performed by CAT (Chemometric Agile Tool) a chemometric software based on R, developed by the Chemistry Group of the Italian Chemical Society [30]. The data matrix and the detailed PCA analysis is available as Supplementary Materials.

2.7. Analytical Determinations

The most promising extract obtained by PUAE and the corresponding glyceric macerate (RNGM), in order to make a comparison, were characterized to evaluate their total phenolic contents (TPC) and their radical scavenging activity (RSA). All the measurements were performed in duplicate and the results are expressed as mean ± standard deviation (SD). Statistical analysis was performed by the Excel Data Analysis Tool (Microsoft Corporation, Seattle, WA, USA).

2.7.1. Determination of The Total Phenolic Compounds (TPC)

The Folin-Ciocalteu spectrophotometric method was applied to estimate the TPC of the *R. nigrum* bud preparations [31]. 0.2 mL of sample appropriately diluted, 1 mL of Folin-Ciocalteu reagent (diluted 1:10 with deionized water), 0.8 mL of aqueous sodium carbonate 7.5% *w*/*v* solution were added in a test tube and vortexed. After an incubation period of 30 min at room temperature in the dark, the absorbance was recorded at 760 nm by an Agilent 8453 UV-Vis spectrophotometer with 1 nm resolution. A calibration curve, using gallic acid as a standard, has been used to evaluate the polyphenolic concentration. The TPC was expressed as milligrams of gallic acid equivalent (GAE) pulled-out from 100 mL of bud extract (mg GAE/100 mL).

2.7.2. Determination of Radical Scavenging Activity (RSA)

The DPPH• assay was applied to evaluate the RSA of the *R. nigrum* bud preparations [32]. Determinations were performed as described in a previous paper [24]. The absorbance at 515 nm was recorded by an Agilent 8453 UV-Vis spectrophotometer with 1 nm resolution. A multilevel calibration with ascorbic acid as standard was used to evaluate the RSA and to express the results as milligrams of ascorbic acid equivalent (AAE) in 100 mL of bud extract (mg AAE/100 mL).

2.8. HPLC Analysis

HPLC methods were used for phytochemical analysis both on *R. nigrum* bud preparations and PUAE extracts. Analysis were focused on flavonols, phenolic acids (benzoic and cinnamic acids), and catechins, as polyphenolic markers with a demonstrated health-promoting activity [33]. Bioactive compounds were identified and quantified by comparison and combination of their retention times and UV spectra with those of authentic standards. The calibration parameters for all the employed analytical standards were previously reported by the authors [22,34]. The total bioactive compound content (TBCC) was determined as sum of the selected and identified markers with health-promoting activities and positive antioxidant effects on human health-status according to "multimarker approach" [35]: phytochemicals were grouped into different bioactive classes in order to evaluate each class contribution to phytocomplex composition. All analyses were triplicated and the results expressed as mg/100 g of fresh weight (FW).

Samples were filtered with circular pre-injection filters (0.45 µm, polytetrafluoroethylene membrane) prior to HPLC-DAD analysis. Chromatographic analysis was carried out using an Agilent 1200 High-Performance Liquid Chromatograph coupled to an Agilent UV-Vis diode array detector (Agilent Technologies, Santa Clara, CA, USA), based on HPLC methods previously validated for fresh fruits, herbal medicines, and other food products [2,22,23]. Chromatographic conditions were set in order to obtain a phytochemical information with a good resolution and a reasonable analysis time.

Bioactive molecule separation was achieved on a Kinetex C18 column (4.6 mm × 150 mm, 5 μm, Phenomenex, Torrance, CA, USA). Different mobile phases were used for bioactive compound characterization and several linear gradients in different slopes were optimized because some compounds were similar in structure with each other in the same chemical class: (1) a solution of 10 mM KH_2PO_4/H_3PO_4 and acetonitrile with a flow rate of 1.5 $mL{\cdot}min^{-1}$ (method A—analysis of cinnamic acids and flavonols); (2) a solution of methanol/water/formic acid (5:95:0.1 *v/v/v*) and a mix of methanol/formic acid (100:0.1 *v/v*) with a flow rate of 0.6 $mL{\cdot}min^{-1}$ (method B—analysis of benzoic acids and catechins). Selected wavelengths were suitable to achieve more specific peaks as well as a smooth baseline after a full scan on the chromatogram from 190 to 400 nm; in particular, UV spectra were recorded at 330 nm (A) and 280 nm (B). Information on used chromatographic methods and selected markers are reported in the Supplementary material.

3. Results and Discussion

3.1. Optimization of The PUAE Experimental Conditions by DoE Using Untargeted Phytochemical Fingerprint

The PUAE conditions have been optimized by a Faced Central Composite Design (CCD), whose results are shown in Table 1. The DoE response variable to be optimized was obtained by an untargeted spectroscopic method combined with chemometrics previously described by the authors [2,20]. Briefly each of the nine extracts, obtained according the experimental plan and spectroscopic analyzed, was described by a vector of 701 UV-Vis absorbances plus 701 fluorescence emissions, as a holistic nontargeted fingerprint 1402-dimensional of the corresponding extract. Since these multivariate vectors of UV-Vis absorptions and fluorescence emissions (701 + 701 variables) of each extract have been proven to be strictly correlated to the whole polyphenolic fraction of the extracts they were combined in a multivariate data matrix: The DoE response matrix. This matrix, composed of nine rows and 1402 columns ($M_{9,1402}$: nine objects corresponding to the nine experiments and 1402 variables which represent the spectral absorptions/emissions), has been elaborated by principal component analysis (PCA), an unsupervised patter recognition technique, in order to extract the useful analytical information and to reduce its dimensionality.

The untargeted polyphenolic phytochemical fingerprints (UV-Vis absorptions and Fluorescence emissions) of each extract obtained by DoE were reported in Figure 1 and compared with the corresponding commercial product RNGM.

Figure 2 shows the score-plot on the first two principal components (PCs), whose explained variance are 90.3% and 6.8%, respectively.

The first PC (PC1) retains all the useful information of the 1402 original variables and thus the corresponding scores were used as the response variable of each experiment in the experimental matrix. The other PCA details are reported in the Supplementary Materials (i.e., score matrix, loading matrix, eigenvalues, explained variance plot). The following model of the CCD has been obtained by applying multiple linear regression to the experimental matrix:

$$Y = -2.6708 - 0.6814X_1\,{*} - 1.2487X_2\,{**} - 0.1010X_1X_2 + 2.6000X_1{}^2\,{**}\ 1.4061X_2{}^2\,{*}$$

* indicates the significance of the coefficients: * = $p < 0.05$, ** = $p < 0.01$.

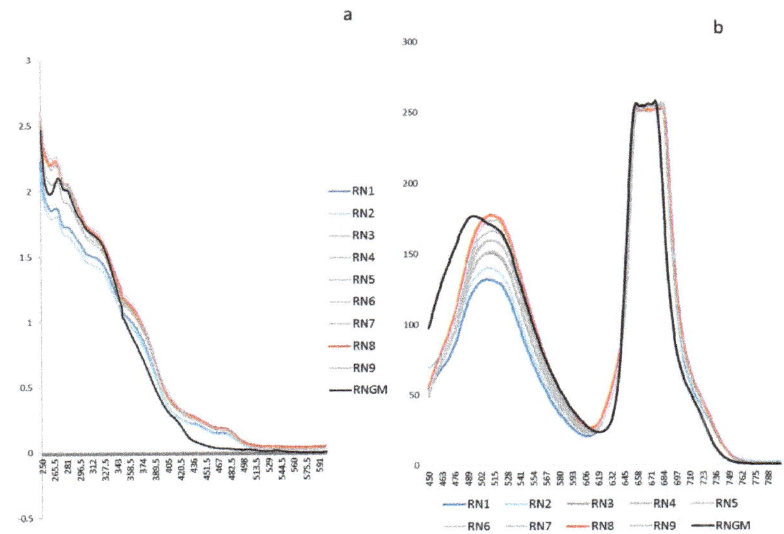

Figure 1. Untargeted spectroscopic fingerprints of the *R. nigrum* bud (RNB) phytocomplex: (**a**) The UV-Vis averaged spectra (250–600 nm) of the nine experiments selected by the DoE and the corresponding *R. nigrum* glyceric macerate (RNGM) tested at the same dilution (1:20 in the extraction solvent); (**b**) The 2D Fluorescence averaged spectral emissions (450–800 nm) of the nine experiments selected by the DoE and the corresponding RNGM tested at the same dilution (1:20 in the extraction solvent).

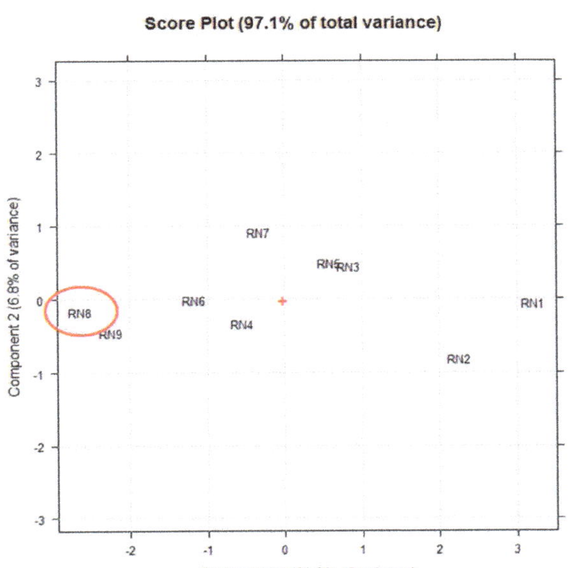

Figure 2. The score plot on the first two principle components (PCs) selected by principle component analysis (PCA) using the vector of UV-Vis (250–600 nm) spectra coupled to fluorescence (450–800 nm) absorptions of each extract (RN1–RN9) as a multivariate untargeted signal. The red + represents the central point of the plot and the red circle highlights the extract obtained in the best experimental conditions.

All the linear and quadratic terms are significant as highlighted in the plot of the coefficients (Figure 3).

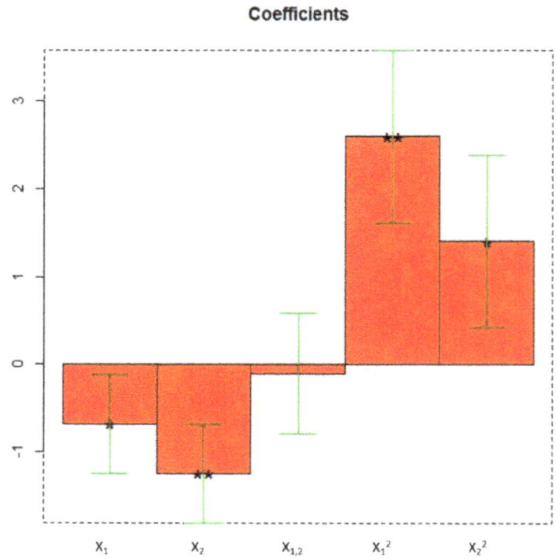

Figure 3. The coefficients of the model of Y (PC1 scores) obtained by the Faced Central Composite Design (X_1: duty cycle; X_2: extraction time). * = $p < 0.05$, ** = $p < 0.01$.

Particularly, the linear term X_2 (** = $p < 0.01$) corresponding to the extraction time, the quadratic term X_1^2 (** = $p < 0.01$), the linear term X_1 (* = $p < 0.05$) corresponding to the duty cycle, and the quadratic term X_2^2 (* = $p < 0.05$) are the statistically significant coefficients. They should be increased or, on the contrary, decreased to improve or to minimize the Y response variable respectively. Experiments whose scores on PC1 are negative (Figure 2) correspond to the highest absorptions/emissions of the phytocomplex as shown in Figure 1, thus they must be decreased. RN8 represents the best experimental conditions and this extract has been analytically characterized and compared with RNGM.

3.2. Analytical Characterization of The Most Promising PUAE Extract (RN8) and the Corresponding RNGM

3.2.1. Determination of the Total Phenolic Compounds (TPC) and the Radical Scavenging Activity (RSA)

RN8, representing the most promising *R. nigrum* extract obtained by PUAE, and the corresponding RNGM, were analytically characterized to evaluate their total phenolic contents (TPC) and their radical scavenging activity (RSA). As reported in Table 2, both the bud preparations presented quantitatively similar RSA values: 1158.58 ± 73.24 mg/100 mL of bud extract for RN8 and 1137.04 ± 38.49 mg/100 mL of bud extract for RNGM, respectively. Regarding TPC, RN8 showed a higher value with respect to RNGM (415.56 ± 5.52 mg/100 mL vs. 276.44 ± 3.85 mg/100 mL). However, the Folin-Ciocalteu assay is a nonspecific method to quantify phenols and polyphenols. In fact, this reagent does not measure only phenols, but can react with some reducing substances (i.e., ascorbic acid) [36]. For this reason, the phenol content could be overestimated and further investigations on the phytocomplex composition should be carried out. Nevertheless, the higher value of RN8 is promising and indicative of almost no oxidative alterations potentially induced by ultrasounds with respect to maceration.

Table 2. Total phenolic compounds (TPC) and radical scavenging activity (RSA) of the most promising *R. nigrum* extract obtained by pulsed ultrasound-assisted extraction (PUAE) (RN8) compared to the corresponding commercial product (RNGM).

Determination		RNGM		RN8	
		Mean Value	SD	Mean Value	SD
TPC	*mg GAE/100 mL bud extract*	276.44	3.85	415.56	5.52
RSA	*mg AAE/100 mL bud extract*	1137.04	38.49	1158.58	73.24

Results are reported as mg/100 mL of bud extract and expressed as mean value ± standard deviation (SD) ($n = 2$). GAE: gallic acid equivalent; AAE: ascorbic acid equivalent.

3.2.2. Targeted Phytochemical Fingerprint

Antioxidant compounds (in particular, polyphenols) may play a critical health-promoting role in humans for disease prevention due to their synergistic or additive biological effects (phytocomplex) that influence human health better than a single molecule or a group of few compounds [37]. In this study, 13 biologically active compounds (grouped into four polyphenolic classes) were selected as markers for fingerprint analysis because they have been described as important health-effective substances in humans [38]. The phytochemical fingerprints of RNGM and RN8 are reported in Table 3.

Table 3. Targeted phytochemical fingerprint by HPLC-DAD of the polyphenolic compounds in the most promising *R. nigrum* extract (RN8) obtained by PUAE compared to the corresponding glyceric macerate (RNGM).

Bioactive Class	Compound	RNGM		RN8	
		Mean Value	SD	Mean Value	SD
		(mg/100 g FW)		(mg/100 g FW)	
Cinnamic acids	*caffeic acid*	22.48	0.04	20.76	0.48
	chlorogenic acid	n.d.	/	n.d.	/
	coumaric acid	5.21	0.15	1.05	0.25
	ferulic acid	n.d.	/	n.d.	/
Flavonols	*hyperoside*	n.d.	/	n.d.	/
	isoquercitrin	n.d.	/	n.d.	/
	quercetin	49.53	0.49	80.14	1.08
	quercitrin	30.86	0.85	48.18	0.94
	rutin	17.25	0.22	20.88	0.48
Benzoic acids	*ellagic acid*	69.66	0.08	75.37	0.30
	gallic acid	0.31	0.09	0.64	0.05
Catechins	*catechin*	95.88	0.26	55.85	2.78
	epicatechin	59.83	0.37	49.08	0.48

Results are reported as mg/100 g of bud fresh weight (FW) and expressed as mean value ± standard deviation (SD) ($n = 3$). * n.d. = not detectable.

Among the analyzed compounds, chlorogenic acid, ferulic acid, hyperoside, and isoquercitrin were not detected. Bioactive compounds were separated and identified via HPLC-DAD. Adding other markers with demonstrated biological activity may be a necessary step for a better identification of the chromatographic pattern in further fingerprint studies together with a mass spectrometry detection of unknown peaks.

In RNGM samples, catechins were the most important bioactive class (44.36%), with flavonols as the second most abundant (27.82%), followed by benzoic acids and cinnamic acids (19.93% and 7.89%, respectively), while in PUAE extracts RN8, the quantitative relationships between catechins and flavonols were reversed: 29.81% for catechins and 42.39% for flavonols, while cinnamic acids (6.20%) and benzoic acids (21.60%) showed percentages similar to the bud macerates (Figure 4).

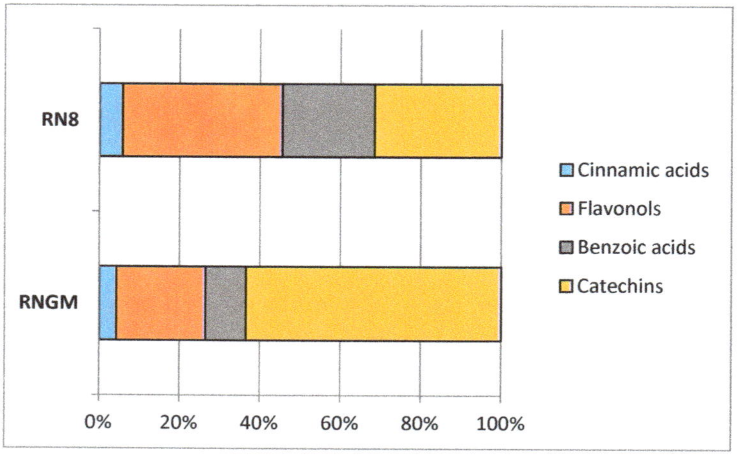

Figure 4. Contribution of each polyphenolic class to the total phytocomplex in RN8 and RNGM analyzed samples.

Figure 5 reports the polyphenolic chromatographic profile of analyzed samples: RNGM and RN8 presented qualitatively and quantitatively similar phenolic patterns. The most important differences were only detected in four compounds: (i) quercetin (48.53 ± 0.49 mg/100 g of bud fresh weight, FW, for RNGM and 80.14 ± 1.08 mg/100 g FW for RN8); (ii) quercitrin (30.86 ± 0.85 mg/100 g FW for RNGM and 48.18 ± 0.94 mg/100 g FW for RN8); (iii) catechin (95.88 ± 0.26 mg/100 g FW for RNGM and 55.85 ± 2.78 mg/100 g FW for RN8); (iv) epicatechin (59.83 ± 0.37 mg/100 g FW for RNGM and 49.08 ± 0.48 mg/100 g FW for RN8). Caffeic acid showed levels slightly higher in RNGM, while ellagic acid presented levels slightly higher in RN8.

Our results highlight that the traditional glyceric macerate and the alternative PUAE extract show similar total polyphenolic levels (and a qualitatively similar chromatographic pattern), but some differences in specific bioactive compounds (in particular, flavonols and catechins) were also detected, due to the different extraction method [39]. For this reason, PUAE yielded an extract rich in biological active molecules with potentially high health-promoting activity, but maybe with a practical use which could be different from traditional bud preparations. In any case, this research is only a preliminary study and further phytochemical, clinical, toxicological, and pharmaceutical in vitro and in vivo tests should be carried out to confirm this preliminary hypothesis.

Cinnamic acids and flavonols (**A**).

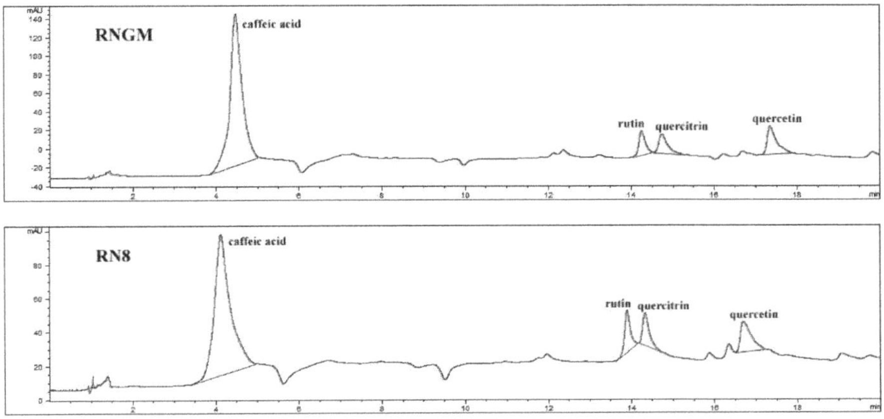

Benzoic acids and catechins (**B**).

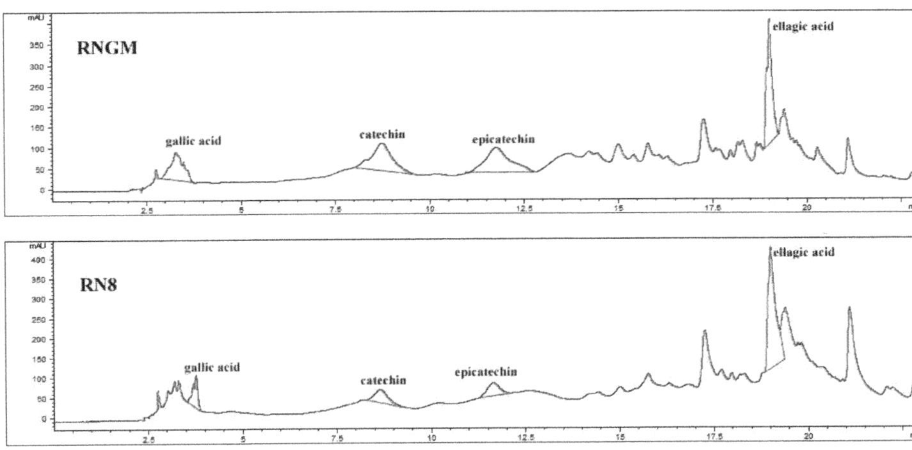

Figure 5. Chromatographic pattern of polyphenolic compounds identified in RN8 and RNGM samples: (**A**) Cinnamic acids and flavonols; (**B**) Benzoic acids and catechins. In the chromatograms, the x-axis represents the signal current intensity (mAU), while the y-axis represents time (min).

4. Conclusions

In this work, PUAE, as an alternative time-saving method to produce *R. nigrum* bud derivatives, was presented according to the green chemistry principles. The unconventional extraction conditions were optimized by DoE at the lab scale using untargeted fingerprints coupled to chemometrics, but this same quick strategy could be analogously applied to transfer this method to an industrial scale. The impact of the UAE with respect to traditional maceration was evaluated in terms of recovering of the total phenolic content, the antiradical scavenging activity and the profiles of the most important bioactive compounds. In particular, PUAE provided the extract named RN8 in a few minutes, compared to the 21 day-long maceration, whose total polyphenolic levels and antiradical scavenging are similar or even slightly increased with respect to RNGM. Furthermore, RN8 presents a qualitatively similar chromatographic pattern, even if some differences in flavonols and catechins were detected. Due to these differences, an LC-MS study of RN8 is mandatory in the near future. Nevertheless, this is a promising preliminary result to provide alternative uses of *Ribes nigrum* bud derivatives using this unconventional time-saving extraction method.

Supplementary Materials: Supplementary materials are available online at http://www.mdpi.com/2304-8158/8/10/466/s1. Table S1: score matrix; Table S2: loading matrix; Table S3: eigenvalue matrix; Table S4: chromatographic conditions. Figure S1: percentage of explained variance plot.

Author Contributions: Conceived the experiments, R.B., P.Z., A.P.; Designed the experiments, F.T. and D.D.; Followed the experiments, R.B. and G.L.B.; Performed the experiments and analyzed the data, P.Z., F.T., D.D. and R.B.; Wrote the paper, F.T.

Funding: This research was funded by an European Union project called FINNOVER (n° 1198), http://www.interreg-finnover.com/.

Conflicts of Interest: The authors declare no conflict of interest.

References

1. FINNOVER Interreg Alcotra Project 2017–2020. Available online: http://www.interreg-finnover.com/ (accessed on 29 July 2019).
2. Turrini, F.; Donno, D.; Boggia, R.; Beccaro, G.L.; Zunin, P.; Leardi, R.; Pittaluga, A.M. An innovative green extraction and re-use strategy to valorize food supplement by-products: Castanea sativa bud preparations as case study. *Food Res. Int.* **2019**, *115*, 276–282. [CrossRef]
3. The EU Food Supplements Directive 2002/46/EC. Available online: http://data.europa.eu/eli/dir/2002/46/oj (accessed on 20 July 2019).
4. Pharmacopée Française. *Codex Medicamentarius Gallicus, Codex Français: Monographie, Préparations Homéopathiques*; Ordre National des Pharmaciens: Paris, France, 1965; Available online: http://ansm.sante.fr/Mediatheque/Publications/Pharmacopee-francaise-Plan-Preambule-index (accessed on 26 July 2019).
5. Donno, D.; Beccaro, G.L.; Cerutti, A.K.; Mellano, M.G.; Bounous, G. Bud Extracts as New Phytochemical Source for Herbal Preparations—Quality Control and Standardization by Analytical Fingerprint. In *Phytochemicals—Isolation, Characterisation and Role in Human Health*, 1st ed.; Rao, A.V., Rao, L.G., Eds.; InTech: Rijeka, Croatia, 2015; pp. 187–218.
6. Donno, D.; Mellano, M.G.; Cerutti, A.K.; Beccaro, G.L. Biomolecules and Natural Medicine Preparations: Analysis of New Sources of Bioactive Compounds from *Ribes* and *Rubus* spp. Buds. *Pharmaceuticals* **2016**, *9*, 7. [CrossRef]
7. Donno, D.; Boggia, R.; Zunin, P.; Cerutti, A.K.; Guido, M.; Mellano, M.G.; Prgomet, Z.; Beccaro, G.L. Phytochemical fingerprint and chemometrics for natural food preparation pattern recognition: An innovative technique in food supplement quality control. *J. Food Sci. Technol.* **2016**, *53*, 1071–1083. [CrossRef]
8. Benvenuti, S.; Pellati, F.; Melegari, M.; Bertelli, D. Polyphenols, anthocyanins, ascorbic acid, and radical scavenging activity of *Rubus*, *Ribes*, and Aronia. *Food Chem. Toxicol.* **2004**, *69*, 164–169. [CrossRef]
9. Paunović, S.M.; Nikolić, M.; Miletić, R.; Mašković, P.; Milinković, M.; Karaklajić-Stajić, Z. Phytochemical Screening and Biological Activity of Extract Berries of Black Currant (*Ribes nigrum* L.). *Erwerbs-Obstbau* **2019**, *61*, 71–78. [CrossRef]
10. Donno, D.; Beccaro, G.L.; Mellano, M.G.; Cerutti, A.K.; Marconi, V.; Bounous, G. Botanicals in *Ribes nigrum* bud-preparations: An analytical fingerprinting to evaluate the bioactive contribution to total phytocomplex. *Pharm. Biol.* **2013**, *51*, 1282–1292. [CrossRef]
11. Chemat, F.; Rombaut, N.; Meullemiestre, A.; Turk, M.; Perino, S.; Fabiano-Tixier, A.S.; Abert-Vian, M. Review of Green Food Processing techniques. Preservation, transformation, and extraction. *Innov. Food Sci. Emerg. Technol.* **2017**, *41*, 357–377. [CrossRef]
12. Chemat, F.; Vian, M.A.; Cravotto, G. Green extraction of natural products: Concept and principles. *Int. J. Mol. Sci.* **2012**, *13*, 8615–8627.
13. Green Chemistry's 12 Principles, United States Environmental Protection Agency. Available online: https://www.epa.gov/greenchemistry/basics-green-chemistry#twelve (accessed on 26 July 2019).
14. Li, Y.; Fabiano-Tixier, A.S.; Tomao, V.; Cravotto, G.; Chemat, F. Green ultrasound-assisted extraction of carotenoids based on the bio-refinery concept using sunflower oil as an alternative solvent. *Ultrason. Sonochem.* **2013**, *20*, 12–18. [CrossRef]

15. Chemat, F.; Zill-e-Huma; Khan, M.K. Applications of ultrasound in food technology: Processing, preservation and extraction. *Ultrason. Sonochem.* **2011**, *18*, 813–835. [CrossRef]
16. Vinatoru, M.; Mason, T.J.; Calinescu, I. Ultrasonically assisted extraction (UAE) and microwave assisted extraction (MAE) of functional compounds from plant materials. *Trends Anal. Chem.* **2017**, *97*, 159–178. [CrossRef]
17. Chemat, F.; Rombaut, N.; Sicaire, A.G.; Meullemiestre, A.; Fabiano-Tixier, A.S.; Abert-Vian, M. Ultrasound assisted extraction of food and natural products. Mechanisms, techniques, combinations, protocols and applications. A review. *Ultrason. Sonochem.* **2017**, *34*, 540–560. [CrossRef]
18. Cravotto, G.; Binello, A.; Orio, L. Green extraction techniques: For high-quality natural products. *Agro Food Ind. Hi Tech* **2011**, *22*, 24–36.
19. Medina-Torres, N.; Ayora-Talavera, T.; Espinosa-Andrews, H.; Sánchez-Contreras, A.; Pacheco, N. Ultrasound Assisted Extraction for the Recovery of Phenolic Compounds from Vegetable Sources. *Agronomy* **2017**, *7*, 47. [CrossRef]
20. Boggia, R.; Turrini, F.; Anselmo, M.; Zunin, P.; Donno, D.; Beccaro, G.L. Feasibility of UV–VIS-Fluorescence Spectroscopy combined with pattern recognition techniques to authenticate a new category of plant food supplements. *J. Food Sci. Tech.* **2017**, *54*, 2422–2432. [CrossRef]
21. Leardi, R. Experimental design in chemistry: A tutorial. *Anal. Chim. Acta* **2009**, *652*, 161–172. [CrossRef]
22. Donno, D.; Mellano, M.; Hassan, S.; De Biaggi, M.; Riondato, I.; Gamba, G.; Giacoma, C.; Beccaro, G.L. Assessing Nutritional Traits and Phytochemical Composition of Artisan Jams Produced in Comoros Islands: Using Indigenous Fruits with High Health-Impact as an Example of Biodiversity Integration and Food Security in Rural Development. *Molecules* **2018**, *23*, 2707. [CrossRef]
23. Donno, D.; Mellano, M.G.; Prgomet, Z.; Beccaro, G.L. Advances in *Ribes* x nidigrolariaRud. Bauer and A. Bauer fruits as potential source of natural molecules: A preliminary study on physico-chemical traits of an underutilized berry. *Sci. Hortic.* **2018**, *237*, 20–27. [CrossRef]
24. Turrini, F.; Zunin, P.; Catena, S.; Villa, C.; Alfei, S.; Boggia, R. Traditional or hydro-diffusion and gravity microwave coupled with ultrasound as green technologies for the valorization of pomegranate external peels. *Food Bioprod. Process.* **2019**, *117*, 30–37. [CrossRef]
25. Boggia, R.; Turrini, F.; Villa, C.; Lacapra, C.; Zunin, P.; Parodi, B. Green extraction from pomegranate marcs for the production of functional foods and cosmetics. *Pharmaceuticals* **2016**, *9*, 63–74. [CrossRef]
26. Turrini, F.; Boggia, R.; Leardi, R.; Borriello, M.; Zunin, P. Optimization of the Ultrasonic-Assisted Extraction of Phenolic Compounds from *Oryza Sativa* L. 'Violet Nori' and Determination of the Antioxidant Properties of its Caryopses and Leaves. *Molecules* **2018**, *23*, 844. [CrossRef]
27. Sadeckà, J.; Tòthovà, J. Fluorescence spectroscopy and chemometrics in the food classification—A review. *Czech J. Food Sci.* **2007**, *25*, 159–173. [CrossRef]
28. Barnes, R.J.; Dhanoa, M.S.; Lister, S.J. Standard normal variate transformation and de-trending of near-infrared diffuse reflectance spectra. *Appl. Spectrosc.* **1989**, *43*, 772–777. [CrossRef]
29. Wold, S.; Esbensen, K.; Geladi, P. Principal component analysis. *Chemom. Intell. Lab. Syst.* **1987**, *2*, 37–52. [CrossRef]
30. Italian Chemical Society. Division of Analytical Chemistry-Group of Chemometrics. *CAT Chemometric Agile Tool*. Available online: http://www.gruppochemiometria.it/index.php/software (accessed on 29 July 2019).
31. Singleton, V.L.; Rossi, J.A., Jr. Colorimetry of total phenolics with phosphomolybdic-phosphotungstic acid reagents. *Am. J. Enol. Viticult.* **1965**, *16*, 144–158.
32. Brand-Williams, W.; Cuvelier, M.E.; Berset, C. Use of a free radical method to evaluate antioxidant activity. *LWT-Food Sci. Technol.* **1995**, *28*, 25–30. [CrossRef]
33. Shahidi, F.; Ambigaipalan, P. Phenolics and polyphenolics in foods, beverages and spices: Antioxidant activity and health effects—A review. *J. Funct. Foods* **2015**, *18*, 820–897. [CrossRef]
34. Donno, D.; Mellano, M.G.; De Biaggi, M.; Riondato, I.; Rakotoniaina, E.N.; Beccaro, G.L. New Findings in *Prunuspadus* L. Fruits as a Source of Natural Compounds: Characterization of Metabolite Profiles and Preliminary Evaluation of Antioxidant Activity. *Molecules* **2018**, *23*, 725. [CrossRef]
35. Mok, D.K.W.; Chau, F.T. Chemical information of Chinese medicines: A challenge to chemist. *Chemometr. Intell. Lab.* **2006**, *82*, 210–217. [CrossRef]
36. Jagota, S.K.; Dani, H.M. A new colorimetric technique for the estimation of vitamin C using Folin phenol reagent. *Anal. Biochem.* **1982**, *127*, 178–182. [CrossRef]

37. Donno, D.; Mellano, M.G.; Cerutti, A.K.; Beccaro, G.L. Nutraceuticals in Alternative and Underutilized Fruits as Functional Food Ingredients: Ancient Species for New Health Needs. In *Alternative and Replacement Foods*, 1st ed.; Holban, A.M., Grumezescu, A.M., Eds.; Academic Press: Cambridge, MA, USA, 2018; pp. 261–282.
38. Balasundram, N.; Sundram, K.; Samman, S. Phenolic compounds in plants and agri-industrial by-products: Antioxidant activity, occurrence, and potential uses. *Food Chem.* **2006**, *99*, 191–203. [CrossRef]
39. Rocchetti, G.; Blasi, F.; Montesano, D.; Ghisoni, S.; Marcotullio, M.C.; Sabatini, S.; Cossignani, L.; Lucini, L. Impact of conventional/non-conventional extraction methods on the untargeted phenolic profile of Moringa oleifera leaves. *Food Res. Int.* **2019**, *115*, 319–327. [CrossRef] [PubMed]

foods

MDPI

Article

Developing and Validating a Method for Separating Flavonoid Isomers in Common Buckwheat Sprouts Using HPLC-PDA

Davin Jang [1], Young Sung Jung [2], Mi-Seon Kim [1], Seung Eel Oh [3], Tae Gyu Nam [3,*] and Dae-Ok Kim [1,2,*]

1 Graduate School of Biotechnology, Kyung Hee University, Yongin 17104, Korea; davin1031@khu.ac.kr (D.J.); miseonkim95@khu.ac.kr (M.-S.K.)
2 Department of Food Science and Biotechnology, Kyung Hee University, Yongin 17104, Korea; chembio@khu.ac.kr
3 Korea Food Research Institute, Wanju 55365, Korea; dr51@kfri.re.kr
* Correspondence: ntg97@kfri.re.kr (T.G.N.); DOKIM05@khu.ac.kr (D.-O.K.); Tel.: +82-31-201-3796 (D.-O.K.); Fax: +82-31-204-8116 (D.-O.K.)

Received: 10 September 2019; Accepted: 1 November 2019; Published: 4 November 2019

check for updates

Abstract: Buckwheat sprouts that are synthesized during the germination process are rich in flavonoids, including orientin, vitexin, rutin, and their isomers (isoorientin, isovitexin, and quercetin-3-*O*-robinobioside, respectively). The purpose of this study was to optimize and validate an analytical method for separating flavonoid isomers in common buckwheat sprout extract (CSE). Factors, such as range, linearity, precision, accuracy, limit of detection, and limit of quantification, were evaluated for each standard using high-performance liquid chromatography (HPLC). On the basis of resolution and symmetry, a column temperature of 40 °C with 0.1% (*v*/*v*) acidic water and acetonitrile as mobile phases, at a flow rate of 1 mL min^{-1} were determined to be the optimal analytical conditions. Calibration curves for orientin, isoorientin, vitexin, isovitexin, and rutin exhibited good linearity with correlation coefficients of 0.9999 over the 6.25–100.00 μg mL^{-1} range. Recovery values of 96.67–103.60% confirmed that the method was accurate for all flavonoids. The relative standard deviations of intra-day repeatability and inter-day reproducibility confirmed method preciseness, with values of less than 5.21% and 5.40%, respectively. The developed method was used to analyze flavonoids in CSE, with isomers satisfactorily separated and simultaneously quantified. We demonstrated that the developed HPLC method can be used to monitor flavonoids in buckwheat sprouts.

Keywords: common buckwheat sprout; flavonoid isomer; quercetin-3-*O*-robinobioside; validation; chromatographic separation

1. Introduction

Buckwheat is a pseudocereal belonging to the *Polygonaceae* family that grows rapidly and is tolerant to cold [1]. Buckwheat is found almost everywhere but is mainly grown in the northern hemisphere [2]. Common buckwheat (*Fagopyrum esculentum* Möench) and tartary buckwheat (*F. tataricum* Gaertner) are the most consumed buckwheat species [1]. Among the edible parts of common buckwheat, the sprouts have attracted considerable attention in recent years [3] as they are considered to be a popular health food and are widely consumed because of their bioactive compounds [4,5]. The sprout-germination process induces the hydrolysis of triglycerides in the seeds and produces the energy required for various biochemical reactions through the tricarboxylic acid cycle [6], and the content of bioactive compounds in the seeds increases through chemical reactions, for example, the flavone glycoside content is known to increase during the germination of common buckwheat [7,8]. Common buckwheat sprout (CS) has been studied for a variety of pro-health benefits, such as its antioxidant capacity [7,9] and anti-inflammatory effects [10,11].

CS has been reported to have more abundant flavone *C*-glucosides than those of the tartary species [12]. The flavone *C*-glucosides in CS are present as orientin and isoorientin, from luteolin as the parent, as well as vitexin and isovitexin, from apigenin as the parent (Figure S1). In addition, rutin is a representative flavonoid present in common buckwheat and CS [12]. A recent report revealed that quercetin-3-*O*-robinobioside (Q3R), which is an isomer of rutin, exists in CS [8]. Q3R is known to be present in cotyledon and immature common buckwheat, rooibos, mature saskatoon fruit, and jujube fruit [8,13–15]. To summarize, the six main flavonoids found in CS exist as three sets of isomer pairs, with differences due to the position and form of the sugar.

Flavonoids and their glycosides can be used as quality control markers for many phytomedicines and medicinal plants [16]. Various analytical techniques have been developed for the separation and detection of flavonoid glycosides, in which the most-widely employed method is reversed-phase high-performance liquid chromatography (HPLC), coupled with photodiode array (PDA) detection and/or mass spectrometry (MS) [14,16,17]. Qualitative analysis of flavonoids is possible using MS even when baseline separation has not been secured. However, structural isomers cannot be distinguished only on the basis of MS/MS information because they have the same molecular weight and similar fragment patterns [16,18,19]. Therefore, flavonoid glycosides need to be separated in order to enable their accurate quantitative analysis in food. Previous studies reported the effects of column temperature, composition of mobile phase, and flow rate on the separation of a flavonoid isomer for qualitative analysis [20,21]. Even with a column temperature of 40 °C and the addition of acid to solvents, orientin and Q3R in CS were not separatable with isoorientin and rutin, respectively [7,9]. Q3R and rutin in mature saskatoon and jujube fruits were not completely separated and were not suitable for quantification [13,22]. Therefore, a suitable HPLC method for quantitatively analyzing flavonoid isomers present in CS is necessary.

In this study, we introduce an HPLC analysis method that simultaneously quantifies two types of flavone-*C*-glycoside isomer and flavonol-*O*-glycoside isomer found in common buckwheat sprout extract (CSE). The HPLC conditions of the developed analytical method, including mobile phase, column temperature, and flow rate, were optimized. The developed method was validated by determining the range, linearity, precision, accuracy, limit of detection (LOD), and limit of quantification (LOQ) for each compound.

2. Materials and Methods

2.1. Chemicals and Reagents

Orientin, isoorientin, vitexin, and isovitexin (all with ≥99% purity) were purchased from Extrasynthese (Genay, France). Rutin hydrate (≥94%), HPLC-grade formic acid (≥98%), and dimethyl sulfoxide (DMSO) were obtained from Sigma Aldrich Co., LLC (St. Louis, MO, USA). HPLC-grade water, methanol, and acetonitrile were purchased from Thermo Fisher Scientific (Waltham, MA, USA).

2.2. CSE Preparation

CS seeds were purchased from First Village Farmer's Union under Hallasan Mountain (Jeju-do, Republic of Korea) in 2018. Seeds were planted for 6 days at 25 °C in a dark growth chamber. Sprouts were harvested and freeze-dried (FD 8518; Ilshin Lab Co., Ltd., Dongducheon, Korea) for 3 days. The CSE was obtained by mixing the dried buckwheat sprouts with 90% (*v*/*v*) aqueous methanol. The mixture was ultrasonicated for 30 min and the supernatant was acquired by centrifugation (2232× *g*) for 10 min and filtered through a 0.45-µm polyvinylidene fluoride syringe filter (Millipore, Billerica, MA, USA). The residue was re-extracted once using the procedure described above. The filtrate was evaporated on a rotary evaporator (N-1000; Eyela, Tokyo, Japan) in a water bath at 40 °C. The extract was stored in a deep freezer (WSM-2700UC; Grand Woosung Inc., Seoul, Korea) at −50 °C and freeze-dried (FD 8518) for 3 days. All experiments were performed in triplicate.

2.3. Flavonoids Analysis by Reversed-Phase HPLC

The 3 sets of isomer pairs were analyzed by modifying an existing method [8]. CSE and its flavonoids were analyzed using HPLC (Alliance e2695; Waters, Milford, MA, USA) with the Empower 3 software (Waters), a PDA detector (2998, Waters), and a ProntoSIL 120-5-C18-ace-EPS column (4.6 × 250 mm, 5.0 μm; Bischoff, Leonberg, Germany), with the flavonoids and extract monitored at 360 nm. The column temperature was set to 40 °C and an injector volume of 5 μL was used. Gradient elution was carried out with 0.1% (*v*/*v*) formic acid in water (solvent A) and acetonitrile (solvent B). All solvents were filtered and degassed. A flow rate of 1.0 mL min^{-1} was used. The following binary mobile-phase linear gradients were used: 100% A at 0 min, 90% A at 4 min, 86% A at 20 min, 84% A at 30 min, 84% A at 36 min, 80% A at 44 min, 80% A at 50 min, 75% A at 54 min, 30% A at 58 min, 30% A at 62 min, 15% A at 66 min, 15% A at 70 min, 100% A at 72 min, and 100% A at 75 min. Flavonoids were identified by comparing their retention time and ultraviolet (UV) spectra with those of their respective standards. The Q3R was identified as in previous studies and was quantified using the calibration curve for rutin because the standard for Q3R is not commercially available [13,22].

2.4. Optimizing the Chromatography Conditions

The conditions for the separation of the flavonoids in CSE include the mobile phase, temperature, and flow rate. Water or 0.1% (*v*/*v*) formic acid in water as solvent A and methanol or acetonitrile as solvent B were used. The column temperatures were set to 20, 30, and 40 °C. The separation of the 3 sets of isomer pairs was evaluated. Flow rates of 0.6, 0.8, and 1.0 mL min^{-1} were used. The column and method used are listed in Section 2.3. The resolution (*Rs*) and symmetry factor were calculated as follows [23]:

$$Rs = 1.18 \times (t_{R2} - t_{R1})/(W_2 + W_1)$$

where $t_{R2} - t_{R1}$ is the difference in retention time and $W_2 + W_1$ is the sum of the peak widths at half of peak height, and,

$$Symmetry\ factor = W_{0.05h}/2f$$

where $W_{0.05h}$ is the peak width at 1/20th the peak height above the peak baseline, and *f* is the distance along the horizontal line drawn from the leading edge of the peak to the vertical line drawn from where the peak dissects a horizontal line drawn at 1/20th of the peak height above the peak baseline ($W_{0.05h}$).

2.5. Method Validation

The validation criteria are based on the guidelines published by the Ministry of Food and Drug Safety [23]. Spectral scans of peaks were performed with the PDA detector (2998, Waters) in order to verify method specificity. The linearity of each calibration curve, range, precision, accuracy, LOD, and LOQ were also evaluated.

2.5.1. Linearity

Stock solutions of standard compounds in CSE were prepared at a concentration of 10 mg mL^{-1} in DMSO and 10% (*v*/*v*) DMSO in methanol was used to dilute the stock solution to the appropriate concentration. The concentration ranges of the standard compounds were appropriately set to include the CSE to be analyzed through preliminary experiments. Linearity ranges were determined by diluting the standard solutions to six different concentrations. The method was found to be linear in 6.25–100.00 μg mL^{-1} range for all standard compounds. Each concentration was analyzed in triplicate. The analytical curve was obtained from the peak area corresponding to each standard compound at six different concentrations. The linearity of each calibration curve is expressed by its correlation coefficient (R).

2.5.2. Accuracy, Precision, and Recovery

Accuracy was evaluated as the percent recovery (%) at three concentrations (25.00, 50.00, and 100.00 μg mL^{-1}) of spiked standard solutions and blanks (dilution solvent). Precision is expressed as the relative standard deviation (RSD) of intra-day repeatability (analysis was performed on the same day ($n = 3$) with the same instrument) and inter-day reproducibility (three different days ($n = 3 \times 3$) using the same instrument).

2.5.3. Limits of Detection and Quantification

LODs and LOQs for the analyte flavonoids were calculated from the standard deviation of the response and the slope of the calibration curve. The standard curves were constructed in the 0.61–20.00 μg mL^{-1}. The LOD was calculated as $3.3 \times \sigma/s$ (σ = standard deviation of the response, s = slope of the standard curve), while $10 \times \sigma/s$ was used for the LOQ.

2.6. Statistical Analysis

Analysis was performed in triplicate and the results obtained are represented as means ± standard deviations. One-way analysis of variance followed by the Tukey's test ($p < 0.05$) was applied to determine the significances of the differences among the means. Tests for statistical significance were performed using IBM SPSS software (Version 23; IBM SPSS Statistics Inc., Armonk, NY, USA).

3. Results and Discussion

3.1. Effects of Elution Conditions on Isomer Separation

The mobile phase, temperature, and flow rate are significant isomer-separation factors in HPLC [24,25]. The data for the various isomers in CSE using different mobile phases are listed in Table 1, at a column temperature of 40 °C and a flow rate of 1.0 mL min^{-1}. *Rs*, a quantitative value that indicates the degree of separation between adjacent components should be greater than 1.5 in order to satisfy the baseline-separation criteria [23]. The data listed in Table 1 are for water as solvent A and an organic solvent as B. Luteolin, apigenin, and quercetin derivatives include orientin and isoorientin, vitexin and isovitexin, and Q3R and rutin, respectively, and the *Rs* values for the luteolin and quercetin derivatives were less than 1.5 when methanol was used as solvent B. We conclude that methanol is not suitable as solvent B. On the other hand, all components were completely separated using acetonitrile, which has a higher elution strength than methanol and is universally used due to its UV cutoff and viscosity [25]. Symmetry factors were calculated in order to determine whether or not proton (hydrogen ion) donor needs to be added to the water that accompanies the acetonitrile. Peaks appear to be more symmetric as the symmetry factor approaches unity [23] and values in the 0.99–1.03 range were observed for all components using acidic water as mobile phases (Table 1). Peaks were not completely separated when methanol was used as a mobile phase, which means that symmetry factors could not be calculated. Therefore, the optimal mobile phases for analyzing CSE are water containing 0.1% (*v*/*v*) formic acid as solvent A and acetonitrile as solvent B.

Table 1. Effect of the mobile-phase composition on the separation of isomers in common buckwheat sprout extract.

Composition of Mobile Phase (Solvent A/Solvent B)	Resolution (*Rs*)			Symmetry Factor					
	Luteolin Derivatives [1]	Apigenin Derivatives [2]	Quercetin Derivatives [3]	Orientin	Isoorientin	Vitexin	Isovitexin	Q3R [4]	Rutin
Water/Methanol	1.23 ± 0.00 [c,5]	3.66 ± 0.03 [c]	0.00 ± 0.00 [c]	n.d. [6]	n.d.	n.d.	n.d.	n.d.	n.d.
Water/Acetonitrile	2.53 ± 0.03 [a]	10.17 ± 0.06 [b]	2.09 ± 0.04 [a]	1.17 ± 0.04	1.18 ± 0.05	0.96 ± 0.01	0.99 ± 0.01	1.12 ± 0.05	1.05 ± 0.02
Acidic water [7]/Methanol	1.19 ± 0.00 [c]	3.71 ± 0.01 [c]	0.00 ± 0.00 [c]	n.d.	n.d.	n.d.	n.d.	n.d.	n.d.
Acidic water/Acetonitrile	1.87 ± 0.00 [b]	10.30 ± 0.03 [a]	1.93 ± 0.02 [b]	1.03 ± 0.02	1.03 ± 0.01	0.99 ± 0.02	0.99 ± 0.01	1.02 ± 0.02	1.00 ± 0.02

[1] Luteolin derivatives, orientin and isoorientin; [2] Apigenin derivatives, vitexin and isovitexin; [3] Quercetin derivatives, Q3R and rutin; [4] Q3R, quercetin-3-*O*-robinobioside; [5] Means with different superscripts in the same column indicate significant differences ($p < 0.05$) by Tukey's test, [6] n.d., not detected; [7] Acidic water is water containing 0.1% (*v*/*v*) formic acid.

The chromatogram shown in Figure 1 reveals the effect of temperature on isomer separation. The calculated *Rs* values are listed in Table S1. The above-mentioned optimal mobile phases were used at a flow rate was 1.0 mL min^{-1}. The *Rs* values for the isomer sets were 1.05 (luteolin derivatives), 10.83 (apigenin derivatives), and 0.00 (quercetin derivatives) at 20 °C (Figure 1A and Table S1). Only vitexin and isovitexin (the apigenin derivatives) satisfied the baseline-separation criterion at 20 °C. The luteolin derivatives and apigenin derivatives exhibited *Rs* values of 1.58 and 9.64 at 30 °C, which satisfy the baseline-separation criterion (Figure 1B and Table S1). The *Rs* value of Q3R and rutin (the quercetin derivatives) at 30 °C was 1.15, which clearly does not satisfy the baseline-separation criterion. In contrast, Figure 1C and Table S1 reveal that all isomer sets were baseline separated at 40 °C, with *Rs* values of 1.87, 10.30, and 1.93 for the luteolin, apigenin, and quercetin derivatives, respectively. The analytes also eluted faster as the column temperature was increased from 20 °C to 40 °C as higher temperatures result in lower mobile-phase viscosities and pressures, which lead to shorter retention times [26]. However, high temperatures do not necessarily guarantee efficient separation [24,27], for example, the *Rs* value of the apigenin derivatives seemed to be independent of the increase in temperature with values of 10.83, 9.64, and 10.30 at 20 °C, 30 °C, and 40 °C, respectively (Figure 1). Therefore, appropriate temperature conditions are very important when developing an analytical method. Based on the above results, all subsequent analyses in this study were carried out at 40 °C.

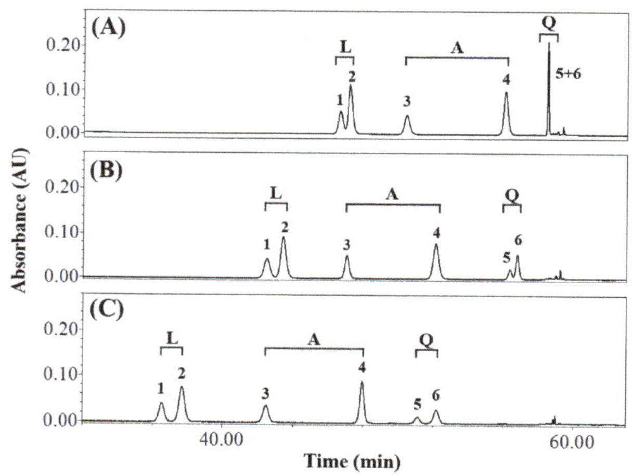

Figure 1. Reversed-phase high-performance liquid chromatography (HPLC) traces of common buckwheat sprout extract acquired at: (**A**) 20 °C, (**B**) 30 °C, and (**C**) 40 °C detected at 360 nm. Peak number: 1, orientin; 2, isoorientin; 3, vitexin; 4, isovitexin; 5, quercetin-3-*O*-robinobioside; 6, rutin. L, luteolin derivatives; A, apigenin derivatives; Q, quercetin derivatives.

The effects of flow rate on elution were evaluated at 0.6, 0.8, and 1.0 mL min^{-1} (Figure 2 and Table S2) using the optimal solvents and temperature. All components were separated except for the quercetin derivatives (*Rs* = 1.44) at a flow rate of 0.6 mL min^{-1} (Table S2). The components eluted faster as the flow rate was increased from 0.6 to 1.0 and the flavonoids eluted at 46.52–60.03 min at 0.6 mL min^{-1}, but at 36.28–52.00 min at 1.0 mL min^{-1}. As a result, a flow rate of 1.0 mL min^{-1} was chosen in subsequent work.

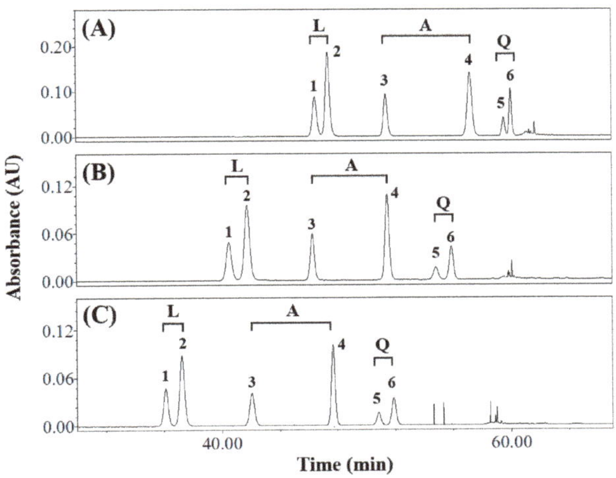

Figure 2. Reversed-phase HPLC traces of common buckwheat sprout extract acquired at flow rates of: (**A**) 0.6 mL min^{-1}, (**B**) 0.8 mL min^{-1}, and (**C**) 1.0 mL min^{-1} detected at 360 nm. Peak number: 1, orientin; 2, isoorientin; 3, vitexin; 4, isovitexin; 5, quercetin-3-*O*-robinobioside; 6, rutin. L, luteolin derivatives; A, apigenin derivatives; Q, quercetin derivatives.

3.2. Isolating the Three Isomer Pairs from CS

The structures of the three sets of isomer pairs are displayed in Figure S1. Orientin and isoorientin are isomers that have luteolin as their parent, in which a glucose unit is attached at the 8-C and 6-C position of the flavonoid A ring, respectively. Likewise, vitexin and isovitexin are isomers with apigenin as the parent. Rutin contains a rutinose (6-*O*-α-L-rhamnosyl-D-glucose) at the C-3 position of the quercetin, while Q3R contains a robinobiose (6-*O*-α-L-rhamnosyl-D-galactose) at the same position. Figure 3A shows the ordering of orientin, isoorientin, vitexin, isovitexin, and rutin standards obtained using the optimal HPLC conditions, which is similar to that of a previous CSE analysis result [8]. The -OH of glucose moiety and the -OH of the flavonoid A-ring can interact to form hydrogen bonds [28], resulting in electron distributions and polarities that depend on the position of glucose bonding to the flavonoid (6-C or 8-C). Therefore, when reversed-phase HPLC is used, the flavone-8-C-glycosides elute faster than the corresponding flavone-6-C-glycosides [29]. The HPLC trace of the CSE also shows that the flavone-8-C-glycosides elute faster than the flavone-6-C-glycosides (Figure 3). The Q3R and rutin isomer sets were also detected in the CSE (Figure 3B).

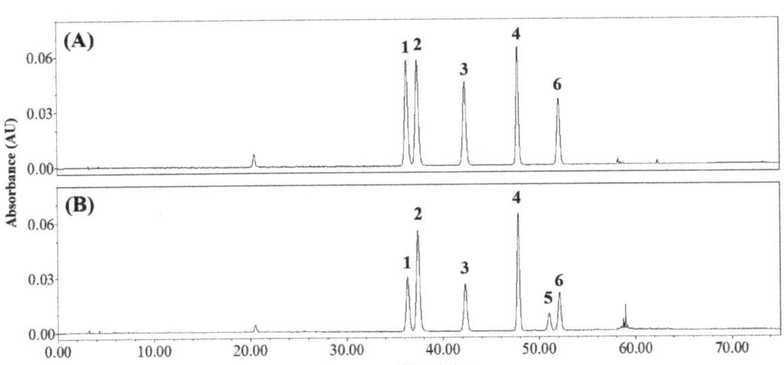

Figure 3. Reversed-phase HPLC traces of (**A**) standards and (**B**) common buckwheat sprout extract. Peak number: 1, orientin; 2, isoorientin; 3, vitexin; 4, isovitexin; 5, quercetin-3-*O*-robinobioside; 6, rutin.

The UV spectra corresponding to the peaks of flavonoids in CSE separated by HPLC-PDA are shown in Figure 4. Band I, which is observed at 300–380 nm, is due to electron transitions in the cinnamoyl group (B-ring of flavonoid backbone), while band II (240–280 nm) corresponds to electron transitions of the benzoyl group (A-ring of flavonoid backbone) [30]. Figure 4A,B show that orientin, isoorientin, vitexin, and isovitexin have the same absorption maximum for bands II (λ_{max} = 267.1 nm). The sugar attached at the 6-C or 8-C position of the benzoyl group does not affect the UV spectra. Orientin and isoorientin exhibit band I maxima at 348.4 nm, while those of vitexin and isovitexin are evident at 334.0 nm (Figure 4A,B). Q3R showed band I maxima similar to rutin, while band II maxima was quite different (Figure 4C). The UV spectra of flavonoids are affected by a variety of functional groups (methyl, glycosyl, and hydroxyl) and other molecular features (2,3 double bond and/or a ketone in the C-ring) [31]. For example, -OH groups coupled to the B-ring systematically cause a red-shift of the λ_{max} of band I [31], with electron delocalization contributing these observations. Consequently, the band I λ_{max} values of orientin and isoorientin (3′,4′-OH) are higher than those of vitexin and isovitexin (3′-OH). Q3R and rutin exhibit higher band I values (λ_{max} = 353.2 nm) than the flavones (λ_{max} = 348.4, 334.0 nm) and the 3-OH present in the flavonol structure reduces π conjugation through stabilization [31].

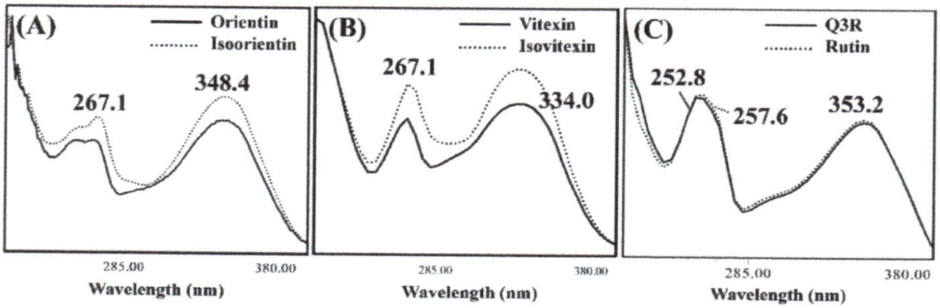

Figure 4. UV spectra of (**A**) orientin and isoorientin, (**B**) vitexin and isovitexin, and (**C**) quercetin-3-*O*-robinobioside (Q3R) and rutin acquired with the HPLC-PDA (photodiode array).

3.3. Validating the Applied Method

3.3.1. Linearity and Range

Calibration curves based on six concentrations were constructed over the 6.25–200.00 μg mL^{-1} range (Table 2). We analyzed five components, however Q3R was not analyzed because no validated standard compound exists. According to guidelines for the validation of analytical procedures for compounds, such as medicines, published by the Ministry of Food and Drug Safety, a response is considered to be linear if the correlation coefficient (R) of the calibration curve exceeds 0.9900 [23]. All calibration curves in this study exhibited good linearities with correlation coefficients over 0.9999, indicating that the responses to the external standards in this study are suitably linear. Therefore, we conclude that a concentration range of 6.25 μg mL^{-1} or higher ensures good linearity when quantifying flavonoids in CSE.

Table 2. Calibration curves constructed from standard flavonoid solutions.

Flavonoid	Range (μg mL^{-1})	Calibration Curve	Correlation Coefficient (R)
Orientin	6.25–200.00	y = 13074.0x + 3111.6	0.9999
Isoorientin	6.25–200.00	y = 13895.0x + 1876.0	0.9999
Vitexin	6.25–200.00	y = 10604.0x + 4006.4	0.9999
Isovitexin	6.25–200.00	y = 11450.0x + 6941.8	0.9999
Rutin	6.25–200.00	y = 7751.5x + 4289.6	0.9999

3.3.2. Precision, Accuracy, and Recovery

Validation results, including accuracy and precision, are listed in Table 3. Accuracy and precision were measured for each component using three spiked base-solvent replicates at concentrations of 25, 50, and 100 μg mL^{-1}. RSDs of retention times were less than 0.2%. Accuracy is an indication of the degree to which the measured value is close to the true value. The relative area of the main component is set to 100% and the amount of analyte is calculated and expressed as a percent recovery. Values of recovery that ranged between 96.67% and 103.60% were measured for the five components (Table 3). All percent recoveries were within in the 95% confidence interval criterion and were highly accurate. A higher concentration of a spiked standard (25,100 μg mL^{-1}) in the blank solvent resulted recovery closer to 100%. We conclude that the method provides values that are close to the true values and is therefore suitable for accurate analysis.

Precision was assessed by the intra-day repeatability and inter-day reproducibility on three different days, and is expressed as a RSD. The RSDs of intra-day repeatability measured three times in a single day were found to be 2.67–5.21% (Table 3). Based on the RSD values of the five flavonoids measured in this study, we showed that the developed analysis method is consistent and precise [32,33].

3.3.3. Limits of Detection and Quantification

The five standard components were analyzed over the 0.61–20.00 μg mL^{-1} range in order to determine LODs and LOQs, which were calculated from the standard deviation of the response and the slope of the calibration curve (Table 3). LOD and LOQ are used to verify the capacity of a method to detect and quantify materials at low concentrations [23]. The results show the concentrations used in the standard curves are appropriate and consistent with previous validation studies [34,35]. These results show that the new method is very sensitive toward the quantification of flavonoids in CSE.

3.4. *Applying the Analytical Method to CSE*

We verified the method using CSE, the results of which are listed in Table 4. The developed HPLC method was used to analyze eight CSEs produced at different batches. All samples were prepared in the same manner (described above). The flavonoids in each extract were detected in the elution order with an increasing retention time as follows: orientin, isoorientin, vitexin, isovitexin, Q3R, and rutin. Except for Q3R, which was detected only in the extract, the t_R values of the other five components are consistent with those listed in Table 3. Q3R was detected at 51.1 min, eluting between isovitexin and rutin. All samples exhibited satisfactory validation data, with the calculated LODs and LOQs listed in Table 3. The flavonoid content in the CSEs was calculated based on the standard curves provided in Table 2. There were differences in content per growth lot, but the results show that the content of 6-*C*-glycosyl flavones is higher than that of the flavone-8-*C*-glycosides in the CSEs, as reported previously [11,12]. The results also show that there was more rutin than Q3R in CSEs, in line with reported results [11]. Similar trends were observed despite differences in sowing times and cultivation environments. Hence, the validated method was successfully applied to various CSEs.

Table 3. Validation data for the simultaneous analysis of flavonoids in common buckwheat sprout extract using HPLC.

Flavonoid	t_R [1] (min)	RSD [2] (%) of t_R	LOD [3] (μg mL^{-1})	LOQ [4] (μg mL^{-1})	Spiking Level (μg mL^{-1})	Accuracy (%)		Precision (RSD; %)	
						Recovery (n = 3)	Confidence Interval (95%)	Intra-Day Repeatability (n = 3)	Inter-Day Reproducibility (n = 3 × 3 Days)
Orientin	36.4	0.13	0.32	0.96	25	97.10	96.91–103.59	3.74	4.04
					50	102.96		3.78	3.13
					100	100.69		2.80	4.30
Isoorientin	37.5	0.14	0.09	0.26	25	97.05	96.87–103.54	3.33	3.56
					50	102.88		4.18	2.85
					100	100.69		2.67	4.35
Vitexin	42.4	0.12	0.26	0.77	25	96.67	96.48–103.83	3.84	4.55
					50	103.09		3.69	3.46
					100	100.70		2.72	4.22
Isovitexin	47.9	0.07	0.29	0.88	25	97.12	96.84–104.20	3.61	4.13
					50	103.60		3.76	3.44
					100	100.83		2.69	4.24
Rutin	52.1	0.10	0.42	1.28	25	97.25	96.97–103.90	5.21	5.40
					50	103.35		4.25	3.57
					100	100.71		2.96	4.03

[1] t_R, retention time. [2] RSD, relative standard deviation. [3] LOD, limit of detection. [4] LOQ, limit of quantification.

Table 4. Application of the developed HPLC method to common buckwheat sprout extracts.

		Orientin	Isoorientin	Vitexin	Isovitexin	Q3R [1]	Rutin
t_R [2] (min)		36.4	37.5	42.4	47.9	51.1	52.1
Content (mg g^{-1} DW [4])	No. 1	6.63 ± 0.17 ^{d,3}	12.63 ± 0.21 ^{a}	5.90 ± 0.11 ^{e}	10.48 ± 0.18 ^{b}	3.50 ± 0.06 ^{f}	9.98 ± 0.21 ^{c}
	No. 2	6.99 ± 0.04 ^{d}	13.37 ± 0.20 ^{a}	6.69 ± 0.32 ^{d}	11.62 ± 0.25 ^{b}	4.12 ± 0.07 ^{e}	10.53 ± 0.23 ^{c}
	No. 3	7.02 ± 0.08 ^{d}	13.48 ± 0.29 ^{a}	6.89 ± 0.23 ^{d}	12.14 ± 0.29 ^{b}	3.69 ± 0.04 ^{e}	10.74 ± 0.44 ^{c}
	No. 4	7.64 ± 0.11 ^{e}	14.11 ± 0.27 ^{b}	9.50 ± 0.06 ^{c}	16.59 ± 0.22 ^{a}	3.42 ± 0.07 ^{f}	8.60 ± 0.08 ^{d}
	No. 5	8.16 ± 0.44 ^{c}	15.24 ± 0.71 ^{a}	7.93 ± 0.24 ^{c}	13.83 ± 0.68 ^{a}	4.56 ± 0.64 ^{d}	12.18 ± 0.53 ^{b}
	No. 6	8.90 ± 0.22 ^{c}	16.67 ± 0.33 ^{a}	9.84 ± 0.25 ^{b}	17.09 ± 0.50 ^{a}	3.65 ± 0.03 ^{d}	9.22 ± 0.11 ^{bc}
	No. 7	7.33 ± 0.20 ^{c}	14.00 ± 0.62 ^{a}	7.29 ± 0.26 ^{c}	12.92 ± 0.44 ^{a}	4.45 ± 0.25 ^{d}	11.74 ± 0.44 ^{b}
	No. 8	6.99 ± 0.77 ^{b}	13.27 ± 1.28 ^{a}	7.26 ± 0.72 ^{b}	13.34 ± 1.02 ^{a}	3.90 ± 0.46 ^{c}	10.70 ± 1.05 ^{b}
	No. 9	8.48 ± 0.26 ^{d}	15.83 ± 0.46 ^{b}	9.65 ± 0.32 ^{c}	16.80 ± 0.50 ^{a}	4.21 ± 0.13 ^{e}	9.84 ± 0.28 ^{c}

[1] Q3R, quercetin-3-*O*-robinobioside. [2] t_R, retention time. [3] Means with different superscripts in the same row indicate significant differences ($p < 0.05$) by Tukey's test. [4] DW, dry weight.

4. Conclusions

The HPLC method developed and optimized in this study is capable of concurrently analyzing flavonoids in CSE. Validation data, namely linearity, precision, percentage of recovery, LOD, and LOQ were found to be adequate for the intended purpose. The method was successfully used to quantify and qualify CSE on the basis of retention times and UV spectra. The developed method was very effective for the analysis of flavone-6-*C*-glycosides and flavone-8-*C*-glycosides in CSE. In addition, the method can also separate stereoisomers of flavonol glycosides bearing different sugars (glucose or galactose). The broad versatility and good efficiency is a valuable feature of this simultaneous analysis method. The utility of the method can be further extended to the simultaneous analysis of flavone and flavonol isomers present in food.

Supplementary Materials: The following are available online at http://www.mdpi.com/2304-8158/8/11/549/s1, Figure S1: Structures of flavonoids in common buckwheat sprouts, Table S1: Resolution of common buckwheat sprout extract analyzed at: (A) 20 °C, (B) 30 °C, and (C) 40 °C, Table S2. Resolution of common buckwheat sprout extract analyzed on flow rate of: (A) 0.6, (B) 0.8, and (C) 1.0 mL min^{-1}.

Author Contributions: Conceptualization, D.-O.K.; Funding acquisition, S.E.O.; Investigation, D.J., M.-S.K., S.E.O., and T.G.N.; Methodology, D.J., Y.S.J., and T.G.N.; Supervision, D.-O.K.; Validation, D.J., Y.S.J., and M.-S.K.; Writing — original draft, D.J., and Y.S.J.; Writing, review, and editing, T.G.N., and D.-O.K.

Funding: This research was supported by the Main Research Program (E0162502-04) of the Korea Food Research Institute funded by the Ministry of Science and ICT.

Conflicts of Interest: The authors declare no conflict of interest.

References

1. Zhang, Z.-L.; Zhou, M.-L.; Tang, Y.; Li, F.-L.; Tang, Y.-X.; Shao, J.-R.; Xue, W.-T.; Wu, Y.-M. Bioactive compounds in functional buckwheat food. *Food Res. Int.* **2012**, *49*, 389–395. [CrossRef]
2. Giménez-Bastida, J.A.; Zieliński, H. Buckwheat as a functional food and its effects on health. *J. Agric. Food Chem.* **2015**, *63*, 7896–7913. [CrossRef] [PubMed]
3. Nam, T.G.; Kim, D.-O.; Eom, S.H. Effects of light sources on major flavonoids and antioxidant activity in common buckwheat sprouts. *Food Sci. Biotechnol.* **2018**, *27*, 169–176. [CrossRef] [PubMed]
4. Kim, S.-L.; Kim, S.-K.; Park, C.-H. Introduction and nutritional evaluation of buckwheat sprouts as a new vegetable. *Food Res. Int.* **2004**, *37*, 319–327. [CrossRef]
5. Park, C.H.; Yeo, H.J.; Park, Y.E.; Chun, S.W.; Chung, Y.S.; Lee, S.Y.; Park, S.U. Influence of chitosan, salicylic acid and jasmonic acid on phenylpropanoid accumulation in germinated buckwheat (*Fagopyrum esculentum* Moench). *Foods* **2019**, *8*, 153. [CrossRef] [PubMed]
6. Lin, L.-Y.; Peng, C.-C.; Yang, Y.-L.; Peng, R.Y. Optimization of bioactive compounds in buckwheat sprouts and their effect on blood cholesterol in hamsters. *J. Agric. Food Chem.* **2008**, *56*, 1216–1223. [CrossRef] [PubMed]
7. Kim, S.-J.; Zaidul, I.S.M.; Suzuki, T.; Mukasa, Y.; Hashimoto, N.; Takigawa, S.; Noda, T.; Matsuura-Endo, C.; Yamauchi, H. Comparison of phenolic compositions between common and tartary buckwheat (*Fagopyrum*) sprouts. *Food Chem.* **2008**, *110*, 814–820. [CrossRef]

8. Nam, T.-G.; Lee, S.M.; Park, J.-H.; Kim, D.-O.; Baek, N.-I.; Eom, S.H. Flavonoid analysis of buckwheat sprouts. *Food Chem.* **2015**, *170*, 97–101. [CrossRef]
9. Liu, C.-L.; Chen, Y.-S.; Yang, J.-H.; Chiang, B.-H. Antioxidant activity of tartary (*Fagopyrum tataricum* (L.) Gaertn.) and common (*Fagopyrum esculentum* Moench) buckwheat sprouts. *J. Agric. Food Chem.* **2008**, *56*, 173–178. [CrossRef]
10. Ishii, S.; Katsumura, T.; Shiozuka, C.; Ooyauchi, K.; Kawasaki, K.; Takigawa, S.; Fukushima, T.; Tokuji, Y.; Kinoshita, M.; Ohnishi, M.; et al. Anti-inflammatory effect of buckwheat sprouts in lipopolysaccharide-activated human colon cancer cells and mice. *Biosci. Biotechnol. Biochem.* **2008**, *72*, 3148–3157. [CrossRef]
11. Nam, T.G.; Lim, T.-G.; Lee, B.H.; Lim, S.; Kang, H.; Eom, S.H.; Yoo, M.; Jang, H.W.; Kim, D.-O. Comparison of anti-inflammatory effects of flavonoid-rich common and tartary buckwheat sprout extracts in lipopolysaccharide-stimulated RAW 264.7 and peritoneal macrophages. *Oxid. Med. Cell. Longev.* **2017**, *2017*, 9658030. [CrossRef] [PubMed]
12. Lee, M.-K.; Park, S.-H.; Kim, S.-J. A time-course study of flavonoids in buckwheats (*Fagopyrum* species). *Korean J. Agric. Sci.* **2011**, *38*, 87–94. [CrossRef]
13. Ozga, J.A.; Saeed, A.; Wismer, W.; Reinecke, D.M. Characterization of cyanidin-and quercetin-derived flavonoids and other phenolics in mature saskatoon fruits (*Amelanchier alnifolia* Nutt.). *J. Agric. Food Chem.* **2007**, *55*, 10414–10424. [CrossRef] [PubMed]
14. Pawlowska, A.M.; Camangi, F.; Bader, A.; Braca, A. Flavonoids of *Zizyphus jujuba* L. and *Zizyphus spina-christi* (L.) Willd (Rhamnaceae) fruits. *Food Chem.* **2009**, *112*, 858–862. [CrossRef]
15. Patel, O.; Muller, C.; Joubert, E.; Louw, J.; Rosenkranz, B.; Awortwe, C. Inhibitory interactions of *Aspalathus linearis* (rooibos) extracts and compounds, aspalathin and Z-2-(β-D-glucopyranosyloxy)-3-phenylpropenoic acid, on cytochromes metabolizing hypoglycemic and hypolipidemic drugs. *Molecules* **2016**, *21*, 1515. [CrossRef]
16. Pereira, C.A.M.; Yariwake, J.H.; McCullagh, M. Distinction of the *C*-glycosylflavone isomer pairs orientin/isoorientin and vitexin/isovitexin using HPLC-MS exact mass measurement and in-source CID. *Phytochem. Anal.* **2005**, *16*, 295–301. [CrossRef]
17. Willems, J.L.; Khamis, M.M.; Saeid, W.M.; Purves, R.W.; Katselis, G.; Low, N.H.; El-Aneed, A. Analysis of a series of chlorogenic acid isomers using differential ion mobility and tandem mass spectrometry. *Anal. Chim. Acta* **2016**, *933*, 164–174. [CrossRef]
18. Geng, P.; Sun, J.; Zhang, M.; Li, X.; Harnly, J.M.; Chen, P. Comprehensive characterization of *C*-glycosyl flavones in wheat (*Triticum aestivum* L.) germ using UPLC-PDA-ESI/HRMS^n and mass defect filtering. *J. Mass Spectrom.* **2016**, *51*, 914–930. [CrossRef]
19. Stalikas, C.D. Extraction, separation, and detection methods for phenolic acids and flavonoids. *J. Sep. Sci.* **2007**, *30*, 3268–3295. [CrossRef]
20. Yáñez, J.A.; Andrews, P.K.; Davies, N.M. Methods of analysis and separation of chiral flavonoids. *J. Chromatogr. B* **2007**, *848*, 159–181. [CrossRef]
21. Xu, F.; Liu, Y.; Zhang, Z.; Yang, C.; Tian, Y. Quasi-MS^n identification of flavanone 7-glycoside isomers in *Da Chengqi Tang* by high performance liquid chromatography-tandem mass spectrometry. *Chin. Med.* **2009**, *4*, 15. [CrossRef] [PubMed]
22. Choi, S.-H.; Ahn, J.-B.; Kim, H.-J.; Im, N.-K.; Kozukue, N.; Levin, C.E.; Friedman, M. Changes in free amino acid, protein, and flavonoid content in jujube (*Ziziphus jujube*) fruit during eight stages of growth and antioxidative and cancer cell inhibitory effects by extracts. *J. Agric. Food Chem.* **2012**, *60*, 10245–10255. [CrossRef] [PubMed]
23. Ministry of Food and Drug Safety. *Analytical Method Guideline about Validation of Drugs and etc.*; Ministry of Food and Drug Safety: Cheongju, Korea, 2015; pp. 1–51. Available online: http://www.nifds.go.kr/brd/m_15/view.do?seq=10030&srchFr=&srchTo=&srchWord=&srchTp=&itm_seq_1=0&itm_seq_2=0&multi_itm_seq=0&company_cd=&company_nm=&page=45 (accessed on 7 May 2019).
24. Xie, R.; Tu, M.; Wu, Y.; Adhikari, S. Improvement in HPLC separation of acetic acid and levulinic acid in the profiling of biomass hydrolysate. *Bioresour. Technol.* **2011**, *102*, 4938–4942. [CrossRef] [PubMed]
25. Welch, C.J.; Brkovic, T.; Schafer, W.; Gong, X. Performance to burn? Re-evaluating the choice of acetonitrile as the platform solvent for analytical HPLC. *Green Chem.* **2009**, *11*, 1232–1238. [CrossRef]

26. Giegold, S.; Teutenberg, T.; Tuerk, J.; Kiffmeyer, T.; Wenclawiak, B. Determination of sulfonamides and trimethoprim using high temperature HPLC with simultaneous temperature and solvent gradient. *J. Sep. Sci.* **2008**, *31*, 3497–3502. [CrossRef] [PubMed]
27. Plumb, R.; Mazzeo, J.R.; Grumbach, E.S.; Rainville, P.; Jones, M.; Wheat, T.; Neue, U.D.; Smith, B.; Johnson, K.A. The application of small porous particles, high temperatures, and high pressures to generate very high resolution LC and LC/MS separations. *J. Sep. Sci.* **2007**, *30*, 1158–1166. [CrossRef]
28. Praveena, R.; Sadasivam, K.; Deepha, V.; Sivakumar, R. Antioxidant potential of orientin: A combined experimental and DFT approach. *J. Mol. Struct.* **2014**, *1061*, 114–123. [CrossRef]
29. Andersen, O.M.; Markham, K.R. Flavonoids: Chemistry, biochemistry, and applicaions. In *Separation and quantification of flavonoids*; Marston, A., Hostettmann, K., Eds.; CRC Press: Boca Raton, FL, USA, 2005; pp. 1–36.
30. Naseem, B.; Sabri, A.; Hasan, A.; Shah, S.S. Interaction of flavonoids within organized molecular assemblies of anionic surfactant. *Colloids Surf. B Biointerface* **2004**, *35*, 7–13. [CrossRef]
31. Anouar, E.H.; Gierschner, J.; Duroux, J.-L.; Trouillas, P. UV/Visible spectra of natural polyphenols: A time-dependent density functional theory study. *Food Chem.* **2012**, *131*, 79–89. [CrossRef]
32. Jovanov, P.; Guzsvány, V.; Lazić, S.; Franko, M.; Sakač, M.; Šarić, L.; Kos, J. Development of HPLC-DAD method for determination of neonicotinoids in honey. *J. Food Compos. Anal.* **2015**, *40*, 106–113. [CrossRef]
33. Lu, Q.-Y.; Lee, R.-P.; Huang, J.; Yang, J.; Henning, S.M.; Hong, X.; Heber, D.; Li, Z. Quantification of bioactive constituents and antioxidant activity of Chinese yellow wine. *J. Food Compos. Anal.* **2015**, *44*, 86–92. [CrossRef]
34. Coelho, E.M.; da Silva Padilha, C.V.; Miskinis, G.A.; de Sá, A.G.B.; Pereira, G.E.; de Azevêdo, L.C.; Lima, M.D.S. Simultaneous analysis of sugars and organic acids in wine and grape juices by HPLC: Method validation and characterization of products from northeast Brazil. *J. Food Compos. Anal.* **2018**, *66*, 160–167. [CrossRef]
35. Calabrò, M.L.; Galtieri, V.; Cutroneo, P.; Tommasini, S.; Ficarra, P.; Ficarra, R. Study of the extraction procedure by experimental design and validation of a LC method for determination of flavonoids in *Citrus bergamia* juice. *J. Pharm. Biomed. Anal.* **2004**, *35*, 349–363. [CrossRef]

foods

MDPI

Article

Changes in Polyphenolics during Storage of Products Prepared with Freeze-Dried Wild Blueberry Powder

Laura Lavefve [1], Cindi Brownmiller [1], Luke Howard [1,*], Donovon Reeves [1], Sean H. Adams [2,3], Jin-Ran Chen [2,3], Eva C. Diaz [2,3,4] and Andy Mauromoustakos [5]

1 Department of Food Science, University of Arkansas, 2650 North Young Avenue, Fayetteville, AR 72704, USA; ldlavefv@uark.edu (L.L.); cbrownm@uark.edu (C.B.); dxr017@uark.edu (D.R.)
2 Department of Pediatrics, University of Arkansas for Medical Sciences, Little Rock, AR 72202, USA; shadams@uams.edu (S.H.A.); chenjinran@uams.edu (J.-R.C.); ECDiazfuentes@uams.edu (E.C.D.)
3 Arkansas Children's Nutrition Center, 15 Children's Way, Little Rock, AR 72202, USA
4 Arkansas Children's Research Institute, 1 Children's Way, Little Rock, AR 72202, USA
5 Agricultural Statistics Lab, 104 Agricultural Annex, University of Arkansas, Fayetteville, AR 72701, USA; amavro@uark.edu
* Correspondence: lukeh@uark.edu; Tel.: +47-9575-29-78; Fax: +47-9575-69-36

Received: 16 March 2020; Accepted: 6 April 2020; Published: 9 April 2020

check for updates

Abstract: Wild blueberry (WBB) powder can be added to the formulation of foods to encourage consumption of health-promoting polyphenolics, but the stability of polyphenolics throughout storage is important. We determined the stability of polyphenolics in five products (ice pop, oatmeal bar, graham cracker cookie, juice, and gummy product) prepared with WBB powder. Samples stored at 21 °C, 4.4 °C, or −20 °C (ice pops only) were analyzed at 0, 2, 4, 6, and 8 weeks for polyphenolic content and percent polymeric color. Total anthocyanins decreased over storage and storage temperatures in all products. However, the ice pop and the refrigerated juice both retained over 90% of their initial total anthocyanin content. The refrigerated oatmeal bar also showed good retention of anthocyanins (86%), but the gummy product retained only 43% and 51% when stored at 4.4 °C or 21 °C, respectively. The lower amount of polyphenolic compounds recovered in the gummies stored at 4.4 °C compared to 21 °C may be attributed to reduced extraction efficiency as a result of gel hardening at refrigerated temperature. Chlorogenic acid and flavonols were generally more stable than anthocyanins throughout storage.

Keywords: anthocyanin; chlorogenic acid; flavonol; polymeric color; storage; wild blueberry

1. Introduction

Lowbush "wild" blueberries (*Vaccinium angustifolium* Ait) are considered a nutrient-rich healthy food, due in large part to their exceptional phenolic content [1,2] and antioxidant activity [2]. Lowbush blueberries are particularly rich in anthocyanins and the anthocyanin profile is complex compared with other fruits [3,4]. They contain five of the six anthocyanidins commonly found in nature (delphinidin, cyanidin, petunidin, peonidin, and malvidin), which can have three different sugar moieties attached (galactose, glucose, and arabinose) as well as acyl groups such as acetyl-, malonyl-, or coumaryl- also attached to the sugar moieties [4,5]. Blueberries are also rich in proanthocyanidins [1,6], chlorogenic acid [7,8], and flavonols [7,9].

Diets rich in blueberries or their polyphenolic-rich extracts have been associated with lower cardiovascular risk, weight gain and metabolic syndrome, and neurological diseases (reviewed in [10]). In addition, studies involving blueberries have identified polyphenolic-derived phenolic acids that improve cell differentiation and proliferation of osteoblasts in vitro and promote bone growth and limit bone loss in rodents [11–13]. These health-promoting effects are due to a myriad of mechanisms associated with blueberry polyphenolics, including prevention of oxidative stress and inflammation, and vaso- and lipid modulation [14–16]. Many human studies reporting positive health outcomes have used freeze-dried wild blueberry (WBB) powder [17–21], which is a natural source of concentrated polyphenolics. However, the freeze-dried WBB powder may be tart or astringent and not always palatable to consume. This can be problematic in feeding trials in children and adults. In our previous work, we developed five food products (gummy, oatmeal bar, graham cracker cookie, juice, and ice pop) prepared with freeze-dried WBB powder that were evaluated for children's acceptability and desire to eat [22]. These results are useful in designing food products as well as menu items that could be used in clinical trials of WBB-rich diets. In addition to evaluating sensory properties, it is important to validate the storage stability of polyphenolics in these products, before use in clinical trials, to ensure that a consistent dose of polyphenolics can be maintained. Blueberry polyphenolics, especially anthocyanins, are unstable in various processed forms such as juices, jams, purees, and canned berries when stored at ambient temperature [23–25]. Additionally, anthocyanins in freeze-dried WBB powder are susceptible to degradation when stored at ambient temperature with a reported half-life of 139 days at 25 °C [26].

The mechanism responsible for loss of anthocyanins during storage is unknown, but anthocyanin losses are commonly accompanied by increased polymeric color values, suggesting that anthocyanins form polymers with proanthocyanidins [27]. In addition to polymerization, many other factors can affect the stability of anthocyanins including exposure to elevated temperatures, light, oxygen, metals, sugars, and ascorbic acid [28]. At present, refrigeration of blueberry products such as jam [24] and juices [29,30] is the best approach to mitigate polyphenolic losses during storage.

This study was undertaken to determine the stability of anthocyanins, flavonols, chlorogenic acid, and percent polymeric color in five blueberry products prepared with freeze-dried WBB powder. Gummy, oatmeal bar, graham cracker cookie, and juice were stored at 21 °C and 4.4 °C and evaluated for anthocyanin, flavonol, and chlorogenic acid content and percent polymeric color over eight weeks of storage. An ice pop product stored at −20 °C was evaluated for its anthocyanin and chlorogenic acid content over eight weeks of storage.

2. Materials and Methods

2.1. Standards and Solvents

HiActive® North American wild blueberry powder was purchased from FutureCeuticals, Inc. (Momence, IL, USA). Rutin, chlorogenic acid, high performance liquid chromatography (HPLC) grade methanol, HPLC grade acetonitrile, potassium metabisulfite, formic acid, acetic acid, chlorogenic acid and rutin were purchased from Sigma-Aldrich (St. Louis, MO, USA). A standard mixture of delphinidin, cyanidin, petunidin, peonidin, pelargonidin, and malvidin glucosides was purchased from Polyphenols (Sandnes, Norway).

2.2. Preparation of Blueberry Products

Samples of juice, ice pop, gummy, oatmeal bar, and graham cracker cookie, each containing 15 g of WBB powder per serving, were prepared and packaged as previously described [22]. One serving of oatmeal bar, ice pop, and graham cracker cookie was equivalent to one piece each (61 g, 121 g and 50 g, respectively), a juice serving was 135 g, and a gummy serving was 7 pieces, or 113 g. The amount of 15 g of WBB powder used in product formulations was calculated and converted from previous animal studies to humans [31]. The graham cracker cookies and oatmeal bars were prepared with minimal thermal treatment. This involved only the use of brief microwave heating to solubilize the ingredients in order to avoid thermal loss of phenolic compounds, but still obtain a ready-to-consume non-baked product. The blueberry juice and ice pop were prepared with an anthocyanin concentrate, previously extracted from the WBB powder [22]. This procedure was used to produce juice and ice pop products with no particulates. The formulation was adjusted with water so the anthocyanin content of the products was equivalent to that found in 15 g of WBB powder per serving. The preparation and processing of the samples for the storage study were performed in two separate experiments, using the same sample of wild freeze-dried blueberries obtained from FutureCeuticals Inc. (Momence, IL, USA). The WBB powder was stored at 15.5 °C for four months between the two experiments. The samples from Experiment 1 were stored at 21 °C and the samples from Experiment 2 were stored at 4.4 °C. The ice pop products prepared in Experiment 1 were stored at −20 °C. Three samples of each packaged product were evaluated at time 0 (immediately after preparation) and after 2, 4, 6, and 8 weeks of storage.

2.3. Extraction of Polyphenolics from Freeze-dried Blueberries and Products

Polyphenolics were extracted by homogenizing 5 g of WBB-containing food product or 1 g of WBB powder in 25 mL of extraction solution containing methanol/water/formic acid (60:37:3 v/v/v), to the smallest particle size using a Euro Turrax T18 Tissuemizer (Tekmar-Dohrman Corp, Mason, OH, USA) for 1 min. Homogenates were centrifuged for 5 min at 10,864 × *g*. The pellet was re-extracted two additional times with 25 mL of extraction solution and centrifuged for 5 min at 10,864 × *g*. The filtrates were pooled and adjusted to 100 mL with extraction solvent in a volumetric flask. Prior to HPLC analysis, 5 mL of extract were dried in a Thermo Savant Speed Vac Plus SC210A (Thermo Fisher Scientific, Waltham, MA, USA) and reconstituted in 1 mL 5% formic acid in water. All samples were passed through 0.45 μm nylon syringe filters (VWR, Radnor, PA, USA) into 1 mL HPLC vials prior to HPLC analysis. The ice pop and juice samples did not undergo extraction due to prior extraction of anthocyanins to make the concentrate used in the formulation but were filtered using the 0.45 μm nylon syringe filters prior to HPLC analysis.

2.4. HPLC Analysis of Anthocyanins and Chlorogenic Acid

Anthocyanins and chlorogenic acid were analyzed by HPLC using the method of Cho and others (2004) [7]. Samples (50 μL) were analyzed using a Waters HPLC system (Waters Corp, Milford, MA, USA) equipped with a model 600 pump, a model 717 Plus autosampler, and a model 996 photodiode array detector. Separation was carried out at room temperature using a 4.6 mm × 250 mm Symmetry C_{18} column (Waters Corp, Milford, MA, USA) preceded by a 3.9 mm × 20 mm Symmetry C_{18} guard column. The mobile phase was a linear gradient of 5% formic acid (A) and methanol (B) from 2% B to 60% B for 60 min at a flow rate of 1 mL/min. The system was equilibrated for 20 min at the initial gradient prior to each injection. Detection wavelengths of 320 nm and 510 nm were used to monitor chlorogenic acid and anthocyanin peaks, respectively. Individual anthocyanin monoglucosides and acylated anthocyanin derivatives were quantified as delphinidin, cyanidin, petunidin, peonidin, and malvidin glucoside equivalents using external calibration curves (3.75, 7.5, 15, 30, 60, 120, 240 mg/L; $R^2 > 0.9977$ for each anthocyanin glucoside) of a mixture of authentic standards (Polyphenols, Sandnes, Norway). Chlorogenic acid was quantified using external calibration curves (4, 8, 16, 32, 64, 128, 256 mg/L; $R^2 = 0.9988$) of an authentic standard (Sigma-Aldrich, St. Louis, MO, USA). Results are expressed as mg of anthocyanin or chlorogenic acid per g of WBB powder.

2.5. HPLC Analysis of Flavonols

Flavonols were analyzed by HPLC using the same HPLC system described above according to the method of Cho et al. (2005) [32]. Separation was performed at room temperature on a 4.6 mm × 250 mm Aqua C_{18} column (Phenomenex, Torrance, CA, USA) preceded by a 3.0 mm × 4.0 mm ODS C_{18} guard column. The mobile phase was a linear gradient of 2% acetic acid (A) and 0.5% acetic acid in water and acetonitrile (50:50 v/v) (B) from 10% B to 55% B in 50 min and from 55% B to 100% B in 10 min at a flow rate of 1 mL/min. The system was equilibrated for 20 min at the initial gradient prior to each injection. A detection wavelength of 360 nm was used to monitor flavonol peaks. Flavonols were quantified as rutin equivalents using an external calibration curve (3.3, 6.6, 13.2, 26.4, 52.8, 105.6, 211.2 mg/L; $R^2 = 0.9999$) of an authentic standard (Sigma-Aldrich, St. Louis, MO, USA), with results expressed as mg of rutin equivalents per g of WBB powder.

2.6. HPLC/ESI-MS Analysis of Polyphenolics

An analytical Hewlett Packard 1100 series HPLC instrument (Palo Alto, CA, USA) equipped with an autosampler, binary HPLC pump, and UV/Vis detector was used. For HPLC/MS analysis, the HPLC apparatus was interfaced to a Bruker model Esquire-LC/MS ion trap mass spectrometer (Billerica, MA, USA). Mass spectral data were collected with the Bruker software (Bruker Co., DataAnalysis version 4.0, Billerica, MA, USA), which also controlled the instrument and collected the signal at 520 nm. Typical conditions for mass spectral analysis conducted in positive-ion electrospray mode for anthocyanins and negative-ion electrospray mode for flavonols included a capillary voltage of 4000 V, a nebulizing pressure of 30.0 psi, a drying gas flow of 9.0 mL/min, and a temperature of 300 °C. Data were collected in full scan mode over a mass range of *m/z* 50–1000 at 1.0 s per cycle. Characteristic ions (*m/z*) were used for peak assignment. For compounds where chemical standards were commercially available, retention times were also used to confirm the identification of components.

2.7. Analysis of Percent Polymeric Color

Percent polymeric color (% PC) of extracts was determined using the spectrophotometric assay of Giusti and Wrolstad (2001) [33]. Sample extracts were diluted with water in order to have an absorbance reading between 0.5 and 1.0 at 512 nm when evaluated by an 8452A Diode Array Spectrophotometer (Hewlett Packard, Palo Alto, CA, USA). For analysis, 0.2 mL of 0.90 M potassium metabisulfite was added to 2.8 mL diluted sample (bisulfite bleached sample) and 0.2 mL of DI water was added to 2.8 mL diluted sample (non-bleached, control sample). After equilibrating for 15 min, but not more than 1 h, samples were evaluated at λ = 700 nm, 512 nm, and 420 nm. Color density was calculated using the control sample according to the following formula:

$$\text{Color Density} = [(A_{420\text{ nm}} - A_{700\text{ nm}}) + (A_{512\text{ nm}} - A_{700\text{ nm}})] \times \text{Dilution Factor} \quad (1)$$

Polymeric color was determined using the bisulfite-bleached sample using the following formula:

$$\text{Polymeric Color} = [(A_{420\text{ nm}} - A_{700\text{ nm}}) + (A_{512\text{ nm}} - A_{700\text{ nm}})] \times \text{Dilution Factor} \quad (2)$$

Percent polymeric color was calculated using the formula:

$$\%\ \text{Polymeric color} = (\text{polymeric color}/\text{color density}) \times 100 \quad (3)$$

2.8. Statistical Analysis

The effect of storage time (0, 2, 4, 6, and 8 weeks) on anthocyanins, flavonols, chlorogenic acid, and % polymeric color in each blueberry product was evaluated using the Fit Model platform of JMP (JMP Pro, version 15, SAS Institute, Cary, NC, USA)), and the percent retention of each compound after 8 weeks of storage was calculated using the fit model equation. The effect of storage temperature on phenolic compounds stability was not evaluated in this study due to the length of time (4 months) the WBB powder was stored between processing the products in Experiment 1 (products stored at 21 °C) and Experiment 2 (products stored at 4.4 °C). During this four-month storage time, the powder stored at 15.5 °C presumably absorbed moisture evident by powder clumping, resulting in different amounts of polyphenolics in the products immediately after processing. Principal component analysis (PCA) was performed with the total and individual anthocyanins variables, using the Multivariate platform in JMP, on the mean value (n = 3) of each sample per time point and using the correlation method. Correlations among total anthocyanins and percent polymeric color were determined by pairwise correlations method in the multivariate platform of JMP.

3. Results

3.1. Identification of Anthocyanins by HPLC/MS

The WBB powder used to prepare the products contained at least 22 anthocyanins (Table 1), which were identified by comparing their mass-to-charge (*m/z*) values and elution orders with previous studies [4,5,34]. Blueberries are unique in that three different sugars (galactose, glucose, arabinose) are commonly attached to the five anthocyanidins (delphinidin, cyanidin, petunidin, peonidin, malvidin) [34]. This was confirmed in our study; however, we were unable to detect peonidin-3-arabinoside using our HPLC method. We were unable to obtain complete separation of all of the anthocyanins present in the extract due to the complexity of the anthocyanin profile. Peak 15 contained two co-eluting compounds, namely cyanidin-3-(6′′-malonyl) galactoside and cyanidin-3-(6′′-acetyl) galactoside, and peak 18 was composed of three co-eluting compounds, namely delphinidin-3-rutinoside, cyanidin-3-(6′′-malonyl) glucoside, and malvidin-3-(6′′-acetyl) galactoside. We were unable to identify peak 17, which appeared to be a delphinidin derivative based on its aglycone *m/z* of 303, but the molecular ion *m/z* value was ambiguous. Many of the anthocyanins were present in acylated form. Two of the cyanidin glycosides (galactoside and glucoside) were acylated with malonic acid, whereas delphinidin, cyanidin, and malvidin galactosides as well as petunidin, peonidin, and malvidin glucosides were acylated with acetic acid moieties.

Table 1. Peak assignments, retention times (RT in min), and mass spectral data of anthocyanins in extract from WBB powder.

Peak	HPLC (RT in min)	Identification	*m/z* [M$^+$]	Fragments
1	24.9	delphinidin-3-galactoside	465	303
2	26.2	delphinidin-3-glucoside	465	303
3	27.1	cyanidin-3-galactoside	449	287
4	28.1	delphinidin-3-arabinoside	435	303
5	28.8	cyanidin-3-glucoside	449	287
6	29.9	petunidin-3-galactoside	479	317
7	30.3	cyanidin-3-arabinoside	419	287
8	31.1	petunidin-3-glucoside	479	317
9	31.8	peonidin-3-galactoside	463	301
10	32.9	petunidin-3-arabinoside	449	317
11	33.2	peonidin-3-glucoside	463	301
12	33.6	malvidin-3-galactoside	493	331
13	35.1	malvidin-3-glucoside	493	331
14	36.6	malvidin-3-arabinoside	463	331
15	37.7	cyanidin-3-(6′′-malonyl) galactoside + cyanidin-3-(6′′-acetyl) galactoside	535 491	287 287
16	40.0	delphinidin-3-(6′′-acetyl) galactoside	507	303
17	41.6	delphinidin derivative	-	303
18	42.3	delphinidin-3-rutinoside malvidin-3-(6′′-acetyl) galactoside cyanidin-3-(6′′-malonyl) glucoside	611 535 535	303 331 287
19	42.7	malvidin-3-(6′′-acetyl) galactoside	535	331
20	43.6	petunidin-3-(6′′-acetyl) glucoside	521	317
21	45.7	peonidin-3-(6′′-acetyl) glucoside	505	301
22	46.6	malvidin-3-(6′′-acetyl) glucoside	535	331

3.2. Identification of Flavonols by HPLC/MS

The WBB powder used to prepare the products contained at least 12 flavonols (Table 2), which were identified by comparing their mass-to-charge (*m/z*) values and elution orders with previous studies [9,32]. The WBB powder contained one syringetin derivative, three myricetin derivatives, and eight quercetin derivatives. We were unable to identify peak 8, which appeared to be a quercetin derivative based on its aglycone *m/z* of 300, but the molecular ion *m/z* of 623 was ambiguous.

Table 2. Peak assignments, retention times (RT in min), and mass spectral data of flavonols in extract from WBB powder.

Peak	HPLC (RT in min)	Identification	*m/z* [M⁻]	Fragments
1	39.2	myricetin-3-galactoside	479	316
2	39.8	myricetin-3-glucoside	479	316
3	43.8	myricetin-3-rhamnoside	463	316
4	44.2	quercetin-3-rutinoside	609	300
5	45.0	quercetin-3-galactoside	463	300
6	45.8	quercetin-3-glucoside	463	300
7	47.0	quercetin-3-glucuronide	477	301
8	48.9	Unknown	623	505, 433, 300
9	50.3	quercetin-3-pentoside	433	300
10	51.0	quercetin-3-rhamnoside	447	300
11	51.1	syringetin-3-galactoside/glucoside	507	344
12	51.8	quercetin-3-(6′′-acetyl) galactoside	505	300

3.3. Stability of Anthocyanins in Blueberry Products during Storage

3.3.1. Ice Pop

The total anthocyanin content of the ice pop over eight weeks of storage at −20 °C is shown in Figure 1. The total amount of anthocyanins significantly decreased with storage time ($p = 0.0056$), but the percent retention remained high with 93% of total anthocyanins retained in the product after eight weeks. Consistent with our results, total anthocyanin content of frozen blueberries was stable over three months of storage at −20 °C [35]. Changes in major individual anthocyanins in the ice pop over eight weeks of storage at −20 °C are shown in Figure S1. Most of the individual anthocyanins did not significantly decrease over storage. For the anthocyanins that decreased during storage, their percent retention after eight weeks remained over 87%: malvidin-3-glucoside (87.3%), malvidin-3-galactoside (94.5%), cyanidin-3-galactoside (91.2%), malvidin-3-(6′′-acetyl) glucoside (91.1%), petunidin-3-glucoside (91.4%).

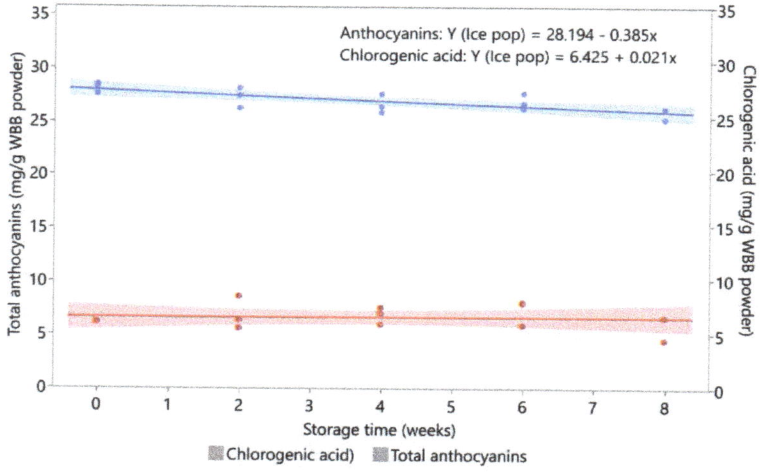

Figure 1. Stability of total anthocyanins and chlorogenic acid in ice pop stored at −20 °C ($n = 3$/time point). Shaded area around lines represents 95% confidence intervals for predicted values.

3.3.2. Oatmeal Bar

The total anthocyanin content of the oatmeal bar decreased with storage time at 4.4 °C ($p = 0.0008$) and 21 °C ($p = 0.0008$). The mean total anthocyanin content of both storage temperatures over eight weeks of storage is shown in Figure 2. After eight weeks of storage, the oatmeal bar retained 86.4% of the total anthocyanins present in the control sample (Day 0) when stored at 4.4 °C and 74.1% when stored at 21 °C. Changes in the major individual anthocyanins in the oatmeal bar over eight weeks of storage are shown in Figure S2. At 4.4 °C, seven individual anthocyanins did not significantly decrease over storage: petunidin-3-(6′′-acetyl) glucoside, petunidin-3-arabinoside, petunidin-3-glucoside, delphinidin-3-(6′′-acetyl) glucoside, cyanidin-3-arabinoside, the two co-eluting compounds (cyanidin-3-(6′′-malonyl) galactoside + cyanidin-3-(6′′-acetyl) galactoside), and the three co-eluting compounds (delphinidin-3-rutinoside + malvidin-3-(6′′-acetyl) galactoside + cyanidin-3-(6′′-malonyl) glucoside). The percent retention values of the anthocyanins over eight weeks of storage at 4.4 °C were all > 75%. When stored at 21 °C, petunidin-3-arabinoside, malvidin-3-(6′′-acetyl) glucoside, the unknown delphinidin derivative, and the two co-eluting compounds (cyanidin-3-(6′′-malonyl) galactoside + cyanidin-3-(6′′-acetyl) galactoside) did not significantly decrease over eight weeks of storage. The percent retention of the other anthocyanins ranged from 68.2% (malvidin-3-glucoside) to 82.8% (petunidin-3-galactoside).

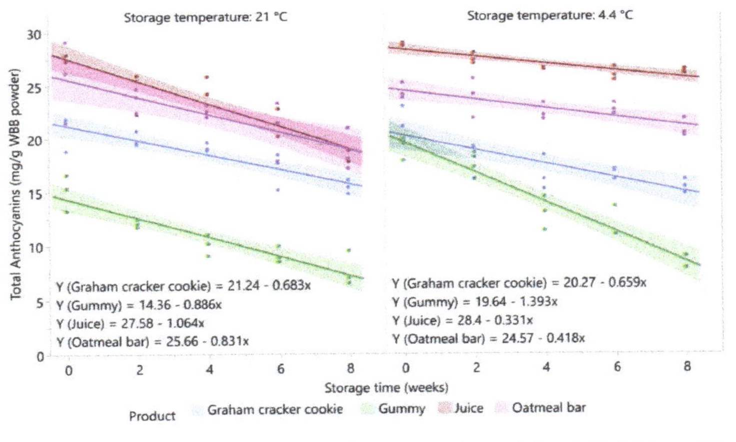

Figure 2. Stability of total anthocyanins in blueberry stored at 21 °C and 4.4 °C (n = 3/time point). Shaded area around lines represents 95% confidence intervals for predicted values.

3.3.3. Graham Cracker Cookie

The total anthocyanin content of the graham cracker cookie decreased with storage time at 4.4 °C ($p = 0.0003$) and 21 °C ($p < 0.0001$). The mean total anthocyanin content of both storage temperatures over eight weeks of storage is shown in Figure 2. After eight weeks of storage, the graham cracker cookie retained about 74% of the total anthocyanins present in the control sample (Day 0) for both storage temperatures. Changes in the major individual anthocyanins in the graham cracker cookie over eight weeks of storage are shown in Figure S3. At both storage temperatures, petunidin-3-arabinoside and the two co-eluting compounds (cyanidin-3-(6′′-malonyl) galactoside + cyanidin-3-(6′′-acetyl) galactoside) did not significantly decrease with storage time. Percent retention of the three delphinidin-3-glycosides (glucoside, galactoside, and arabinoside) was >80% at 4.4 °C. However, the unidentified delphinidin derivative showed only 32.9% retention, while all the other compounds showed retentions >50%. When stored at 21 °C, delphinidin-3-arabinoside and petunidin-3-galactoside showed >80% retentions, while the unknown delphinidin derivative was the only compound showing <50% retention after eight weeks of storage (47.1%).

3.3.4. Juice

The total anthocyanin content of the juice decreased with storage time for each storage temperature ($p < 0.0001$). The total anthocyanin content of juice stored at 4.4 °C and 21 °C is shown in Figure 2. After eight weeks of storage, the juice stored at 4.4 °C retained 90.7% of total anthocyanins compared with control samples (Day 0), whereas the juice stored at 21 °C retained 69.1%. Concentrations of anthocyanins are known to readily decline during storage of blueberry juice at ambient temperature [23,36], but refrigeration is an effective treatment to ameliorate anthocyanin losses [29,37,38]. Changes in the major individual anthocyanins in the juice stored at 4.4 °C and 21 °C over eight weeks of storage are shown in Figure S4. At 4.4 °C, peonidin-3-galactoside, cyanidin-3-arabinoside, malvidin-3-galactoside, malvidin-3-glucoside, and malvidin-3-(6′′-acetyl) galactoside remained stable over the eight weeks of storage. At 4.4 °C, all anthocyanins showed >50% retention, with the minimal percent retention being 57.7% for the unknown delphinidin derivative. This compound, however, did not significantly decrease over storage at 21 °C, along with the two co-eluting anthocyanins (cyanidin-3-(6′′-malonyl) galactoside + cyanidin-3-(6′′-acetyl) galactoside). Besides these two compounds, the percent retention of anthocyanins at 21 °C ranged from 59% (malvidin-3-(6′′-acetyl) glucoside) to 75.5% (petunidin-3-glucoside).

3.3.5. Gummy Product

The total anthocyanin content of the gummy product decreased with storage time for each storage temperature ($p < 0.0001$). The total anthocyanin content of the gummy product stored at 4.4 °C and 21 °C is shown in Figure 2. After eight weeks of storage, the gummy product stored at 4.4 °C and 21 °C retained 43.2% and 50.6%, respectively, of their original total anthocyanin content (Day 0). Consistent with our findings, levels of total anthocyanins declined in gelatin gels prepared with grape pomace extract over 24 weeks of storage at 21 °C, with losses most pronounced in gels exposed to neon light [39]. Maier et al. (2009) [39] also reported similar retention of total anthocyanins in gels stored for 24 weeks at 6 °C and 24 °C. The lower amount of anthocyanins recovered in the gummies stored at 4.4 °C compared to the same product stored at 21 °C may be explained by reduced extraction efficiency due to the hardening of the gel at low temperature, as opposed to degradation late during storage. Changes in the major individual anthocyanins in the gummy product stored at 4.4 °C and 21 °C over eight weeks of storage are shown in Figure S5. At 4.4 °C, all the individual anthocyanins decreased with storage time with retentions <50%, except for the two co-eluting anthocyanins (cyanidin-3-(6′′-malonyl) galactoside + cyanidin-3-(6′′-acetyl) galactoside) (60.9%) and malvidin-3-glucoside (52.1%). The percent retentions of the rest of the anthocyanins at this storage temperature ranged from 29.3% to 49.2%. When stored at 21 °C, two anthocyanins did not significantly decrease with time, namely the unknown delphinidin derivative and the two co-eluting compounds (cyanidin-3-(6′′-malonyl) galactoside + cyanidin-3-(6′′-acetyl) galactoside). For the rest of the anthocyanins, percent retentions ranged from 40% (cyanidin-3-glucoside and cyanidin-3-galactoside) to 71% (malvidin-3-(6′′-acetyl) glucoside).

In all the products, the individual anthocyanin loss did not appear to be impacted by the anthocyanidin structure or the type of sugar moiety attached (data not shown).

3.3.6. Product Comparison of Anthocyanin Composition over Eight Weeks of Storage

The distribution of the products according to their individual anthocyanin profile as affected by storage time (0, 2, 4, 6, and 8 weeks) can be visualized on a PCA scores plot (Figure 3). The first principal component (PC1) explained 83.9% of the variation with all the individual anthocyanins being positively loaded on PC1. Therefore, PC1 represents the amount of individual anthocyanins. The juice and ice pop samples had high scores on PC1 (except for the juice samples stored at 21 °C for 6 and 8 weeks). The oatmeal bar samples also had positive scores on PC1 for the earlier storage times, whereas the oatmeal bar samples stored at 21 °C for eight weeks were the only oatmeal sample to have a negative score. Except for the control samples (Day 0), all the graham cracker cookie samples had negative scores on PC1, regardless of the storage temperature. Finally, all gummy samples (for both storage at 4.4 °C and 21 °C) had negative scores on PC1, with scores becoming smaller with storage time. The PCA figure confirmed higher values of anthocyanins in the juice and ice pop samples, as well as, to a lesser extent, the oatmeal bars. The graham cracker cookie and gummy samples did not demonstrate high values for anthocyanins, with a clear loss of anthocyanins with storage time for the gummy samples.

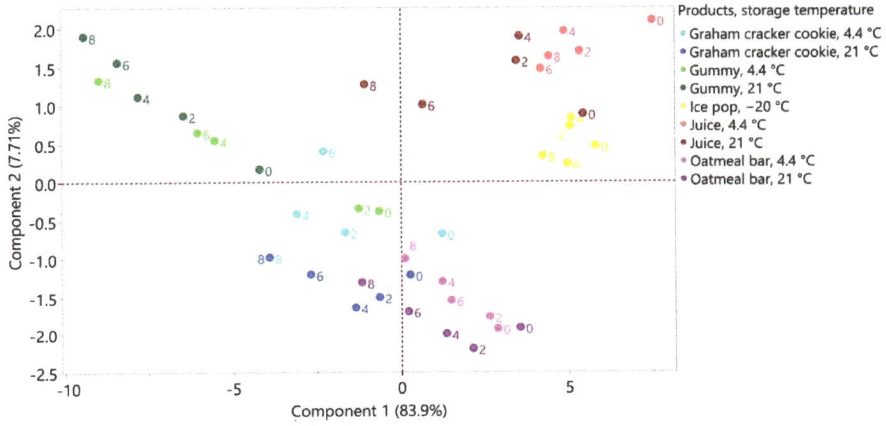

Figure 3. Principal component analysis (PCA) scores plot of blueberry products stored at −20 °C (ice pop), 4.4 °C and 21 °C (graham cracker cookie, gummy, juice, oatmeal bar) for 0, 2, 4, 6, and 8 weeks (n = 3/time point).

3.4. Changes in Percent Polymeric Color in Blueberry Products during Storage

Changes in percent polymeric color (% PC) values in oatmeal bar, graham cracker cookie, juice, and gummy product stored at 4.4 °C and 21 °C over eight weeks of storage are shown in Figure 4.

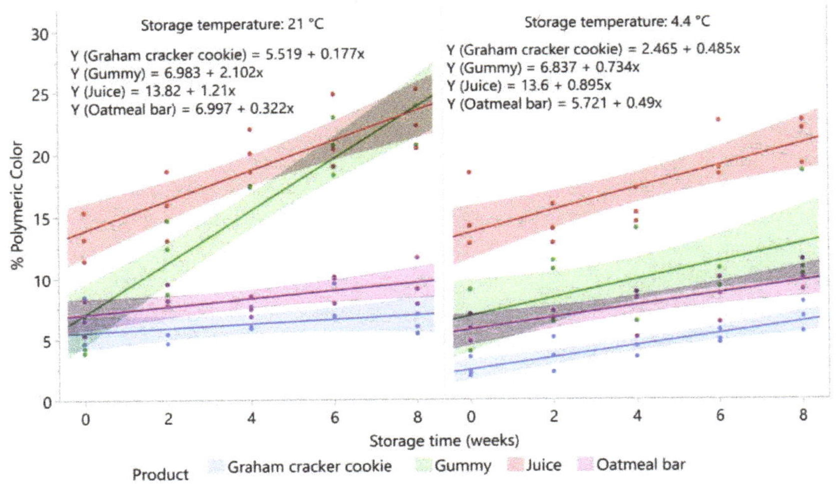

Figure 4. Stability of percent polymeric color in blueberry stored at 21 °C and 4.4 °C (n = 3/time point). Shaded area around lines represents 95% confidence intervals for predicted values.

3.4.1. Oatmeal Bar

Percent polymeric color values of oatmeal bar samples increased over storage at 4.4 °C ($p = 0.0012$) and 21 °C ($p = 0.0173$). For samples stored at 4.4 °C, % PC values increased from 7% at day 0 to 9.6% at eight weeks, and from 5.7% at day 0 to 9.6% at eight weeks in samples stored at 21 °C. Percent polymeric color values showed significant inverse correlations to levels of total anthocyanins at 4.4 °C ($r_{xy} = -0.65$; $p = 0.0081$) and at 21 °C ($r_{xy} = -0.64$; $p = 0.0103$) during storage.

3.4.2. Graham Cracker Cookie

Percent polymeric color values of graham cracker cookie samples significantly increased over storage at 4.4 °C ($p < 0.0001$), but not at 21 °C ($p = 0.2075$). For samples stored at 4.4 °C, % PC values increased from 2.5% at day 0 to 6.3% at eight weeks, but when stored at 21 °C, % PC values remained at 6.2% on average during eight weeks of storage. Percent polymeric color values showed significant inverse correlations to levels of total anthocyanins at 4.4 °C ($r_{xy} = -0.56$; $p = 0.0298$) and 21 °C ($r_{xy} = -0.58$; $p = 0.0237$) during storage.

3.4.3. Juice

Percent polymeric color values of juice samples significantly increased over storage at 4.4 °C ($p = 0.0008$) and 21 °C ($p < 0.0001$). For samples stored at 4.4 °C, % PC values increased from 13.6% at day 0 to 20.8% at eight weeks, and the values increased from 13.8% at day 0 to 23.5% at eight weeks when stored at 21 °C. Percent polymeric color values showed significant inverse correlations to levels of total anthocyanins at 4.4 °C ($r_{xy} = -0.58$; $p = 0.0239$), and more strongly at 21 °C ($r_{xy} = -0.71$; $p = 0.0029$) during storage.

3.4.4. Gummy Product

Percent polymeric color values of gummy samples significantly increased over storage at 4.4 °C ($p = 0.0229$) and 21 °C ($p < 0.0001$). For samples stored at 4.4 °C, % PC values increased from 6.8% at day 0 to 12.7% at eight weeks, and in samples stored at 21 °C, % PC values increased from 6.9% at day 0 to 23.8% at eight weeks of storage. Percent polymeric color values showed a moderate inverse correlation to levels of total anthocyanins at 4.4 °C ($r_{xy} = -0.563$; $P = 0.036$), but a much greater inverse correlation at 21 °C ($r_{xy} = -0.93$; $p < 0.0001$) during storage.

Percent polymeric color values typically show an inverse correlation with total anthocyanins during storage of blueberry products [23,24], and inverse correlations with each individual anthocyanins in all the products and storage temperature (data not shown). Higher percent polymeric color values indicate that a higher percentage of anthocyanins are resistant to bleaching in the presence of potassium metabisulfite. Since the sulfonic acid adduct attaches at C4 on the middle heterocyclic ring, it is thought that anthocyanin–procyanidin polymers are formed via a direct condensation reaction, resulting in a C4–C8 anthocyanin–procyanidin linkage as the major polymers formed in blueberries during storage. Hence, it is possible that declines in anthocyanins during storage of the blueberry products are not true losses due to degradation, but the conversion of monomeric anthocyanins to anthocyanin–procyanidin polymers. Anthocyanins can be degraded via a hydration reaction, where the flavylium ion is converted to a hemiketal structure, which is rapidly converted to *cis*-chalcone, which slowly arranges to a *trans*-chalcone structure [40]. The *trans*-chalcone structure is highly unstable and rapidly degrades to hydroxybenzoic acid derivatives [40]. However, we do not consider that this reaction was responsible for anthocyanin losses in the blueberry products over storage since we did not observe an increase in phenolic acid derivatives in our HPLC chromatograms at 280 nm (data not shown).

3.5. Stability of Chlorogenic Acid during Storage

The stability of chlorogenic acid in the four blueberry products stored at 4.4 °C and 21 °C is shown in Figure 5. Chlorogenic acid was stable in all products over storage regardless of storage temperature, except for the juice and oatmeal bar stored at 4.4 °C, where levels significantly decreased ($p = 0.0249$ and $p = 0.0133$, respectively). At 4.4 °C, the chlorogenic acid content decreased from 4.3 to 3.6 mg/g WBB powder in the juice and from 3.0 to 2.6 mg/g WBB powder in the oatmeal bar. At 21 °C, chlorogenic acid in the juice showed a slight increasing trend; however, this change was not statistically significant ($p = 0.6496$). Chlorogenic acid was also stable in the ice pop over eight weeks of storage at −20 °C ($p = 0.8332$), with an average value of 6.5 mg/g WBB powder over storage (Figure 1). Initial levels of chlorogenic acid were higher in all products stored at 21 °C compared with 4.4 °C storage, which may be due to the variation in processing the two sets of samples for the storage study, or possible the degradation of chlorogenic acid in the WBB powder used to prepare the products. The WBB powder used to prepare samples for the refrigerated storage study was stored at 15.5 °C for three months prior to preparing the samples. Blueberries contain polyphenol oxidase, which can readily oxidize chlorogenic acid [41]. Chlorogenic acid was previously found to be stable in blueberry juice, puree, and canned berries stored for six months at 25 °C [23], but blueberry jams lost 27% of chlorogenic acid over six months of storage at 25 °C [24].

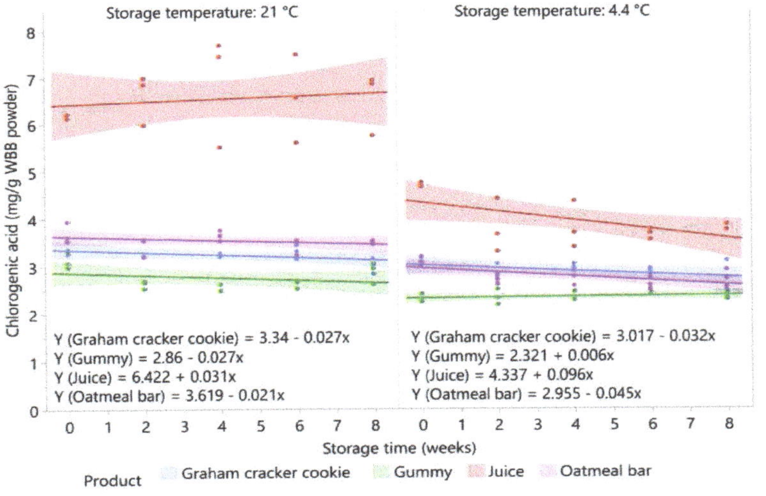

Figure 5. Stability of chlorogenic acid in blueberry stored at 21 °C and 4.4 °C (n = 3/time point). Shaded area around lines represents 95% confidence intervals for predicted values.

3.6. Stability of Flavonols in Blueberry Products during Storage

The stability of total flavonols in the five blueberry products is shown in Figure 6. Total flavonol levels in the oatmeal bar stored at both temperatures and in the gummy product and graham cracker cookie stored at 4.4 °C were stable over eight weeks of storage as well as the juice samples stored at 4.4 °C from two to eight weeks ($p > 0.05$). Flavonols in the juice stored at 21°C were also stable over time despite an upward trend, but the increase was not significant ($p = 0.1935$). Consistent with our findings, total flavonol concentrations were found to be relatively stable (<15% losses) in blueberry jam, juice, puree, and canned berries over six months of storage at 25 °C [24,42].

However, total flavonol content significantly decreased over storage in the gummy product ($p = 0.0085$) and graham cracker cookie ($p = 0.0237$) stored at 21 °C. Total flavonol levels declined by 45.7% and 28.5%, respectively, in these products over eight weeks of storage. In the gummy product, the most marked loss occurred from six to eight weeks of storage. Moisture loss during late storage presumably led to hardening of the gummies, resulting in incomplete extraction of the flavonols.

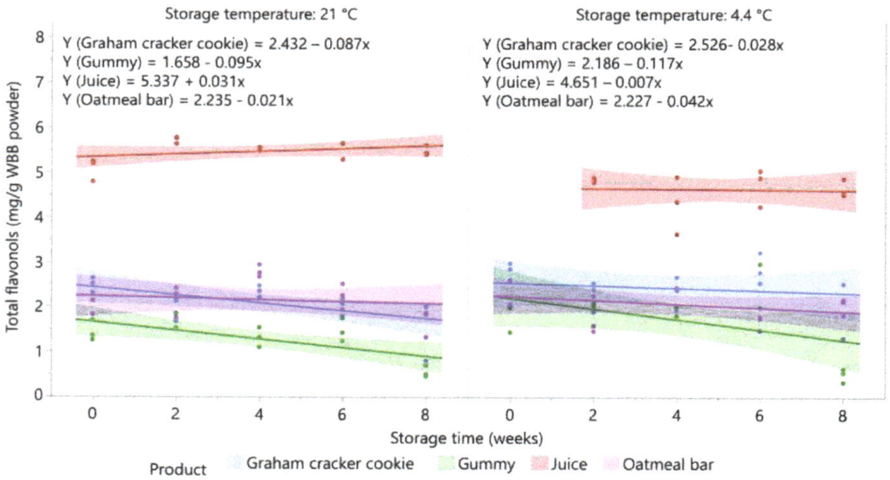

Figure 6. Stability of total flavonols in blueberry stored at 21 °C and 4.4 °C (n = 3/time point). Shaded area around lines represents 95% confidence intervals for predicted values.

Individual flavonols in oatmeal bar and juice were all stable for both storage temperature as well as in graham cracker cookie stored at 4.4 °C (data not shown). In the graham cracker cookie stored at 21 °C, myricetin-3-galactoside was the most unstable flavonol showing 57.8% retention after eight weeks ($p = 0.038$). Quercetin-3-rutinoside ($p = 0.0225$), quercetin-3-galactoside ($p = 0.0195$), quercetin-3-glucoside ($p = 0.037$), quercetin-3-rhamnoside ($p = 0.0058$), quercetin-3-(6′′-acetyl) galactoside ($p = 0.0195$), and syringetin-3-galactoside/glucoside ($p = 0.0237$) were all retained at levels from 61.1% to 70.1%. Four individual flavonols significantly decreased in the gummy product stored at 4.4 °C. The percent retention of quercetin-3-glucuronide ($p = 0.0012$) and quercetin-3-rhamnoside ($p = 0.0016$) were 33% and 38.5%, respectively. In the gummy product stored at 21 °C, only two flavonols (myricetin-3-glucoside and the syringetin derivative) showed no significant decrease over eight weeks of storage. All other flavonols in the gummy product stored at 21 °C were adversely affected by storage time with retentions ranging between 45.2 and 60.1% after eight weeks.

4. Conclusions

The stability of polyphenolics over eight weeks of storage in food products made with WBB powder varied according to product type. Polyphenolic compounds from the ice pop, oatmeal bar, and juice were shown to be stable over storage and are good candidates for further use in applications in which stored food items are to be used for delivery of significant amounts of polyphenolics (e.g., controlled feeding trials). The other food items may also be used for these applications, but the relative retention of bioactive polyphenolics as outlined herein should be taken into account when dose and delivery are designed. The gummy product showed relatively poor retention of anthocyanins and flavonols after eight weeks of storage, which may be due to extraction issues rather than true losses. In summary, incorporating WBB powder into food products in which key molecules remain intact during storage can improve the consumption of blueberry phytochemicals. The effect of the product matrix on bioavailability of retained polyphenolic compounds needs further investigation.

Supplementary Materials: The following are available online at http://www.mdpi.com/2304-8158/9/4/466/s1, Figure S1: Individual anthocyanin content of blueberry ice pop stored at −20 °C over eight weeks. Bars represent standard error of the mean (n = 3/time point). Anthocyanidins: Cyd = cyanidin, Dpd = delphinidin, Mvd = malvidin, Pnd = peonidin, Ptd = petunidin. Sugar moiety: ara = arabinoside, gal = galactose, glc = glucose. Figure S2: Individual anthocyanin content of blueberry oatmeal bar stored at 4.4 and 21 °C over eight weeks. Bars represent standard error of the mean (n = 3/time point). Anthocyanidins: Cyd = cyanidin, Dpd = delphinidin, Mvd = malvidin,

Pnd = peonidin, Ptd = petunidin. Sugar moiety: ara = arabinoside, gal = galactose, glc = glucose. Figure S3: Individual anthocyanin content of blueberry graham cracker cookie stored at 4.4 and 21 °C over eight weeks. Bars represent standard error of the mean (n = 3/time point). Anthocyanidins: Cyd = cyanidin, Dpd = delphinidin, Mvd = malvidin, Pnd = peonidin, Ptd = petunidin. Sugar moiety: ara = arabinoside, gal = galactose, glc = glucose. Figure S4: Individual anthocyanin content of blueberry juice stored at 4.4 and 21 °C over eight weeks. Bars represent standard error of the mean (n = 3/time point). Anthocyanidins: Cyd = cyanidin, Dpd = delphinidin, Mvd = malvidin, Pnd = peonidin, Ptd = petunidin. Sugar moiety: ara = arabinoside, gal = galactose, glc = glucose. Figure S5: Individual anthocyanin content of blueberry gummy stored at 4.4 and 21 °C over eight weeks. Bars represent standard error of the mean (n = 3/time point). Anthocyanidins: Cyd = cyanidin, Dpd = delphinidin, Mvd = malvidin, Pnd = peonidin, Ptd = petunidin. Sugar moiety: ara = arabinoside, gal = galactose, glc = glucose.

Author Contributions: Conceptualization, L.H., S.H.A., J.-R.C., and E.C.D.; Methodology, L.H., C.B., and S.H.A.; Formal analysis, L.H., L.L., C.B., D.R., and A.M.; Investigation: L.H., C.B., and D.R.; Resources, L.H.; Writing—original draft, L.H., and L.L.; Writing—review and editing, L.H., L.L., S.H.A., J.-R.C., E.C.D., and A.M.; Supervision, L.H., and S.H.A.; Project administration, L.H., and S.H.A.; Funding acquisition, L.H., S.H.A., J.-R.C., and E.C.D. All authors have read and agreed to the published version of the manuscript.

Funding: This study was supported by the Arkansas Children's Nutrition Center (USDA-Agricultural Research Service Project 6026-51000-012-06S), Little Rock, AR, USA. Eva C. Diaz is partially supported by the Arkansas Center for Advancing Pediatric Therapeutics funded by the National Institutes of Health (Award # 8UG1OD024945).

Conflicts of Interest: The authors declare no conflict of interest.

References

1. Esposito, D.; Chen, A.; Grace, M.H.; Komarnytsky, S.; Lila, M.A. Inhibitory effects of wild blueberry anthocyanins and other flavonoids on biomarkers of acute and chronic inflammation in vitro. *J. Agric. Food Chem.* **2014**, *62*, 7022–7028. [CrossRef] [PubMed]
2. Prior, R.L.; Cao, G.; Martin, A.; Sofic, E.; McEwen, J.; O'Brien, C.; Lischner, N.; Ehlenfeldt, M.; Kalt, W.; Krewer, G.; et al. Antioxidant Capacity As Influenced by Total Phenolic and Anthocyanin Content, Maturity, and Variety of Vaccinium Species. *J. Agric. Food Chem.* **1998**, *46*, 2686–2693. [CrossRef]
3. Böhm, H.G. Mazza und E. Miniati: Anthocyanins in Fruits, Vegetables and Grains. 362 Seiten, zahlr. Abb. und Tab. CRC Press, Boca Raton, Ann Arbor, London, Tokyo 1993. Preis: 144. *Food Nahr.* **1994**, *38*, 343. [CrossRef]
4. Kalt, W.; McDonald, J.E.; Ricker, R.D.; Liu, X. Anthocyanin content and profile within and among blueberry species. *Can. J. Plant Sci.* **1999**, *79*, 617–623. [CrossRef]
5. Gao, L.; Mazza, G. Quantitation and Distribution of Simple and Acylated Anthocyanins and Other Phenolics in Blueberries. *J. Food Sci.* **1994**, *59*, 1057–1059. [CrossRef]
6. Prior, R.L.; Lazarus, S.A.; Cao, G.; Muccitelli, H.; Hammerstone, J.F. Identification of Procyanidins and Anthocyanins in Blueberries and Cranberries (*Vaccinium* Spp.) Using High-Performance Liquid Chromatography/Mass Spectrometry. *J. Agric. Food Chem.* **2001**, *49*, 1270–1276. [CrossRef]
7. Cho, M.J.; Howard, L.R.; Prior, R.L.; Clark, J.R. Flavonoid glycosides and antioxidant capacity of various blackberry, blueberry and red grape genotypes determined by high-performance liquid chromatography/mass spectrometry. *J. Sci. Food Agric.* **2004**, *84*, 1771–1782. [CrossRef]
8. Kalt, W.; McDonald, J.E. Chemical Composition of Lowbush Blueberry Cultivars. *J. Am. Soc. Hortic. Sci.* **1996**, *121*, 142–146. [CrossRef]
9. Vrhovsek, U.; Masuero, D.; Palmieri, L.; Mattivi, F. Identification and quantification of flavonol glycosides in cultivated blueberry cultivars. *J. Food Compos. Anal.* **2012**, *25*, 9–16. [CrossRef]
10. Kalt, W.; Cassidy, A.; Howard, L.R.; Krikorian, R.; Stull, A.J.; Tremblay, F.; Zamora-Ros, R. Recent Research on the Health Benefits of Blueberries and Their Anthocyanins. *Adv. Nutr.* **2019**, *11*, 224–236. [CrossRef]
11. Chen, J.R.; Lazarenko, O.P.; Wu, X.; Kang, J.; Blackburn, M.L.; Shankar, K.; Badger, T.M.; Ronis, M.J.J. Dietary-induced serum phenolic acids promote bone growth via p38 MAPK/β-catenin canonical Wnt signaling. *J. Bone Miner. Res. Off. J. Am. Soc. Bone Miner. Res.* **2010**, *25*, 2399–2411. [CrossRef] [PubMed]
12. Chen, J.R.; Lazarenko, O.P.; Zhang, J.; Blackburn, M.L.; Ronis, M.J.J.; Badger, T.M. Diet-derived phenolic acids regulate osteoblast and adipocyte lineage commitment and differentiation in young mice. *J. Bone Miner. Res. Off. J. Am. Soc. Bone Miner. Res.* **2014**, *29*, 1043–1053. [CrossRef] [PubMed]
13. Chen, J.R.; Wankhade, U.D.; Alund, A.W.; Blackburn, M.L.; Shankar, K.; Lazarenko, O.P. 3-(3-Hydroxyphenyl)-Propionic Acid (PPA) Suppresses Osteoblastic Cell Senescence to Promote Bone Accretion in Mice. *JBMR Plus* **2019**, *3*, e10201. [CrossRef] [PubMed]

14. Johnson, S.A.; Figueroa, A.; Navaei, N.; Wong, A.; Kalfon, R.; Ormsbee, L.T.; Feresin, R.G.; Elam, M.L.; Hooshmand, S.; Payton, M.E.; et al. Daily blueberry consumption improves blood pressure and arterial stiffness in postmenopausal women with pre- and stage 1-hypertension: A randomized, double-blind, placebo-controlled clinical trial. *J. Acad. Nutr. Diet.* **2015**, *115*, 369–377. [CrossRef]
15. Kalea, A.Z.; Clark, K.; Schuschke, D.A.; Klimis-Zacas, D.J. Vascular reactivity is affected by dietary consumption of wild blueberries in the Sprague-Dawley rat. *J. Med. Food* **2009**, *12*, 21–28. [CrossRef]
16. Rodriguez-Mateos, A.; Rendeiro, C.; Bergillos-Meca, T.; Tabatabaee, S.; George, T.W.; Heiss, C.; Spencer, J.P. Intake and time dependence of blueberry flavonoid-induced improvements in vascular function: A randomized, controlled, double-blind, crossover intervention study with mechanistic insights into biological activity. *Am. J. Clin. Nutr.* **2013**, *98*, 1179–1191. [CrossRef]
17. Bell, L.; Lamport, D.J.; Butler, L.T.; Williams, C.M. A study of glycaemic effects following acute anthocyanin-rich blueberry supplementation in healthy young adults. *Food Funct.* **2017**, *8*, 3104–3110. [CrossRef]
18. Riso, P.; Klimis-Zacas, D.; Del Bo', C.; Martini, D.; Campolo, J.; Vendrame, S.; Møller, P.; Loft, S.; De Maria, R.; Porrini, M. Effect of a wild blueberry (*Vaccinium angustifolium*) drink intervention on markers of oxidative stress, inflammation and endothelial function in humans with cardiovascular risk factors. *Eur. J. Nutr.* **2013**, *52*, 949–961. [CrossRef]
19. Rodriguez-Mateos, A.; Feliciano, R.P.; Cifuentes-Gomez, T.; Spencer, J.P.E. Bioavailability of wild blueberry (poly)phenols at different levels of intake. *J. Berry Res.* **2016**, *6*, 137–148. [CrossRef]
20. Whyte, A.R.; Schafer, G.; Williams, C.M. The effect of cognitive demand on performance of an executive function task following wild blueberry supplementation in 7 to 10 years old children. *Food Funct.* **2017**, *8*, 4129–4138. [CrossRef]
21. Zhong, S.; Sandhu, A.; Edirisinghe, I.; Burton-Freeman, B. Characterization of Wild Blueberry Polyphenols Bioavailability and Kinetic Profile in Plasma over 24-h Period in Human Subjects. *Mol. Nutr. Food Res.* **2017**, *61*, 1700405. [CrossRef] [PubMed]
22. Seo, H.-S.; Adams, S.H.; Howard, L.R.; Brownmiller, C.; Hogan, V.; Chen, J.-R.; Pramudya, R.C. Children's liking and wanting of foods vary over multiple bites/sips of consumption: A case study of foods containing wild blueberry powder in the amounts targeted to deliver bioactive phytonutrients for children. *Food Res. Int.* **2020**, *131*, 108981. [CrossRef] [PubMed]
23. Brownmiller, C.; Howard, L.R.; Prior, R.L. Processing and storage effects on monomeric anthocyanins, percent polymeric color, and antioxidant capacity of processed blueberry products. *J. Food Sci.* **2008**, *73*, H72–H79. [CrossRef] [PubMed]
24. Howard, L.R.; Castrodale, C.; Brownmiller, C.; Mauromoustakos, A. Jam Processing and Storage Effects on Blueberry Polyphenolics and Antioxidant Capacity. *J. Agric. Food Chem.* **2010**, *58*, 4022–4029. [CrossRef] [PubMed]
25. Skrede, G.; Wrolstad, R.E.; Durst, R.W. Changes in Anthocyanins and Polyphenolics During Juice Processing of Highbush Blueberries (*Vaccinium corymbosum* L.). *J. Food Sci.* **2000**, *65*, 357–364. [CrossRef]
26. Fracassetti, D.; Del Bo', C.; Simonetti, P.; Gardana, C.; Klimis-Zacas, D.; Ciappellano, S. Effect of Time and storage temperature on anthocyanin decay and antioxidant activity in wild blueberry (*Vaccinium angustifolium*) powder. *J. Agric. Food Chem.* **2013**, *61*, 2999–3005. [CrossRef] [PubMed]
27. Howard, L.R.; Prior, R.L.; Liyanage, R.; Lay, J.O. Processing and Storage Effect on Berry Polyphenols: Challenges and Implications for Bioactive Properties. *J. Agric. Food Chem.* **2012**, *60*, 6678–6693. [CrossRef]
28. Skrede, G.; Wrolstad, R.E. Flavonoids berries and grapes. In *Functional Foods: Biochemical and Processing Aspects*; CRC Press: Boca Raton, FL, USA, 2012; Volume 2, pp. 71–133.
29. Howard, L.R.; Brownmiller, C.; Mauromoustakos, A.; Prior, R.L. Improved stability of blueberry juice anthocyanins by acidification and refrigeration. *J. Berry Res.* **2016**, *6*, 189–201. [CrossRef]
30. Simard, R.E.; Bourzeix, M.; Herdia, N. Factors influencing color degradation in blueberry juice. *Lebensm. Wiss. Technol.* **1982**, *15*, 177–180.
31. Zhang, J.; Lazarenko, O.P.; Kang, J.; Blackburn, M.L.; Ronis, M.J.J.; Badger, T.M.; Chen, J.R. Feeding blueberry diets to young rats dose-dependently inhibits bone resorption through suppression of RANKL in stromal cells. *PLoS ONE* **2013**, *8*, e70438. [CrossRef]

32. Cho, M.J.; Howard, L.R.; Prior, R.L.; Clark, J.R. Flavonol glycosides and antioxidant capacity of various blackberry and blueberry genotypes determined by high-performance liquid chromatography/mass spectrometry. *J. Sci. Food Agric.* **2005**, *85*, 2149–2158. [CrossRef]
33. Giusti, M.M.; Wrolstad, R.E. Characterization and Measurement of Anthocyanins by UV-Visible Spectroscopy. *Curr. Protoc. Food Anal. Chem.* **2001**, *00*, F1.2.1–F1.2.13. [CrossRef]
34. Wu, X.; Prior, R.L. Systematic identification and characterization of anthocyanins by HPLC-ESI-MS/MS in common foods in the United States: Fruits and berries. *J. Agric. Food Chem.* **2004**, *53*, 2589–2599. [CrossRef] [PubMed]
35. Lohachoompol, V.; Srzednicki, G.; Craske, J. The Change of Total Anthocyanins in Blueberries and Their Antioxidant Effect After Drying and Freezing. *J. Biomed. Biotechnol.* **2004**, *2004*, 248–252. [CrossRef] [PubMed]
36. Syamaladevi, R.M.; Andrews, P.K.; Davies, N.M.; Walters, T.; Sablani, S.S. Storage effects on anthocyanins, phenolics and antioxidant activity of thermally processed conventional and organic blueberries. *J. Sci. Food Agric.* **2012**, *92*, 916–924. [CrossRef]
37. Reque, P.M.; Steffens, R.S.; Jablonski, A.; Flôres, S.H.; Rios, A.D.O.; de Jong, E.V. Cold storage of blueberry (*Vaccinium* spp.) fruits and juice: Anthocyanin stability and antioxidant activity. *J. Food Compos. Anal.* **2014**, *33*, 111–116. [CrossRef]
38. Srivastava, A.; Akoh, C.C.; Yi, W.; Fischer, J.; Krewer, G. Effect of storage conditions on the biological activity of phenolic compounds of blueberry extract packed in glass bottles. *J. Agric. Food Chem.* **2007**, *55*, 2705–2713. [CrossRef]
39. Maier, T.; Fromm, M.; Schieber, A.; Kammerer, D.R.; Carle, R. Process and storage stability of anthocyanins and non-anthocyanin phenolics in pectin and gelatin gels enriched with grape pomace extracts. *Eur. Food Res. Technol. Z. Für Lebensm. Unters. Forsch. Heidelb.* **2009**, *229*, 949–960. [CrossRef]
40. Brouillard, R. Chemical structure of anthocyanins. In *Anthocyanins as Food Colors*; Academic Press: New York, NY, USA, 1982; pp. 1–40.
41. Kader, F.; Rovel, B.; Girardin, M.; Metche, M. Mechanism of Browning in Fresh Highbush Blueberry Fruit (*Vaccinium corymbosum* L). Role of Blueberry Polyphenol Oxidase, Chlorogenic Acid and Anthocyanins. *J. Sci. Food Agric.* **1997**, *74*, 31–34. [CrossRef]
42. Brownmiller, C.; Howard, L.R.; Prior, R.L. Processing and storage effects on blueberry (*Vaccinium corymbosum* L.) polyphenolics. *Acta Hort.* **2009**, *841*, 347–354. [CrossRef]

foods

Article

Validation of Rapid Enzymatic Quantification of Acetic Acid in Vinegar on Automated Spectrophotometric System

Irene Dini [1,*], Ritamaria Di Lorenzo [1], Antonello Senatore [2], Daniele Coppola [2] and Sonia Laneri [1]

1 Department of Pharmacy, University of Naples Federico II, Via Domenico Montesano 49, 80131 Napoli, Italy; ritamaria.dilorenzo@unina.it (R.D.L.); srlaneri@unina.it (S.L.)

2 Lcm Laboratory for Product Analysis of Chamber of Commerce, Corso Meridionale, 80143 Napoli, Italy; a.senatore@si-impresa.na.camcom.it (A.S.); d.coppola@si-impresa.na.camcom.it (D.C.)

* Correspondence: irdini@unina.it

Received: 9 May 2020; Accepted: 4 June 2020; Published: 9 June 2020

Abstract: Vinegar is produced from the fermentation of agricultural materials and diluted acetic acid (diluted with water to 4–30% by volume) via sequential ethanol and acetic acid fermentation. The concentration of acetic acid must be measured during vinegar production. A Community method for analyzing acetic acid in vinegar is a non-specific method based on the assumption that the total acid concentration of the vinegar is attributable to the acetic acid. It consists of titration with a strong base in the presence of an indicator. This test is laborious and has a time-consuming character. In this work, a highly specific automated enzymatic method was validated, for the first time, to quantify the acetic acid in the wine vinegar, in terms of linearity, precision, repeatability, and uncertainty measurement. The results were compared to the Community method of analysis. Regression coefficient ≅ 1 and the normal distribution of residuals in the ANOVA test confirmed the method's linearity. LLOD (0.946 ppm) and LLOQ (2.00 ppm) defined the method's sensitivity. The results of the tested and the Community methods, linearly distributed in the Shapiro–Wilk test, confirmed the method's repeatability. The few anomalous data in the Huber test were due to random errors. The high selectivity of the enzymatic method, which exclusively measures acetic acid concentration, determined the significant differences between the two tests, examined in the accuracy determination. The enzymatic method can be considered applicable since its precision and uncertainty were lower than the Community method values (relative percentage deviations = 10%). The enzymatic method compared to the Community method reduces the analysis time and the risk of errors due to operators (avoid pipetting errors and wrong calculations), minimizes solvent and the sample consumption and guarantees assay quality through method standardization.

Keywords: vinegar; automatized method; quantification

1. Introduction

In the European Member States, products obtained by the fermentation of agricultural materials or by the dilution with water of acetic acid are marketed under the name "vinegar" [1]. According to the raw material used in production, there are many types of vinegar: wine, cider, fruit, malt, malt distillate, spirit, cereal, honey, and whey vinegar. Wine vinegar is widely used as a seasoning, food preservative, and acidifier. Traditional production needs maturation for a long time in the wood to obtain a high acetic degree. Two stages of fermentation lead to the production of wine vinegar. In the first step, the yeasts, generally Saccharomyces, convert the fermentable sugars into ethanol. In the second phase, the bacteria oxidize the ethanol to acetic acid [2]. In Italy, three types of wine vinegar are produced: white wine vinegar, red wine vinegar, and balsamic vinegar. The latter is obtained from fresh grapes, concentrated by a slow heating process (to 1/3 of its original volume), fermented by yeasts (*Zygosaccaharomyces*) and bacteria (*Gluconobacter*), and subsequently refined in wooden barrels (25 years) [3]. Acetic acid is monitored during the acetic fermentation process. A Community method for the analysis of acetic acid consists of direct titration with sodium hydroxide in the presence of phenolphthalein [4]. This method is not selective; it determines the total acidity of the vinegar and attributes it to the content of acetic acid. Titration is an analytical methodology that uses color to measure the quantity of substances. The visual identification of the endpoint can lead to quantification errors. Therefore, accuracy, sampling frequency, and time expenditure are difficulties generally associated with manual titrations. The alternative methods proposed to determine acetic acid in vinegar are spectrophotometry with a fiber optic sensor [5], a titration system with colorimetry (λ480 nm) [6] or an ATR-FT-IR detector, a chemometric test [7], capillary electrophoresis or ion exclusion chromatography with conductimetric detection, [8,9] and liquid chromatography and gas chromatography [10,11]. In this work, we propose the validation of an automated enzymatic method to identify and quantify acetic acid in vinegar. The automated analyzer was designed to disperse the reagents and samples in the cuvette, incubate the samples at a controlled temperature, read the absorbance in the UV-visible spectrum, and calculate the concentrations of the selected molecules using a calibration curve. The highly selective enzymatic reaction allows the detection of acetic acid in spectrum fields without interference. Following regulatory requirements, the validation of the method is essential to establish data traceability and avoid incorrect quantification, which could have economic consequences and damage the reputation of the laboratories. The validation exercise is expensive and time-consuming. It would be desirable for the scientific community to spend more time validating advanced analytical methods for food quality control [12] to eliminate test repetitions and avoid wasting time.

2. Materials and Methods

2.1. Reagents

Enzytec acetic acid Cod. E2580 was purchased from R-Biopharm AG (Darmstadt, Germany). Distilled water was purchased from Sigma-Aldrich (Milan, Italy). Potassium hydrogen phthalate and ethanol were purchased from Carlo-Erba (Milan, Italy).

2.2. Samples Preparation

Three commercial vinegar types were tested: white, red, and balsamic wine vinegar. Samples were diluted 1:125 before analyses.

2.3. Apparatus

The analyzer iCubio iMagic M9 was used and run with full automation for the enzymatic assay for acetic acid determination. It automatically pipetted reagents and samples into the cuvette, allowed incubation at a controlled temperature, read absorbance at the specific wavelength, and calculated the concentration of the analytes with a calibration curve. The parameters used in the automated photometric systems were temperature, 37 °C; wavelengths, 340 nm and 415 nm (bichromatic); and optical path, 1 cm.

2.4. Reference Procedure

Commercial vinegar samples were analyzed by titration to determine the acetic acid content following the Community reference method [12]. A NaOH solution, normalized with potassium hydrogen phthalate (ACS), was gradually added to 5 mL of the vinegar solution. Complete neutralization was indicated by color changes in 2% phenolphthalein solutions in ethanol. Triplicate analyses were carried out. The average of the three analyses was used as the reference value. The mean standard error (pooled standard deviation divided by the average acidity content) was 0.32%.

2.5. Enzymatic Method Determination of Acetic Acid Content

The method reported in the kit instruction (Enzytec acetic acid) was followed. The Enzytec fluid Acid combination Standard (ID-No 5460, 3 × 3 mL) was used to calibrate the automated photometric systems.

2.6. Spectrophotometric Method Validation Parameters

The linearity, precision, sensitivity, and measurement uncertainty were determined.

In the linearity assessment, a matrix-match calibration curve was plotted at 0, 0.06, 0.13, 0.25, and 0.50 g/L for acetic acid.

The method precision was tested, performing ten analyses of the same sample

$$\text{Reproducibility} = \frac{\text{Standard deviation of analyzed samples}}{\text{Standard deviation of reference samples}} \quad (1)$$

Normality, by the Shapiro–Wilk test [13], and the presence of anomalous data, by the Huber test [14], were evaluated to define the method's precision.

The method sensitivity was evaluated by LLOQ (limit of quantification: LOQ = 10σS) and LLOD (limit of detection: LOD = 3.3σS) determinations, where σ is the relative standard deviation and S is the slope of the standard curve.

The method accuracy was determined, performing ten analyses with both methods (Community [4] and enzymatic), determining the residual distribution by the Saphiro–Wilk test, and controlling for anomalous data by *t*-tests.

Type A and B uncertainties were measured following the EURACHEM/CITAC guide [15]. Type A was estimated from 10 repeated readings of the same sample.

$$\text{U Type A} = \sqrt{\frac{\text{variance}}{\text{Degrees of freedom}}} \quad (2)$$

Type B was determined with a metrology approach.

U(t) is the uncertainty associated with 20 mL pipette use U(t). It was obtained considering a certificate of calibration (0.016 mL) and repeatability (0.00050 mL).

U(p) is the uncertainty associated with 10 mL pipette use. It was obtained considering a certificate of calibration (0.096 mL) and repeatability (0.00020 mL).

U(ct) is the uncertainty associated with the calibration curve. It was obtained for the standard, which was measured at three concentrations in triplicate.

$$\text{U(ct)}\ \text{S}\frac{x/y}{b} * \sqrt{1/\text{n} + 1/\text{m}} \quad (3)$$

S = standard deviation of the residual

n = points used for the calibration line

m = readings taken for each sample

U(mr) is the uncertainty associated with a standard preparation.

U(bt) is the uncertainty associated with balances. It was determined considering a certificate of calibration (0.00060 g), repeatability (0.000029 g), and stability (0.000032 g).

U(m) is the uncertainty associated with the use of a 100 mL flask. It was obtained considering a certificate of calibration (0.01 mL) and repeatability (0.00030 mL).

U(k) is the uncertainty associated with the use of a 250 mL flask. It was obtained considering a certificate of calibration (0.025 mL) and repeatability (0.00040 mL).

Compound uncertainty: $\sqrt{(U\ Type\ A)^2 + (U\ Type\ B)^2}$

The accuracy was tested by Student's t test:

$$\text{Accuracy} = \frac{|\overline{X}_{\text{Community}} - \overline{X}|}{\sqrt{S_r^2 + U_{\text{Community}}^2}} \leq t_{p_\nu} \quad (4)$$

$\overline{X}$ = Community method value
$\overline{X}s_{y/x}$ = medium repeatability values
S_r^2 = standard deviation^2
U_{CRM}^2 = reference material uncertainty^2
Uncertainty ($t_{p\nu}$) = k × σr.

2.7. Statistical Analysis

The data were analyzed using the software Statistica Version 7.0 (StatSoft, Hamburg, Germany). The normality of the data was verified by applying the Shapiro–Wilk W-test. A p-value of >0.05 indicates a normal distribution. The non-parametric Huber test determined the outliers.

3. Results

3.1. Method Linearity

The method linearity, by regression coefficient determination (Figure S1), and residuals distribution, by ANOVA tests (Figure S2), were evaluated. $R^2 = 0.99$; the residuals were normally distributed.

3.2. Method Sensitivity

The method detection limit was tested by repeated analyses of blank samples. The LLOD and LLOQ were derived from the regression curve. The quantitation limit (LLOD) of an individual analytical procedure is the lowest amount of acetic acid in a sample that can be quantified with suitable precision and accuracy. The lower limit of detection was 0.0063 g/L. The detection limits (LLOQ) were determined as the concentration giving a peak height three times the background noise. The Lower Limit of Quantitation was 0.0253 g/mL. The LLOQ dilution factor was used to determine the lower end of the measuring range. It was obtained by dividing the (read concentration × 10) / (weight × rate) (0.098 g/100 mL). The last point of the calibration curve line was the upper end of the measuring range (0.40 g/100 mL).

3.3. Measurement Uncertainty

Type A uncertainty due to method repeatability was 0.005 (Table S1). Type B uncertainties—due to method repeatability and associated with the standard preparation, the calibration curve, the balances, the flasks (100 mL and 250 mL), and the pipettes (10 mL and 20 mL)—were less than 10% of the results (Table S2).

3.4. Method Precision

Method repeatability was tested, making ten analyses on the same sample of each type of vinegar and comparing the results with those obtained with the Community method. The Shapiro-Wilk test showed that the data were linearly distributed (Figure S4). The Huber test excluded the presence of anomalous data (Figures S3–S5). The reproducibility of the three kinds of vinegar were red wine vinegar = 0.55, white wine vinegar = 0.69, and balsamic wine vinegar = 0.66. The limits of method repeatability were: upper limit = 0.548 and lower limit = 1.480, considering the ratio between the standard deviation (Sr) of the enzymatic method and the repeatability standard deviation of the reference method (σr), satisfied for nine degrees of freedom (Table S6).

3.5. Accuracy Test

Accuracy was determined, making ten analyses with both methods (Community and enzymatic), and determining significant differences between groups by Student's t test (Figures S4–S6). We cannot reject the null hypothesis that there is no difference between means when $p < 0.05$.

4. Discussion

Automated analyzers are modern instrumentation for routine analytical analysis since they reduce staff errors due to tiredness or a lack of technicality, improve safety, and decrease the amount of reagents, the cost, and the time of analysis. Traditional methods of analysis are struggling to survive in technology. Innovation in technologies brings significant opportunities but also carries risks for society. The validation of new analytical procedures is a developed approach that responds to evolving markets. It is a verification process that checks whether the analytical method achieves predetermined results. The validation of an analytical procedure is used both before its first use and throughout its life, to continually monitor its performance and any critical issues. In food, analysis is indispensable to make available reliable and accurate results with known uncertainty. Therefore, the methods used in analytical laboratories need an accurate validation process to ensure their validity (ISO/IEC, 2005) [16]. Validation is performed to define the linearity, precision, accuracy, and repeatability of the method based on the matrix, the working field, and the uncertainty due to the instrumentation and environmental conditions. Furthermore, it is possible to verify the results by comparing them with those obtained with reference analytical methods. In this work, an enzymatic determination of acetic acid in three different types of vinegar (red wine vinegar, white wine vinegar, and balsamic vinegar) was carried out on automated photometric systems. The method was based on acetate kinase, an enzyme capable of reacting with acetic acid and adenosine-5′-triphosphate, giving acetylphosphate and adenosine-5′-diphosphate (ADP). Acetylphosphate is converted into acetyl-CoA plus phosphate by coenzyme A (CoA) and phosphotransacetylase. ADP reacts with D-glucose through an ADP-dependent exokinase to produce D-glucose-6-phosphate. The latter, in the presence of glucose-6-phosphate dehydrogenase, reacts with NAD^+, turning into D-glucono-δ-lactone-6-phosphate and $NADH^+$ H^+. The concentration of NADH, proportional to the concentration of acetic acid, is determined spectrophotometrically, according to AOAC instructions (AOAC 2012) [17]. The test uses a kit containing ready-to-use reagents and standards. The analytical problem consisted of adapting the procedures proposed by the industry for the wine matrix to the vinegar matrix and the validation of the analytical procedures. The concentration of the solution influences the spectrophotometric reading. The vinegar samples were diluted 125 times to be able to read the absorbance at the desired wavelength. Any change requires an evaluation of the method's performance. The validation of the method, comparative tests with standard methods, and co-validation between laboratories are the possible strategies to achieve this goal [18]. In this case, the method was validated in terms of the linearity, precision, repeatability, measurement of uncertainty, and accuracy. The linearity of the method was demonstrated by the regression coefficient (1) and a residual diagram (straight line) in the ANOVA test. The ANOVA test describes the difference in the standard deviations of the values obtained from the reference compared to the expected values. The reliability of the test depends on the normal distribution of residues with a 95% confidence level. The sensitivity of the method was defined by deriving the LLOD (0.946 ppm) and LLOQ (2.00 ppm) from the regression curve and determining the measuring range (100 ppm ≤ measuring range ≤ 500 ppm). The precision of the method was determined, confirming its reproducibility and repeatability. Repeatability produces the minimum precision value. It was obtained, making ten analyses of the same sample in short intervals of time, by the same operator, in the same laboratory, with the same method and equipment and comparing the repeatability type difference of the method (sr) with that of the reference method (sr). The Shapiro–Wilk test confirmed the linearity between the results obtained by the spectrophotometric method and those obtained by the reference method. The null hypothesis (H0) states that two elements or series of elements are normally distributed. This hypothesis is satisfied if the p-value is higher than 0.05, as happens for the types of analysis carried out (Figure S4). Successively, the absence of anomaly data was tested by the Huber test (Tables S3–S5). The Huber test is a robust statistical method to identify outliers, which may invalidate the resulting analysis, by first fitting most of the data and flagging data points. The presence of very few anomalous data was attributed to random errors. Finally, random and systematic uncertainties were detected to establish method accuracy. Random errors are caused by unpredictable changes in the experiment due to environmental conditions and measuring instruments. The errors due to the instrument or its data-handling system, or the instrument being wrongly used by the experimenter cause systematic errors. In this work, the uncertainties due to method repeatability and associated with the standard preparation, the calibration curve, the balances, the flasks (100 mL and 250 mL), and the pipettes (10 mL and 20 mL) were determined to be irrelevant since they were less than 10% of the results. Finally, the accuracy of the enzymatic method was evaluated. The accuracy showed significant differences between the two populations of data. The enzymatic method underestimated the results because of systematic errors. The percentage difference was calculated, and it was observed that the systematic errors were independent of the matrix but were influenced by the measuring range (five samples had waste results around 10). The differences in measurements between the two tests were due to the high selectivity of the enzymatic method, which exclusively measured the concentration of acetic acid and non-specificity of the Community method, which attributed to the concentration of acetic acid all of the acids present in the sample. Consequently, the underestimation of the enzymatic method was expected. Therefore, the method could be considered applicable since the relative percentage deviations compared to the values obtained with the official method are around 10%.

5. Conclusions

An enzymatic method on an automated spectrophotometric system was used for the quantification of acetic acid in vinegar. The enzymatic method met many validation requirements (linearity, sensitivity, and uncertainties) but not accuracy. However, the method can be considered applicable as the precision and uncertainty are lower than those of the Community method (deviations are around 10%). The validation of instrumental analytical methods not yet used as conventional methods should find more space in the scientific literature. By transferring knowledge on operating methods, the repetition of long and laborious validation processes that involve high costs and waste of time will be avoided.

Supplementary Materials: The following are available online at http://www.mdpi.com/2304-8158/9/6/761/s1, Figure S1: Standard calibration curve, Figure S2: Normal residual probability, Figure S3: Shapiro-Wilk test (5% significance levels), Figure S4: Balsamic wine vinegar accuracy statistical analyses, Figure S5: Red wine vinegar accuracy statistical analyses, Figure S6: White wine vinegar accuracy statistical analyses, Table S1: Data considering calculating type A uncertainty, Table S2: Type B: Systematic uncertainty estimates, Table S3: Red wine vinegar data, Table S4: Withe wine vinegar data, Table S5: Balsamic wine vinegar data, Table S6: Upper and lower limit of repeatability.

Author Contributions: I.D., data curation and writing—review and editing; R.D.L., investigation; A.S., investigation; D.C., investigation.; S.L., investigation. All authors have read and agreed to the published version of the manuscript.

Funding: This research received no external funding.

Conflicts of Interest: The authors declare no conflict of interest.

References

1. Codex Alimentarius Commission. *Draft European Regional Standard for Vinegar*; World Health Organization: Geneva, Switzerland, 1987.
2. Adams, M.R. Vinegar. In *Microbiology of Fermented Food*; Wood, J.B., Ed.; Blackie Academic and Professional: London, UK, 1998; pp. 1–44.

3. Benedetti, B.; Ferrari, S.; Federzoni, G.; Mascolo, A.; Amati, A. Modena e gli aceti speciali. In *Consorzio Tutela Aceto Balsamico di Modena, L'Aceto balsamico di Modena*; Della Camera di Commercio, Industria, Artigianato, Artigianato, Agricoltura di Modena e della Banca Popolare dell'Emilia Romagna: Modena, Italy, 1996; pp. 32–46.
4. *Official Methods of Analysis of the Association of Official Analytical Chemists (AOAC) International*, 17th ed.; AOAC International: Rockville, MD, USA, 2000.
5. Kurauchi, Y.; Ogata, T.; Egashira, N.; Ohga, K. Fiber-optic sensor with a dye-modified chitosan/poly(vinyl alcohol) cladding for the determination of organic acids. *Anal. Sci.* **1996**, *12*, 55–59. [CrossRef]
6. Lenghor, N.; Jakmunee, J.; Vilen, M.; Sara, R.; Christian, G.D.; Grudpan, K. Sequential injection redox or acid–base titration for determination of ascorbic acid or acetic acid. *Talanta* **2002**, *58*, 1139–1144. [CrossRef]
7. Moros, J.; Inon, F.A.; Garrigues, S.; de la Guardia, A. Determination of vinegar acidity by attenuated total reflectance infrared measurements through the use of second-order absorbance-pH matrices and parallel factor analysis. *Talanta* **2008**, *74*, 632–641. [CrossRef] [PubMed]
8. Castro, R.; Moreno, M.V.G.; Natera, R.; Garcia-Rowe, F.; Hernandez, M.J.; Barroso, C.G. Comparative analysis of the organic acid content of vinegar by capillary electrophoresis and ion-exclusion chromatography with conductimetric detection. *Chromatographia* **2002**, *56*, 57–60. [CrossRef]
9. Kitamura, N.; Kotani, A.; Bunseki Kagaku, K.F. A disposable voltammetric cell for determining the titratable acidity in wood-vinegars. *Bunseki Kagaku* **2004**, *53*, 1097–1100. [CrossRef]
10. Cocchi, M.; Lambertini, P.; Manzini, D.; Marchetti, A.; Ulrici, A. Determination of carboxylic acids in vinegars and in aceto Balsamico Tradizionale di Modena by HPLC and GC methods. *J. Agric. Food Chem.* **2002**, *50*, 5255–5261. [CrossRef] [PubMed]
11. Xie, W.; Chai, X. Determination of Total Acid Content in Vinegars by Reaction-Based Headspace Gas Chromatography. *Food Anal. Methods* **2017**, *10*, 419–423. [CrossRef]
12. Dini, I.; Seccia, S.; Senatore, A.; Coppola, D.; Morelli, E. Development and Validation of an Analytical Method for Total Polyphenols Quantification in Extra Virgin Olive Oils. *Food Anal. Methods* **2020**, *13*, 457–464. [CrossRef]
13. Shapiro, S.S.; Wilk, M.B. An analysis of variance test for normality (complete samples). *Biometrika* **1965**, *52*, 591–611. [CrossRef]
14. Huber, P.J. *Robust Statistics*; Wiley: New York, NY, USA, 1981.
15. European Medicines Agency. *EMEA/CHMP/EWP/531305/2008, Concept Paper/Recommendations on the Need for a Guideline on the Validation of Bioanalytical Methods*; European Medicines Agency: London, UK, 2008.
16. Dai, S.Y. Producing Quality Laboratory Data: A Systems Approach. *J. Regul. Sci.* **2016**, *2*, 19–21.
17. AOAC. *Appendix F: Guidelines for Standard Method Performance Requirements (SMPR)*; Official Methods of Analysis of the Association of Official Analytical Chemists (AOAC) International: Rockville, MD, USA, 2012.
18. Martins, M.T.; Paim, C.S.; Steppe, M. Development of a dissolution test for lamotrigine in tablet form using an ultraviolet method. *Braz. J. Pharm. Sci.* **2010**, *46*, 179–186. [CrossRef]

foods

MDPI

Article

Evaluation of Garlic Landraces from Foggia Province (Puglia Region; Italy)

Anna Bonasia [1], Giulia Conversa [1,*], Corrado Lazzizera [1], Pasqua Loizzo [2], Giuseppe Gambacorta [2] and Antonio Elia [1]

[1] Department of the Science of Agriculture, Food and Environment (SAFE), University of Foggia, 71100 Foggia, Italy; anna.bonasia@unifg.it (A.B.); corrado.lazzizera@unifg.it (C.L.); antonio.elia@unifg.it (A.E.)

[2] Department of Soil, Plant and Food Sciences, University of Bari, 70126 Bari, Italy; pasqualoizzo@gmail.com (P.L.); giuseppe.gambacorta@uniba.it (G.G.)

* Correspondence: giulia.conversa@unifg.it

Received: 12 June 2020; Accepted: 26 June 2020; Published: 29 June 2020

check for updates

Abstract: Interest in local landraces has unfortunately decreased over, the last decades, in which they have been continuously subjected to a high genetic erosion in favour of new modern varieties. Within the Puglia region (S-E Italy), Foggia province was found to be the richest in vegetable landraces. In the present study, six garlic landraces collected from this area have been assessed for their chemical composition (minerals, organic acids, free sugars, volatile, and phenolic compounds) along with their main morpho-biometrical traits. A commercial genotype was also considered as a reference standard. The landraces show a large variability, but in general high morphological standards, high levels of cations and phenols, and low levels of volatile-(S)-compounds in comparison with the commercial genotype and the literature values. 'Aglio di Peschici' and 'Aglio Rosso di Monteleone di Puglia' are very rich in minerals and phenols (mainly ferulic acid and iso-rhamnetin). This increase in knowledge on the chemical properties of these garlic landraces could represent a tool for encouraging the consumption of a food product. At the same time, the consumption of these landraces would stimulate their cultivation and could highly contribute to protection against the risk of erosion of agro-biodiversity by their in situ/on-farm conservation.

Keywords: *Allium sativum* L.; agro-biodiversity; local varieties; bulb morphology; phenols; volatile compounds

---

1. Introduction

Garlic (*Allium sativum* L.) is one of the most important bulb vegetables and is mainly used as a spice or flavouring agent for foods. It is used in several types of products such as garlic oil, powder, salt, paste, and flakes.

Garlic has a high nutritive value as a rich source of minerals, carbohydrates, proteins, and vitamins. It also contains numerous bioactive molecules such as organic sulphur-containing (S-compounds) compounds and phenolic compounds [1–3].

The organic S-compounds (present in intact garlic cloves are γ-glutamyl cysteine and cysteine sulfoxides, including deoxyallin and mainly alliin [4]. When garlic is mechanically damaged, the vacuolar enzymes (allinase) convert alliin to allicin. Unstable allicin undergoes non-enzymatic degradation to form a variety of volatile S-compounds—ajoene, vinyl-dithiins, and sulphides as mono-, di-, and tri-sulphides [4]. Numerous studies have reported that the organic S-compounds are responsible for the biological activities revealed in different pathological situations [1,2,5].

Garlic is also characterized by phenolic compounds and the main group consists of phenolic acids (mainly as caffeic acids), followed by flavonoids (mainly as quercetin) [6–8]. A lot of research has been conducted to assess the dietary role of phenolic compounds, their characteristics, metabolic pathways, and biological effects, affirming their capacity to scavenge Reactive Oxygen Species (ROS) and to treat a variety of diseases including heart disease and cancer [2,6].

The Puglia region (South-East Italy) in the centre of the Mediterranean basin has a rich agro-biodiversity, with particular reference to landraces (also known as 'local varieties' and 'farmer's varieties'). Landraces are variable, identifiable populations, usually known by a local name, lacking 'formal' crop improvement, and characterized by a marked adaptation to specific environmental conditions of the area of cultivation [9].

Unfortunately, the regional agro-biodiversity of vegetable crops has been partially lost due to several factors. The modernization and intensification of agriculture have resulted in vast agro-biodiversity loss, since farmers grow modern varieties and hybrids and tend to abandon traditional landraces.

As a result of a regional Project (https://biodiversitapuglia.it//), aimed at collecting and identifying vegetable resources at risk of genetic erosion, the province of Foggia was found to be the richest in landraces among the five provinces of the region [10]. In the *Alliaceae* botanical family the most frequent species found was garlic (*Allium sativum* L.). To stimulate the cultivation of landraces from the farmers ('keepers'), their consumption has to be promoted. To do this, the identification and the highlighting of the chemical properties of the landraces compared with the 'commercial' varieties may represent an added value of the product that could push the preference of consumers.

In this view, the present study describes for the first time the chemical properties of six landraces collected in the province of Foggia, along with their main morpho-biometric traits.

2. Materials and Methods

2.1. Description of Collecting Sites, Plant Materials, and Sampling

Six landraces of garlic (*Allium sativum* L.) were directly collected from local growers which had very small farms in marginal areas, in the province of Foggia, in the North of Puglia region (Italy), which lies geographically between 41°54′44′′N and 41°02′27′′N in latitude and 14°55′58′′E and 14°02′27′′E in longitude. All the interested growers had very small farms in marginal areas and all of them adopted low input agronomic practices such as minimum soil tillage, hand weed control and scarce or no use of irrigation, pesticides or fertilizers.

Details of the collected landraces and the collecting sites are presented in Table 1. Bulbs of 'Spanish white garlic' ('Palmieri Import', Afragola (Na); Origin: Spain), widely available on the European and national market, was used as a reference standard. Pictures of the examined genotypes are shown in Figure S1.

Samples of bulbs (3 ± 0.5 kg per each genotype) were well mixed to obtain three independent replicates, each consisting of 15–40 bulbs. Morphological and biometrical measurements on cloves and bulbs were carried out on fresh material, dealing with 7–15 bulbs for each replicate. Chemical analyses of peeled cloves were performed in triplicate for each replicate, after being freeze-dried (CoolSafe SCANVAC, Labo0Gene, Allerød, Denmark), powdered, packed in hermetic jars and then stored in the dark at −18 ± 1 °C until the analyses were carried out.

Table 1. List of garlic genotypes and description of collecting areas.

Genotype	Acronym	Collection Sites	Landscape Unit [1]	Latitude (N)	Longitude (E)	Altitude (m, asl)	Yearly Mean Temperature [2] (°C)	Yearly Mean Rainfall [2] (mm)
Spanish white garlic (Commercial genotype)	'CG'	Grocery store						
Aglio dei Cortigli (Landrace)	'Cortigli'	Vico del Gargano (Municipality) 'Cortigli' locality	Gargano (mountain)	41.8960	15.9596	445	11.9 [3]	721 [3]
Aglio di Peschici (Landrace)	'Peschici'	Peschici (Municipality)	Gargano (Adriatic Sea coastal plain)	41.94711	16.0098	12	16.6	459
Aglio Rosso di Monteleone di Puglia (Landrace)	'Monteleone'	Monteleone di Puglia (Municipality)	Daunian Apennine (mountain)	41.16768	15.2418	800	11.7	572
Aglio di Anzano di Puglia (Landrace)	'Anzano'	Anzano di Puglia (Municipality)	Daunian Apennine (mountain)	41.12126	15.2914	760	11.6	571
Aglio Bianco di Panni (Landrace)	'Panni'	Panni (Municipality)	Daunian Apennine (mountain)	41.22080	15.2762	728	12.1	558
Aglio Durevole di Panni (Landrace)	'PanniD'	Panni (Municipality)	Daunian Apennine (mountain)	41.22080	15.2762	729	12.1	558

[1] As reported in Piano Paesaggistico Territoriale Regionale—PPTR—www.paesaggiopuglia.it/pptr. [2] As reported in https://it.climate-data.org/. [3] As reported in http://my.meteonetwork.it/station/.

2.2. Appearance and Morpho-Biometrical Measurements

The description of the morphological features for each genotype was done following the descriptors 7.1.11 (bulb shape), 7.1.20 (bulb structure type), 7.1.16.1 (bulb outer skin colour), and 7.1.16.2 (skin colour of the clove), provided by the International Plant Genetic Resources Institute (IPGRI) guidelines [11]. The main biometrical traits for each genotype were assessed. A dry matter concentration of the peeled cloves was calculated after freeze-drying the material.

Measurements of colour on external bulbs (with dry tunics) and cloves (with thin skin) were also performed by images analysis using an image acquisition station (Immagini & Computer, Bareggio, Milano, Italy), equipped with four white lamps, camera NIKON D5200, and software Image Pro Plus 7.0 (Media Cybernetics, Inc., Rockville, MD, USA). Colour indices were based on the CIELAB scale 1976: L*, indicating lightness/darkness, ranging from 0 (black) to 100 (white) value in a grey scale; a* reflecting greenish (if negative) to reddish (if positive) tonality; b* indicating bluish (if negative) to yellowish (if positive) tonality. In addition, the derived parameters, hue angle (h°), and chroma (C*), indicating respectively the hue and the vividness/dullness, were evaluated.

2.3. Chemical Measurements

2.3.1. Standards and Reagents

HPLC-grade methanol, metaphosphoric, formic, methanesulfonic acid, acetonitrile, ethanol were purchased from Sigma-Aldrich (St. Louis, MO, USA). NaOH, sodium carbonate, and sodium bicarbonate were purchased from Merck KGaA (Darmstadt, Germany).

Standards for phenolic (caffeic acid, caftaric acid, quercetin 3-ß-d-glucoside, quercetin, rutin, hyperoside) and volatile (di-allyl-sulphide-DAS; di-allyl-disulphide-DADS; di-allyl-trisulphide-DATS, di-methyl-sulphide, crotonaldehyde, pelargonaldehyde) compounds were purchased from Sigma-Aldrich (MO, USA).

Glucose, fructose, and sucrose standards were purchased from Chem Service (West Chester, USA); all the other standards were purchased from Sigma-Aldrich (MO, USA). Ultrapure water (18.2 MΩ cm resistivity at 25 °C) was obtained from a water purification system Milli Q (Millipore, Germany).

2.3.2. Minerals

Ashes were determined by a muffle furnace according to the AOAC method 923.03.

Inorganic ions were analyzed by ion chromatography (Dionex ICS 3000; Dionex-ThermoFisher Scientific, Waltham, MA, USA). Inorganic cations were extracted from lyophilised samples (1 g), previously ashed (in a muffle furnace at 550 °C for 6 h) and acid digested (20 mL of 1 mol L^{-1} HCl in boiling water for 30 min), before injection into the ion chromatography system. For inorganic anions, the lyophilized samples (0.5 g) were extracted with 50 mL of an eluent solution (3.5 mM sodium-carbonate and 1.0 mM sodium-bicarbonate) in a shaking water bath at room temperature for 30 min. The mixture was filtered through a Whatman no. 2 paper. The filtrates were filtered again through a 0.22 μm Millipore filter, before injection into the ion chromatography system.

The ion chromatography system was equipped with: an isocratic pump; conductivity detector; and a model AS-DV auto-sampler (ThermoFisher Scientific, Waltham, MA, USA); a self-generating ERS-500 suppressor (4 mm), a Dionex Ion-Pac AS23 analytical column (4 × 250 mm, particle size 6 μm), and an eluent solution (3.5 mM sodium-carbonate and 1.0 mM sodium-bicarbonate) at a flow rate of 1 mL min^{-1} (Dionex-ThermoFisher Scientific, Waltham, MA, USA) (specifically for anion analysis); a self-generating DRS-600 suppressor (4 mm), a Dionex IonPack CS12A analytical column (4 × 250 mm, 5 μm), and an eluent solution (20 mM methanesulfonic acid) at a flow rate of 1 mL min^{-1} (Dionex-ThermoFisher Scientific, Waltham, MA, USA) (specifically for cation analysis).

2.3.3. Starch and Simple Carbohydrates

Total starch was analyzed using the Megazyme Total Starch Assay kit (AOAC Method 996.11 and AACC Method 76.13; Megazyme International Ireland Ltd., Wicklow, Ireland) based on the enzymatic hydrolysis method, following McCleary et al. [12]. Lyophilized samples (0.03 g) were pre-extracted with 10 mL of 80% ethanol to remove free glucose; the complete starch hydrolization to glucose was achieved by a combined action of α-amylase and amyloglucosidase; the reaction took place by heating up (80 °C) the samples in the presence of an 8 mL sodium acetate buffer (1.2 M) at pH 3.8 and 0.1 mL of thermostable α-amylase (3000 U/mL) for 15 min. After cooling to 50 °C, 0.1 mL of amyloglucosidase (3300 U/mL) was added which catalyzes complete hydrolysis of the maltodextrins to glucose within 30 min.

The starch concentration was determined as liberated glucose. Then, the mixture was centrifuged at 13,000 rpm (10 min); the supernatant was collected and filtered through a membrane filter (0.45 μm).

Samples were analyzed according to Rohrer [13] using the ICS 3000 System (Dionex-ThermoFisher Scientific, Waltham, MA, USA) and high-performance anion exchange chromatography with pulsed amperometric detection (ED50; Dionex-ThermoFisher Scientific, Waltham, MA, USA), equipped with a Carbopac PA-1 column (CarboPac PA1 Analytical, 4 × 250 mm; particle size 10 μm) (Dionex-ThermoFisher Scientific, Waltham, MA, USA), maintained at 30 °C. Glucose was eluted with NaOH (150 mM) at a flow rate of 1.0 mL min^{-1} for 15 min. Glucose was identified by a comparison of the retention times with the standard. Peak areas were analyzed using the Dionex Chromeleon software (version 6.80, Dionex-ThermoFisher Scientific, Waltham, MA USA).

Total simple carbohydrates were extracted from 15–30 mg of a lyophilized sample and adding 15 mL of ultrapure water and using shaking water baths (Foss, Padova, Italy) for 45 min at room temperature. Then, the mixture was centrifuged at 13,000 rpm (10 min); the supernatant was collected, filtered through a 0.45 μm membrane filter, and analyzed as previously described.

Carbohydrates were identified by a comparison of the retention times with those of sugar standards. Peak areas were analyzed using the Dionex Chromeleon software (version 6.80, Dionex-ThermoFisher Scientific, Waltham, MA, USA).

2.3.4. Organic Acids

The lyophilized sample (0.3 g) was placed in a 50 mL tube and added to 20 mL metaphosphoric acid (0.1%). The sample was mixed in a shaking water bath at room temperature for 15 min, then centrifuged (12,000 rpm, 4 °C for 15 min) (SR16L, ThermoFisher Scientific, Waltham, MA, USA). The supernatant was collected, filtered, and stored at 4 °C until analyzed according to González-Castro et al. [14] with some modification.

Organic acids were separated by the ICS 3000 HPLC System (Dionex-ThermoFisher Scientific, Waltham, MA, USA) equipped with: An isocratic pump, a 10 μL injection loop, an AS-DV auto-sampler, Hydro-RP 80A column (250 × 4.60 mm) (Phenomenex Inc., Castel Maggiore, BO, Italy), maintained at 30 °C combined with a UV-visible detector (RLSC Diode Array Detector, Dionex-ThermoFisher Scientific, Waltham, MA, USA), set to a wavelength of 210 nm and the Chromeleon version 6.50 software.

The eluent consisted of 100 mM Na_2SO_4 at pH 2.6 adjusted with methanesulfonic acid at a flow rate of 1 mL min^{-1}. Individual organic acids were identified by comparing retention times with those of available standards.

2.3.5. Volatile Compounds

Volatile compounds were extracted by the HS-SPME technique using a DVB-CAR-PDMS fiber (Supelco, Bellefonte, Pa., USA) according to Gambacorta et al. [15], with some modifications. The freeze-dried powder sample (0.5 g) and 10 μL of internal standard (2-heptanone) in water at 8.15 μg mL^{-1} were placed in a 12 mL screw-cap vial, tightly capped with a PTFE-silicon septum, and conditioned for 10 min at 50 °C. Then, the fiber was introduced into the headspace of the vial for 30 min, removed, and inserted into the gas-chromatography injection port. Desorption of volatiles from the fiber took place in a split mode (1:20 ratio) at 220 °C. The separation of volatile compounds was performed by a Trace 1300 gas chromatograph (Dionex-ThermoFisher Scientific, Waltham, MA, USA) equipped with a VF-WAXms capillary column (Agilent, Santa Clara, CA, USA), 20 m length × 0.10 mm ID x 0.1 μm film, and coupled with an ISQ single quadrupole mass spectrometer (ThermoFisher Scientific, Waltham, MA, USA). The chromatographic conditions were: Oven, 50 °C (0.1 min) to 180 °C at 13 °C min^{-1}, to 220 °C at 18 °C min^{-1}, held for 3 min; detector, source temperature 250 °C; transfer line temperature 250 °C; carrier gas, helium at a constant flow of 0.4 mL min^{-1}. The impact energy was 70 eV. Data were acquired using the full-scan mode in the range of 33 to 280 m/z at an acquisition rate of 7.2 Hz.

Volatiles were identified by comparing the experimental spectra with those obtained by the available pure standard compounds and with those reported in the NIST Library [16] and quantified using relative areas that related the 3-pentanone as an internal standard. The acquisition and processing of peaks were carried out using the Xcalibur v 2.0 software (ThermoFisher Scientific, Waltham, MA, USA).

2.3.6. Phenolic Compounds

The lyophilized sample (0.05 g) was placed in a 2 mL Eppendorf tube and added to 1 mL of 80% methanol in water. The sample was mixed for 1 min, sonicated for 5 min, and then centrifuged (4000 g 4 °C for 20 min). The clear supernatant was diluted 1:1 with an acetonitrile:water (10:90, vol/vol) solvent mixture containing 0.1% formic, and filtered using re-generated cellulose filters of 0.22 μm pore diameter. The analysis was performed according to Pasqualone et al. [17] with some modification using the UHPLC Dionex Ultimate 3000 RS system (quaternary pump, auto-sampler, column oven, and PDA), coupled by the HESI-II probe with the LTQ Velos Pro ion trap mass spectrometer (Dionex-ThermoFisher Scientific, Waltham, MA, USA). The separation of compounds was performed on a Hypersil GOLD aQ C18 column, 100 mm in length, 2.1 mm ID, and 1.9 μm particle size (Waters, Milford, MA, USA) maintained at 30 °C. A binary mobile phase was used: (A) 0.1% formic acid in water and (B) 0.1% formic acid in acetonitrile, at a constant flow of 0.2 mL/min. The gradient program of solvent B was as follows: 0–30 min from 10% to 70%, 30–33 min isocratic 70%, 33–33.1 min from 70% to 10%. The MS conditions were: Capillary temperature 320 °C; source heater temperature 280 °C; nebulizer gas N_2; sheath gas flow 30 arbitrary units; auxiliary gas flow 15 arbitrary units; capillary voltage—2.8 kV, S-Lens RF Level 60%. Data were acquired in a negative ionization mode.

Phenolic compounds were identified by comparing elution times, molecular ions, MS/MS fragmentation patterns of the experimental spectra with those obtained by the available pure standard compounds or by tentative methods using reported data from the literature. Calibration curves were created to obtain quantification results and were based on the UV signal of each available standard. When no commercial standard was available, a similar compound from the same phenolic group was used as a standard.

2.4. Statistical Analysis

A one-way analysis of variance was performed using the Statistical Analysis Software (SAS, Cary, NC, USA). The least significant difference (LSD) test ($p = 0.05$) was used to establish differences between means.

For a visual analysis of data, the Principal Component Analysis (PCA) was performed using the PAST3 Software (http://folk.uio.no/ohammer/past) [18] on mean standardized ((x-mean)/standard deviation) data. The data matrix considered all genotypes with relative replications. To avoid the presence of highly correlated variables, data were subjected, before the PCA, to the analysis of correlation. The variables which presented a correlation coefficient higher than or equal to 0.8 were grouped, and only one of them was considered in the data matrix.

3. Results and Discussion

Refer to Table 1 for acronyms of garlic genotypes.

3.1. Bulb Appearance and Biometrical Traits

According to IPGRI descriptors, the Commercial Genotype (CG) used in this study was grouped as a 'flat globe' type with a 'regular multi-cloved radial' bulb structure. Among the landraces, the bulb shape ranged from 'broad oval' for 'Panni', 'PanniD', and 'Monteleone', to 'flat globe' for 'Anzano' and 'Cortigli' and to 'globe' for 'Peschici'. All the landraces studied had an 'irregular' bulb structure, except for 'Peschici', which was in the 'regular two-fan group' (Table S1 and Figure S1).

Concerning the colour of bulbs (with dry tunics) and cloves (with thin skin) (Table S1), all landraces (except for 'Monteleone') were characterized by an evident yellow component (+b*) and a very small greenish component (slightly −a*), with a resulting hue and saturation C* values allowing colour identification as creamy. Bulbs and cloves of the 'Peschici' landrace distinctively presented a more vivid colour as indicated by the highest C* values. The instrumental data supported the reddish tonality of cloves of the 'Monteleone' landrace by the highest +a*, the lowest −b*, along with the highest h° values.

A high variability was observed in biometrical traits of both bulbs and cloves (Table S2). The Principal Component Analysis (PCA), carried out on biometrical data, showed that the first two PCs explained 81% of the total variability, attributing 59.1% to PC1 and 21.9% to PC2 (Figure 1). PC1 discriminates the landraces from the Panni Municipality, clustered on the left-hand side of the axis. 'PanniD' was separated from all the other genotypes for the highest concentration in dry matter (DM). The lower moisture concentration of its bulbs is indicative of a higher resistance to post-harvest deterioration. This characteristic may explain the prolonged shelf-life attributed to 'PanniD', known by its local name '*Aglio Durevole di Panni*' meaning a long storage life. PCA also discriminates 'Peschici' from the other genotypes for the highest number of cloves per bulb (Number-c) (highly and positively correlated to PC1) and by the lowest weight of cloves (W-c) (highly and positively correlated to PC2).

This high variability in morpho-biometrical traits in the examined landraces is, of course, the result of the long process of (un)conscious selection made by farmers and of their continued evolution in a certain eco-geographical area [9]. Although garlic is propagated asexually, using the cloves from the previous growing season, there is a great diversity in morphological (*agronomic*) characters, even between genotypes grown in the same areas for a long time. This is probably mostly a consequence of the accumulation of natural mutations and their evolving over time due to the selection pressures and cultural data in the minds of local farmers [19,20].

The variability of the morpho-biometrical traits of garlic genotypes has been widely reported [19,21]. The biometrical traits of the collected landraces are of a high standard in terms of bulb and clove weight in comparison with other Italian landraces [19,22,23] and also in comparison with the tested CG. The DM accumulation of the cloves of collected landraces was also in line with the values reported in various landraces such as those grown in Greece [24] and several genotypes grown in Spain [25].

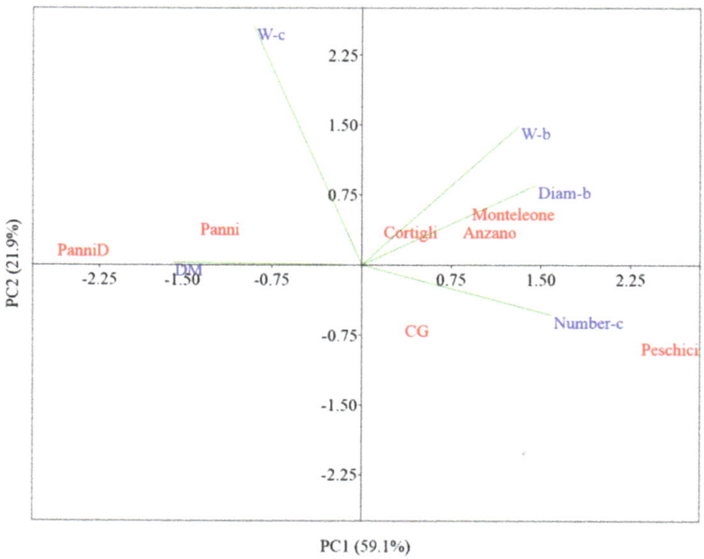

Figure 1. Principal component analysis bi-plot (PC1 vs. PC2) showing the spatial distribution of the main biometrical traits as affected by genotypes. The considered biometrical traits were: Weight of cloves (W-c), number of cloves per bulb (Number-c), weight of bulbs (W-b), diameter of bulbs (Diam-b), concentration of dry matter (DM).

3.2. Mineral Concentration

The concentration of ashes, cations, and anions is reported in Table 2.

Table 2. Concentration of ashes and inorganic ions (g kg^{-1} dw) in bulbs of garlic genotypes.

Genotype [1]	Ashes	Inorganic Cations					Inorganic Anions				
		Total	Na	K	Mg	Ca	Total	NO_3	Cl	PO_4	SO_4
'CG'	34.1 e[3]	8.93 d	0.74 a	7.81 d	0.06 d	0.32 c	1.609 c[3]	0.000 c	0.707 b	0.200 c	0.702 b
'Cortigli'	55.0 b	15.9 b	0.68 a	14.5 b	0.14 a	0.58 b	1.963 b	0.094 b	0.618 c	0.419 b	0.833 a
'Peschici'	70.0 a	22.0 a	0.76 a	20.5 a	0.11 b	0.55 b	2.766 a	0.140 a	1.130 a	0.633 a	0.864 a
'Monteleone'	68.7 a	21.7 a	0.89 a	20.4 a	0.11 b	0.37 c	1.452 c	0.081 b	0.627 bc	0.366 b	0.379 d
'Anzano'	49.4 c	13.8 c	0.72 a	12.0 c	0.07 cd	1.09 a	1.651 c	0.143 a	0.602 c	0.380 b	0.527 c
'Panni'	42.7 d	14.2 c	0.44 b	13.1 c	0.09 bc	0.61 b	1.351 d	0.030 c	0.588 c	0.225 c	0.508 c
'PanniD'	23.7 f	5.90e	0.30 b	5.22 e	0.12 ab	0.26 c	0.732 e	0.014 c	0.327 d	0.161 c	0.231 e
Significance [2]	***	***	***	***	***	***	***	***	***	***	***

[1] Refer to Table 1 for acronyms of garlic genotypes. [2] *** Significant at $p \leq 0.001$. [3] Different letters within the column indicate significant differences at $p = 0.05$.

The content of ash and cations ranged from the highest values in 'Peschici' and 'Monteleone' to the lowest in 'PanniD'. On average, landraces were richer than the CG in ash (51.6 g kg^{-1} dry weight—dw) and cations (15.6 g kg^{-1} dw), particularly K. 'Cortigli' along with 'PanniD' had the highest Mg concentration, whereas 'Anzano' had the highest Ca concentration. Regarding anions, the total concentration in the landraces (1.6 g kg^{-1} dw, on average) was in line with that of the CG; 'Peschici' showed the highest values of all the individual anions, followed by 'Cortigli' (SO_4, NO_3, PO_4). Among the examined landraces, 'PanniD' was the poorest in all anions and cations, except for Mg.

The genotype and/or environment have been cited to be the main determinants in mineral and ash concentration of garlic bulbs [3,23,24]. Apart from the genotypic characteristics, the high incidence of ash and minerals in 'Peschici' in comparison with all the other landraces could also be explained as an effect associated with a higher DM concentration, due to the specific pedo-climatic conditions of this site such as the sandy soils with a limited use of irrigation, and the dry climate (a coastal area characterized by low rainfall and warmer temperatures) (Table 1).

The amount of ash concentration in our landraces (2.0 g 100 g^{-1} fresh weight—fw, on average) was slightly higher than the values reported in Greek garlic landraces [24] and slightly lower than in Indian garlic bulbs [26]. The K levels in our landraces (562 mg 100 g^{-1} fw, on average) can be considered in line with those reported in official food composition databases, such as that of the European Food Safety Authority (EFSA) (579 mg 100 g^{-1} fw) and of the Italian National Centre of Agriculture (INRAN-CREA) (600 mg 100 g^{-1} fw), and higher than those reported by the United States Standard References (USDA) (401 mg 100 g^{-1} fw). The levels of K in our landraces were similar to those found in Greek [24] and in Sicilian (Italy) [23] landraces, while levels of Ca and Mg were significantly lower.

3.3. Starch and Simple Carbohydrates

The values of starch and simple carbohydrates are reported in Table 3.

Table 3. Concentration of starch, simple carbohydrates, and organic acids (g kg^{-1} dw) in bulbs of garlic genotypes.

Genotype [1]	Starch	Simple Carbohydrates				Organic Acids					
		Total	Glucose	Sucrose	Fructose	Total	Oxalic	Citric	Pyruvic	Malic	Ascorbic
'CG'	2.14 b[3]	6.22 a	0.18 a	5.59 b	0.45 a	32.4 ab[3]	2.94 bc	26.5 a	1.96 c	0.94 e	<0.001 c
'Cortigli'	0.83 de	1.28 b	0.19 a	0.84 c	0.25 ab	19.6 d	4.00 b	9.52 c	3.10 c	2.91 de	0.02 ab
'Peschici'	1.30 cd	0.54 b	0.06 a	0.33 c	0.15 b	19.1 d	4.45 b	7.51 c	3.88 bc	3.26 cde	0.02 a
'Monteleone'	1.78 bc	0.57 b	0.07 a	0.29 c	0.21 ab	40.7 a	9.09 a	18.7 b	7.09 a	5.77 bc	0.03 a
'Anzano'	3.06 a	0.47 b	0.10 a	0.15 c	0.22 ab	17.8 d	4.25 b	6.60 c	3.19 c	3.72 cd	0.02 a
'Panni'	1.23 d	7.38 a	0.10 a	7.13 a	0.15 b	30.2 bc	3.32 bc	13.3 bc	6.38 ab	7.20 ab	0.01 bc
'PanniD'	0.46 e	0.50 b	0.07 a	0.37 c	0.06 b	22.5 cd	1.88 c	8.62 c	3.17 c	8.85 a	0.01 bc
Significance [2]	***	***	ns	***	**	***	***	***	***	***	***

[1] Refer to Table 1 for acronyms of garlic genotypes. [2] Significance—ns and **, *** not significant or significant at $p \leq 0.01$ and $p \leq 0.001$, respectively. [3] Different letters within the column indicate significant differences at $p = 0.05$.

Genotypes differed distinctively in the amount of total carbohydrates and individual soluble sugars, except for glucose (0.11 g kg^{-1} dw, on average). 'Panni', as well as the CG, showed the highest level of total simple carbohydrates, mainly due to the contribution of sucrose. The CG had the highest value of fructose, followed by 'Monteleone' and 'Anzano'. Since extensive studies on several vegetables have correlated total simple sugars with perceived sweetness [27], among the examined landraces 'Panni' had the highest sweet flavour perception.

On average, the total carbohydrate content in our landraces (1.79 g kg^{-1} dw) was lower than that of the CG (Table 2). Total sugars, sucrose, glucose, and fructose in our landraces were lower than those reported in other landraces from North-Central Italy [21] and Greece [24] and in several genotypes grown in Spain [25]. The value of total sugars in our landraces was also lower than that reported in the INRAN-CREA (8.4 g 100 g^{-1} fw) or the USDA (1.0 g 100 g^{-1} fw) databases.

On average, among the studied landraces, the greatest fraction of simple carbohydrates was represented by sucrose (85%), followed by fructose (10%) and glucose (5%) (Table 2). Sucrose is reported as the main carbohydrate in genotypes collected from various Italian regions [22,28], Spain [25] and Greece [24]. In rare cases, fructose [25] or glucose [29] has been reported as the main sugar in garlic.

Concerning the starch concentration (Table 2), 'Anzano', followed by CG, was the richest and 'PanniD' the poorest. On average, the starch content in our landraces (1.44 g kg^{-1} dw; 0.054 g 100 g^{-1} fw) was lower than that of the CG and other genotypes [26,29]. Although starch is generally the most widespread carbohydrate reserve in the plant kingdom, in the examined landraces starch was detected in very low concentrations. Similarly, in other Italian local landraces, only traces of starch have been identified (<0.06 g 100 g^{-1} fw), while fructans have been detected in a larger proportion [10]. Thus, it is possible to suppose that the storage function in the *Allium* spp. bulb tissue could be attributed mainly to fructans.

3.4. Organic Acids

The organic acid concentrations of the studied garlic genotypes are presented in Table 3. 'Monteleone' along with the CG showed the highest total organic acid concentrations. In the former, a high content of citric, oxalic, and pyruvic acids was detected, while the CG distinctively showed the highest amount of citric acid and the lowest amount of malic and ascorbic acids. 'Cortigli', 'Peschici', 'Anzano', and 'PanniD' had the lowest total concentrations of organic acids, and 'Panni' had the lowest values of all the individual organic acids, except for malate.

On average, the total organic acid contents in our landraces (25.0 g kg^{-1} dw; 1.0 g 100 g^{-1} fw) were lower than that in the CG (Table 2), and in other genotypes such as those from Greece [24] (2.79 g 100 g^{-1} fw) and from Latvia (3.87 g 100 g^{-1} dw) [30].

In the examined landraces, the organic acid profile was represented by 43% citric, 21% malic, 18% oxalic, and 18% pyruvic acid (Table 2). The organic acid composition of our landraces varied greatly compared with other literature data. A wider composition of organic acids was found in garlic samples from Latvia by Priecina and Karklina [30] and in Italian garlic varieties by Ritota et al. [28], additionally reporting the presence of fumaric, formic, succinic, quinic, salicylic, and tartaric acids, probably due to a different assay implemented.

In the collected landraces, and particularly in 'Monteleone', 'Peschici', and 'Anzano', ascorbic acid (the main biologically active form of vitamin C) was distinctively higher than in the CG (Table 2). However, the averaged values were lower than those reported for some Italian varieties by Fratianni et al. [6] and those reported in the INRAN-CREA (5 mg 100 g^{-1} fw) and in the USDA (31.2 mg 100 g^{-1} fw) food composition databases.

3.5. Volatile Compounds

The volatile compound concentration in the studied garlic genotypes is reported in Table 4.

The precursors of the volatile organic S-compounds in garlic are γ-glutamyl cysteine and cysteine sulfoxides, including deoxyallin and mainly alliin (S-allyl-cysteine-S-oxide) [4], an odourless derivative of cysteine. This latter is enzymatically hydrolyzed (allinase or alliin lyase, E.C.4.4.1.4) into a mixture of both volatile and non-volatile S-containing compounds, after the breakage of the tissue caused by cutting, mastication, and cooking [31]. The volatile S-containing compounds include thiosulphinates, which are very unstable and are transformed into compounds belonging to Sulphides-di-allyl-sulphide (DAS), di-allyl-disulphide (DADS), di-allyl-trisulphide (DATS) or into compounds belonging to 'Vinyldithiins' (cyclic sulfur-containing compounds) or into others [31]. All of them are responsible for the typical flavour of garlic and the protective effects against cardiovascular diseases [1,2,32].

In the current study, the main fraction of volatile substances was the S-compounds (94% for landraces; 79% for the CG), with the non-sulphur compounds (including aldehydes, and 'others') representing only a minor fraction (6% for landraces; 21% for the CG). Among the examined genotypes, a total of 28 compounds were identified in the volatile fraction, 16 of them were S-compounds, while the other compounds were hydrocarbons, alcohols, and ethers (Figure S2).

The total volatile concentration in CG bulbs was distinctively higher than those of landraces (15.0 vs. 2.2 mg kg^{-1} dw, on average), mainly due to the amount of S-compounds (37.8 vs. 2.1 mg kg^{-1} dw, on average). Among the non-sulphur volatile compounds, a lower concentration of total aldehydes (0.04 mg kg^{-1} dw, on average) and 'other' compounds (0.10 mg kg^{-1} dw, on average) was found in the examined landraces. In particular, the landraces were completely devoid of two out of four of the determined aldehydes—'2-butenal-2-methyl' and '4-heptenal'.

Di-methyl-sulphide, trans-propenyl-methyl-di-sulphide, mercaptoacetic acid, 3-vinyl-4H-1, 2-dithiin, and Allicin (S-oxo-di-allyl-di-sulphide) (among the S-compounds), and 4-heptenal-enantic aldehyde (among aldehydes), were detected in very low amounts (<0.01 unit).

Except for the 1-propanethiol S-compound, the genotype distinctively affected the concentration and the profile of volatile compounds.

Table 4. Profile of volatile compounds (mg kg^{-1} dw) in bulbs of garlic genotypes.

RT [1]	Volatile Compounds	MW [2]	*m/z* ions	Genotype [3]							Significance [4]
				CG	Cortigli	Peschici	Monteleone	Anzano	Panni	PanniD	
	total			*14.98 a* [5]	*1.72 b*	*3.35 b*	*1.50 b*	*2.11 b*	*1.81 b*	*2.69 b*	***
	S-compounds			*11.77 a*	*1.64 b*	*3.12 b*	*1.43 b*	*2.00 b*	*1.70 b*	*2.48 b*	***
1.43	di-methyl-**sulphide** [6]	62	62,47,35	0.044 a	0.001 b	0.000 b	0.001 b	0.002 b	0.000 b	0.000 b	*
1.71	1-propanethiol	76	76,47,43	0.149 a	0.030 a	0.030 a	0.016 a	0.034 a	0.037 a	0.045 a	ns
2.03	propylene-**sulphide**	74	41,74,46	0.219 a	0.006 b	0.006 b	0.007 b	0.015 b	0.009 b	0.022 b	***
3.57	di-allyl-**sulphide** (DAS) [6]	114	45,41,73,39	0.316 a	0.009 b	0.023 b	0.014 b	0.010 b	0.012 b	0.021 b	***
4.86	methyl-2-propenyl-**disulphide**	120	120,41,45	0.906 a	0.017 b	0.068 b	0.016 b	0.039 b	0.008 b	0.007 b	***
4.91	trans-propenyl-methyl-**disulphide**	120	73,120,45	0.038 a	0.001 b	0.009 ab	0.001 b	0.006 b	0.001 b	0.005 b	*
5.89	di-methyl-**trisulphide**	126	126,45,79	0.187 a	0.001 b	0.021 b	0.001 b	0.005 b	0.000 b	0.000 b	***
6.69	mercaptoacetic acid	92	47,45,92	0.000 d	0.002 abc	0.001 abcd	0.001b cd	0.000 cd	0.003 a	0.003 ab	*
6.86	di-allyl-**disulphide** (DADS) [6]	146	41,81,39	0.929 a	0.304 c	0.478 b	0.335 c	0.369 bc	0.281 c	0.402 bc	***
7.93	methyl-allyl-**trisulphide**	152	87,73,45	0.522a	0.054 cd	0.306 b	0.039 d	0.142 c	0.017 d	0.013 d	***
9.27	3-vinyl-1,2-dithiin	144	45,144,97	2.203 a	0.255 b	0.424 b	0.201 b	0.303 b	0.286 b	0.439 b	*
9.49	2-vinyl-1,3-dithiane	146	74,72,45	0.222 a	0.024 b	0.037 b	0.020 b	0.051 b	0.021 b	0.046 b	***
9.73	di-allyl-**trisulphide** (DATS) [6]	178	73,41,113	1.219 a	0.521 c	1.138 ab	0.461 c	0.528 c	0.575 bc	0.771 abc	*
10.20	3-vinyl-1,2-dithiacyclohex-5-ene	144	72,71,144	4.766 a	0.389 b	0.555 b	0.305 b	0.457 b	0.435 b	0.682 b	***
10.52	3-vinyl-4H-1,2-dithiin	144	72,71,144	0.000 d	0.008 bc	0.002 cd	0.004 bcd	0.021 a	0.010 b	0.017 a	***
10.56	Allicin (S-oxo-di-allyl-**disulphide**)	162	41,45,72	0.049 a	0.011 b	0.022 b	0.007 b	0.019 b	0.009 b	0.009 b	***
	Non-sulphur compounds			*3.22 a*	*0.09 b*	*0.24 b*	*0.07 b*	*0.10 b*	*0.10 b*	*0.31 b*	***
	Aldehydes			*1.01 a*	*0.01 b*	*0.11 b*	*0.00 b*	*0.01 b*	*0.02 b*	*0.08 b*	***
2.71	2-butenal (crotonaldehyde) [6]	70	41,39,70	0.564 a	0.009 b	0.107 b	0.000 b	0.013 b	0.017 b	0.072 b	***
3.05	2-butenal-2-methyl (2-methyl-2-pentenoic aldehyde)	84	55,84,29,27	0.248 a	0.000 b	0.000 b	0.000 b	0.000 b	0.000 b	0.000 b	**
3.79	4-heptenal (enantic aldehyde)	112	68,67,55	0.062 a	0.000 b	0.000 b	0.000 b	0.000 b	0.000 b	0.000b	***
5.93	nonanal (pelargonaldehyde) [6]	142	57,98,43,56	0.185 a	0.000 b	0.002 b	0.001 b	0.001 b	0.001 b	0.004 b	***
	Others			*2.20 a*	*0.08 b*	*0.13 b*	*0.07 b*	*0.09 b*	*0.08 b*	*0.14 b*	***
1.28	Cyclopropane	42	42,41,39,27	0.482 a	0.023 b	0.034 b	0.019 b	0.033 b	0.024 b	0.037 b	***
1.38	heptane (di-propyl-methane)	100	43,41,29,57	0.220 a	0.000 b	0.000 b	0.000 b	0.000 b	0.000 b	0.000 b	***
3.20	2-propen-1-ol (allyl-alcohol)	58	57,29,31	0.438 a	0.021 b	0.037 b	0.019 b	0.029 b	0.023 b	0.032 b	***
5.11	Pentadecane	212	57,43,71	0.124 a	0.002 b	0.001 b	0.004 b	0.001 b	0.005 b	0.004 b	*
5.33	di-iso-deciyl-ether	298	43,57,41	0.230 a	0.003 b	0.008 b	0.003 b	0.002 b	0.003 b	0.009 b	***
6.07	tetradecane	198	57,43,71	0.083 a	0.003 b	0.015 b	0.002 b	0.000 b	0.002 b	0.010 b	**
6.29	benzene,m-ditert-butyl	190	175,57,41	0.077 a	0.004 b	0.008 b	0.003 b	0.000 b	0.004 b	0.011 b	**
7.33	1,4-dihydro-2,3-benzoxathin-3-oxide	168	104,103,105	0.551 a	0.020 b	0.025 b	0.015 b	0.026 b	0.022 b	0.033 b	***

[1] Retention time (min). [2] Molecular weight. [3] Refer to Table 1 for acronyms of garlic genotypes. [4] Significance—ns and *, **, *** not significant or significant at $p \leq 0.05$, $p \leq 0.01$, $p \leq 0.001$, respectively. [5] Different letters within the row indicate significant differences at $p = 0.05$. [6] Comparison with standards.

A large variation in the amount of (S)-volatile compounds was observed between the GC and the collected landraces (15.0 vs. 2.2 mg kg^{-1} dw—total volatile compounds; 12.0 vs. 2.0 mg kg^{-1} dw—S-compounds). Apart from the genotype, other factors such as 'environment' may affect volatile compound concentrations, as widely proved in the specific literature on garlic [1,33]. The large difference in concentration of (S)-volatile compounds between the CG and landraces could be imputable to different pre- and/or post-harvest management. It is well known that (S) fertilization is strictly correlated to the accumulation of garlic volatile S-compounds [34] and of organo-sulfur-compounds [35], precursors of volatile S-compounds [36]. The collected landraces might have accumulated lower (S) volatile compounds, probably due to the poor S availability in the marginal soils where these plants are collected, also characterized by a low organic matter content and no fertilizer application (in normal growing conditions various fertilizers are used containing significant amounts of sulfur, thus improving S availability). The different post-harvest storage time between the CG and landraces could have also affected their volatile (S)-compound contents. Several studies report that the concentrations of volatile [37] and non-volatile [38] S-compounds are higher in stored than in fresh material. The total amount of volatile (S)-compounds was lower than that found in other studies [4,39]. The allicin in our landraces, representing less than 1% of the total volatile compounds and of the S-compounds, was slightly lower than that found in five endemic Italian varieties [6] and much lower than that found in red garlic from Argentinian germplasm [40] and in Croatian garlic [8].

The qualitative profile of the volatile component in the current study was instead in line with that found in numerous studies reported in the garlic literature [4,32,37,39,41,42]. Genotype and/or 'environment' are also the main determinants of the composition of volatile compounds [1,33].

In the current study, in the examined landraces the five most abundant S-compounds (representing 93% of the S-compounds) were 'di-allyl-trisulphide' (DATS) (0.67 mg kg^{-1} dw; 32% of S-compounds), followed by 3-vinyl-1,2-dithyocyclohex-5-en' (0.47 mg kg^{-1} dw; 23% of S-compounds), di-allyl-disulphide (DADS) (0.36 mg kg^{-1} dw; 18%), 3-vinyl-1,2-dithiin (0.32 mg kg^{-1} dw; 15%), and methyl-allyl-tri-sulphide (0.10 mg kg^{-1} dw; 5%). The quantitative analysis in the CG showed that five compounds represented 85% of the total S-compounds: 3-vinyl-1,2-dithyocyclohex-5-ene (40% of the total), 3-vinyl-1,2-dithiin (18%), DATS (10%), and DADS (8%) and methyl-2-propenyl-di-sulphide (8%).

The sulphide S-compounds were prevalent in the collected landraces, while the S-compounds belonging to the 'Vinyldithiins' family were prevalent in the CG. Although initially controversial, the S-compounds belonging to the 'Vinyldithiin' family are now considered the major components of fresh garlic and some of them have been found to have a very high flavour dilution factor (FD ≥ 1) indicating an intense pungent odour [32,43]. Accordingly, the CG should have a more pungent aroma than the collected landraces. It can be not excluded that the examined landraces were selected over the time as a conscious and intentional aim by growers to reduce the volatile (S)-compounds, responsible for breath and sweat smelling of garlic, which can last for days. Thus, landraces showed this drawback to a more limited extent.

3.6. Phenolic Compounds

Phenolic compounds are reported in Table 5. In the examined genotypes, the phenolic acids (six compounds) represented most of the total phenols (83% on average), and among them, ferulic acid (524 mg kg^{-1} dw, on average) was the most abundant, followed by the caffeic acid derivates—caffeic acid-O-hexoside-1 (211 mg kg^{-1} dw, on average) and caffeic acid-O-hexoside-2 (203 mg kg^{-1} dw, on average). The flavonoid component (six compounds), represented 16% of the total phenols, and among them, iso-rhamnetin (120 mg kg^{-1} dw, on average) and rutoside-1 (60 mg kg^{-1} dw, on average) were the most abundant. Less than 0.3% of total phenols were unknown compounds. A 3-hydroxy-methyl phenol was found only in the CG. Among the flavonoids, quercetol (quercetin), patuletin, hyperoside (hyperin), iso-quercitrin (quercetin 3-β-D-glucoside), and coumaroyl-quinic acid among the phenolic acids were detected at a very low rate (<0.5 unit).

Table 5. Phenolic compounds (mg kg^{-1} dw) in the bulb of garlic genotypes.

Phenols	RT (min)	UV Max (nm)	[M-H]⁻	m/z ions	Genotype [1]							Significance [2]
					CG	Cortigli	Peschici	Monteleone	Anzano	Panni	PanniD	
Caffeic acid-O-hexoside-1 [5]	1.58	379	341	179, 135	404.1 a[3]	186.9 a	181.5 a	141.8 a	169.7 a	192.8 a	205.3 a	ns
Caffeic acid-O-hexoside-2 [5]	1.90	379	341	179, 135	119.6 b	220.0 a	231.8 a	187.8 a	196.0 a	230.3 a	235.2 a	*
Caffeic acid [4]	2.18	323	179	135, 107	6.81 d	10.8 c	5.61 d	14.3 b	14.4 b	10.7 c	17.8 a	***
Ferulic acid [5]	2.74	320	195	177, 89	87.6 d	518.6 bc	851.4 a	790.9 ab	611.6 abc	405.6 c	404.8 c	**
Caftaric acid [4]	3.49	308	311	179	1.87 a	14.7 a	12.2 a	30.2 a	29.0 a	30.5 a	0.00 a	ns
Coumaroylquinic acid [5]	11.33	306	337	191, 163	0.00 b	0.17 b	0.73 b	0.00 b	0.68 a	0.00 b	0.00 b	***
Rutoside-1 (Rutin) [4]	9.21	354	609	609, 300	0.00 b	68.3 ab	91.2 a	102.1 a	68.1 ab	38.3 ab	50.3 ab	*
Iso-rhamnetin [6]	10.05	374	315	165	0.00 b	137.8 a	157.9 a	142.3 a	139.7 a	126.7 a	132.3 a	***
Hyperoside (Hyperin) [4]	10.20	355	463	463, 300	0.00 b	0.00 b	0.20 b	0.00 b	0.00 b	0.22 b	1.09 a	**
Iso-quercitrin (Quercetin 3-ß-d-glucoside) [4]	10.77	355	463	463, 300	0.00 b	0.17 b	0.92 ab	1.39 a	0.00 b	0.00 b	0.00 b	*
Patuletin [6]	12.97	374	331	151	0.00 b	0.00 b	0.00 b	0.20 ab	0.43 a	0.00 b	0.00 b	*
Quercetol (Quercetin) [4]	14.22	374	301	151, 301	0.49 a	0.00 c	0.00	0.00 c	0.00 c	0.00 c	0.22 b	***
3-Hydroxy-methyl phenol [5]	20.03	280	123	123	1.94 a	0.00 b	0.00 b	0.00 b	0.00 b	0.00 b	0.00 b	***
Sum of un-identified phenols					4.2 c	4.4 c	3.0 cd	1.0 e	6.7 b	2.0 de	8.8 a	***
Phenolic acids					526.5 c	954.3 ab	1193 a	1128 ab	1030.3 ab	881.2 b	888.8 b	**
Flavonoids					0.49 d	203.8 bc	248.0 ab	271.1 a	207.5 abc	155.7 c	189.7 bc	***
Total phenols					532.8 c	1162 ab	1444 a	1400 a	1,245 ab	1039 b	1087 b	***

[1] Refer to Table 1 for acronyms of garlic genotypes. [2] Significance—ns, *, **, and *** not significant or significant at $p \leq 0.05$, $p \leq 0.01$, and $p \leq 0.001$, respectively. [3] Different letters within the row of samples indicate significant differences at $p = 0.05$. [4] Comparison with standards. [5] Comparison with the caffeic acid standard. [6] Comparison with the quercetol standard.

Except for caftaric and caffeic acid-O-hexoside-1 acids, the genotype distinctively affected the concentrations and the profile of phenols.

On average, the phenolic concentration in our landraces was much higher than in the CG (1230 vs. 533 mg kg^{-1} dw), due to the high contribution of both components. The CG showed a very small amount of flavonoids (only quercetol).

The high level of phenols in all landraces in comparison with the CG, could be explained as an effect associated with the genotype, but also with the environment [1], the pedo-climatic conditions of the growing site and the technical practices adopted by growers. It is well known that the biosynthesis of phenols/flavonoids is upregulated in response to a wide range of abiotic stresses/factors, ranging from nutrient depletion to cold and drought stress [44,45], aiming to effectively counter the stress-induced oxidative damage. Thus, apart from the genotypic characteristics, the incidence of phenols in the landraces could be related to harsh growing conditions in terms of climate and/or to the scarce availability of nutrients/water, since no or low inputs (water, nutrients, chemicals) were provided by the landrace growers.

The levels of phenolic compounds in our landraces were also substantially greater than that reported for Polish garlic (11.5 mg kg^{-1} dw) [7] and particularly in phenolic acids (0.1–252.1 mg kg^{-1} dw) [46], as well as for the local garlic from Namhae-gun (Korea) both in total phenolic acids (17.9 mg kg^{-1} dw) and in flavonoids (29.9 mg kg^{-1} dw) [47].

The antioxidant, antibacterial, anti-inflammatory, anti-proliferative, and chemopreventive properties of phenolics and flavonoids in vegetables and Allium spp. [2,48] are well known, thus the contribution of the observed landraces to the protection and preservation of human health should be emphasized due to their being very rich in these compounds.

Among the examined landraces, 'Peschici' and 'Monteleone', followed by 'Cortigli' and 'Anzano' showed the highest total phenol concentrations. The most abundant compounds were ferulic acid along with caffeic acid-O-hexoside-2 (phenolic acids) and iso-rhamnetin and rutinoside-1 (flavonoid compounds). In this study, we found significantly higher amounts of ferulic acid both in comparison with the tested GC and also compared with those found in several genotypes in Spain –3.5 mg kg^{-1} dw in Almeria [49]; 0.9–7.3 mg kg^{-1} dw in Andalusia [50]; and in Poland –0.3 mg kg^{-1} dw [7]. Ferulic acid, conferring rigidity to cell walls [51], has been identified as being involved in resistance against thrips [52], stem borers in maize and cotton, and cereal aphids and midges, and in defence against different fungi [53]. Therefore, the higher presence of this compound might be linked to a defence mechanism of landraces to biotic factors.

Hydroxycinnamic acids as well as ferulic and caffeic acids have been reported, similar to many phenols, as antioxidants since they are reactive toward free radicals as a reactive oxygen species [54]. Caffeic acids, found in coffee, fruit, and vegetables such as garlic, are a well-known pharmacological antioxidant with anti-mutagenic activities and anti-inflammatory and immune-modulatory effects. They are also an anti-wrinkle agent and inhibitor of carcinogenesis, as reported by Kim et al. [47]. Moreover, according to the findings of Kallel et al. [55], ferulic and caffeic acids may be the main compounds responsible for the antimicrobial effect of the crude garlic extract.

Concerning iso-rhamnetin, it was identified as a new flavonoid glycoside in *Allium neapolitanum* and evaluated for its anti-aggregation human platelet activity by Carotenuto et al. [56]. Recently, pre-treatment with iso-rhamnetin, extracted from sea buckthorn (*Hippophae rhamnoides* L.), has been shown to play a protective role against acute fulminant hepatitis in mice [57].

'Peschici' and 'Monteleone' appreciably highlighted a high amount of rutoside-1, known as rutin (a flavonol glycoside between quercetin and disaccharide rutinose), which has been found to have important pharmacological effects, such as in the prevention of ulcerative colitis [58]. Both 'Peschici' and 'Monteleone' also showed the highest amounts of iso-quercitrin (also named quercetin 3-β-D-glucoside), even if identified at low concentrations, this quercetin glucoside, which can be isolated from several *Allium* spp. (Chinese onion, garlic, onion, and Welsh onion), has been shown to have anti-proliferative potential in various cancer cell lines [48].

'Anzano' along with 'Monteleone' were the only two genotypes containing patuletin, a compound also detected on *Allium* spp. from Romania (A. *obliquum* L.; A. *senescens* L. subsp. *montanum* (Fries) Holub; A. *schoenoprasum* L. subsp. *Schoenoprasum*) [59]. Isolated in *Urtica urens* L., patuletin has been shown to possess an antioxidant activity and free radical scavenging effects in rats treated with aflatoxin B1, a risk factor for hepatocellular carcinoma [60]. Furthermore, its anti-proliferative, necrotic, and apoptotic activity in tumour cell lines has been claimed recently by Alvarado-Sansininea et al. [61].

Among the examined landraces, those from Panni Municipality were the poorest in total phenolic compounds but contained a good level of caffeic acid-O-hexoside-1 acid (phenolic acid) and iso-rhamnetin and rutoside-1 (flavonoids). Among the investigated genotypes, 'PanniD' showed the highest value of hyperoside (hyperin) and it was the only landrace which accumulated a considerable amount of quercetol (quercetin). Quercetol (quercetin) has important functional benefits, including an anti-inflammatory activity, an anti-histamine effect, allergy medication, as well as anticancer and antivirus activities. It has also been claimed to regulate blood pressure in hypertensive subjects [62]. The hyperoside (hyperin) flavonoid has also been detected in other common Italian varieties of garlic [6]. Its ROS scavenging activity is well known and are the protective effects for PC12 cells against induced-cytotoxicity [63]. This flavonoid is also a candidate as a therapeutic agent for the treatment of vascular inflammatory diseases in humans and in mice endothelial cells [64].

3.7. Principal Component Analysis (PCA)

The results of the PCA using the chemical parameters showed that the first two PCs explained approximately 70% of the total variability, attributing 42% to PC1 and 26% to PC2 (Figure 2).

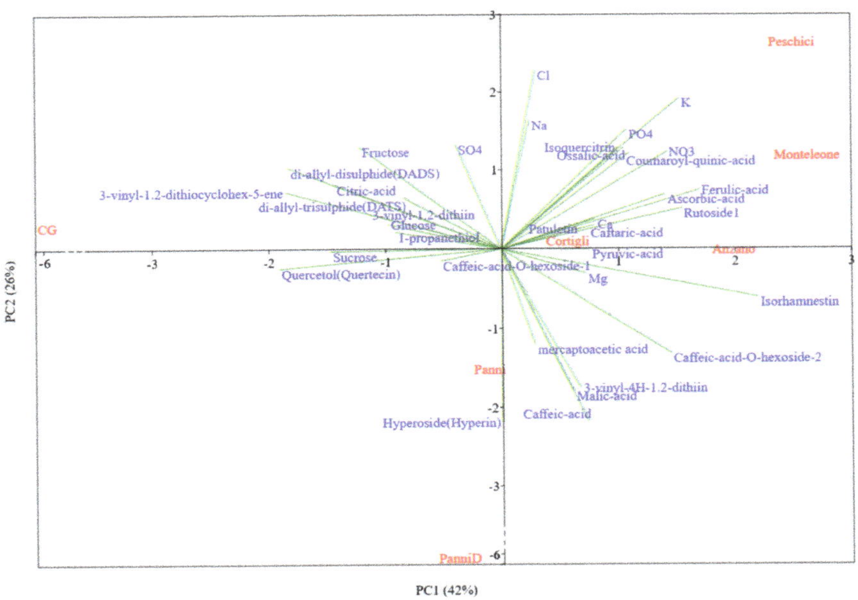

Figure 2. Principal component analysis bi-plot (PC1 vs. PC2) showing the spatial distribution of the main chemical traits as affected by genotypes. Refer to Table 1 for the acronyms of garlic genotypes. Following a previous analysis of correlation, 3-vinyl-1.2-dithiocyclohex-5-ene is representative of a group of seven volatile S-compounds (di-methyl-sulphide, propylene-sulphide, methyl-2-propenyl-disulphide, trans-propenyl-methyl-disulphide, di-methyl-trisulphide, methyl-allyl-trisulphide, and Allicin-S-oxo-di-allyl-di-sulphide); di-allyl-disulphide (DADS) is representative of a group of two volatile S-compounds (di-allyl-sulphide-DAS and 2-vinyl-1.3-dithiane); Isorhamnetin is representative of 3-hydroxy-methyl-phenol.

The group of 'Anzano', 'Monteleone', 'Cortigli', and 'Peschici' landraces separated from the 'PanniD', 'Panni' landraces (low side of PC2) and also from the CG (left side of PC1).

The former group clustered for the concentration of phenolic compounds (Isorhamnetin, Ferulic acid Caffeic-acid-O-hexoside-2, and Rutoside1) (highly and positively correlated to PC1) and for the high concentration of anions and cations (highly and positively correlated to PC2). 'Peschici' and 'Monteleone' set in an extreme position of this quadrant, mainly characterized by the highest mineral (K) concentration and phenols, as also underlined by Anova (Tables 2 and 5). 'Anzano' and 'Cortigli' clustered quite in a cloud-graphical space, suggesting a substantial homogeneity in chemical composition.

Conversely, the CG tended to cluster separately on the left side of PC1 for the general negative correlation with phenolic compounds and for the positive correlation with the volatile S-compounds (di-allyl-disulphide-DADS and 3-vinyl-1,2-dithiocyclohex-5-ene), with a phenolic compound as Quercetol (Quertecin) (exclusively present in GC, Table 5), citric acid and sucrose.

The PC2 axis separated the two Panni Municipality landraces from all the other landraces mainly for the general negative correlation with all the minerals and phenols, except for Hyperoside (Hyperin), exclusively present in 'PanniD' (Table 5).

4. Conclusions

The observed diversity in the morpho-biometrical and chemical traits among the examined landraces can be explained by considering the long process of adaptation to the specific 'environment' (pedo-climatic properties and less intensive agronomic practices), along with the constant selection pressure performed by farmers over a long period. These factors ensured a 'non-homologation' of the final product and specificity of the morpho-biometrical and chemical characteristics of a single genotype. Findings highlighted that the landraces collected from the province of Foggia maintain high biometrical (*productive*) standard features for the market, in line with those of the examined commercial genotype used as a reference standard.

The examined landraces were able to accumulate nutrients and phytochemicals, thus showing interesting features for the human diet. They exhibited high total cations (mainly K^+) and phenols (ferulic acid and derivates of caffeic, among phenolic acids, iso-rhamnetin and rutinoside-1, among flavonoids).

By increasing the knowledge of the properties of these local garlic landraces, this work could represent a key tool for the promotion of their consumption. At the same time, the consumption could encourage their cultivation, thus contributing to the protection of this agro-biodiversity at a high risk of genetic erosion by promoting their in situ/on-farm conservation.

Supplementary Materials: The following are available online at http://www.mdpi.com/2304-8158/9/7/850/s1. Figure S1: Photos of the bulbs of the studied garlic genotypes. Figure S2: Examples of HS-SPME chromatograms of volatile compounds. Table S1: Morphological traits of bulbs and cloves of garlic genotypes. Table S2: Biometric traits of bulbs and cloves of garlic genotypes.

Author Contributions: Conceptualization, A.B., G.C., and A.E.; data curation, A.B.; formal analysis, G.G., P.L., and C.L.; funding acquisition, A.E., and G.C.; supervision, A.B., G.C., and G.G.; validation, A.B. and G.C.; writing—original draft, A.B.; writing—review and editing, A.B., G.C., and A.E. All authors have read and agreed to the published version of the manuscript.

Funding: This research was funded by the Regione Puglia Administration under the "Rural Development Programme (RDP) 2014-2020"—Project "Biodiversity of vegetable crops in Puglia (BiodiverSO)". Measure 10, Sub measure 10.2, Operation 1 "Programme for the conservation and the valorisation of the genetic resources in agriculture".

Acknowledgments: The authors are grateful to Paolo La Rotonda for the valuable help in the collection of garlic samples and to all the growers who kindly donated their bulbs.

Conflicts of Interest: The authors declare no conflict of interest.

References

1. Martins, N.S.; Petropoulos, S.; Ferreira, I.C.F.R. Chemical composition and bioactive compounds of garlic (*Allium sativum* L.) as affected by pre- and post-harvest conditions: A review. *Food Chem.* **2016**, *211*, 41–50. [CrossRef] [PubMed]
2. Moutia, M.; Habti, N.; Badou, A. In vitro and in vivo immunomodulator activities of *Allium sativum* L. *Evid. Based Complementary Altern. Med.* **2018**. [CrossRef] [PubMed]
3. D'Archivio, A.A.; Foschi, M.; Aloia, R.; Maggi, M.A.; Rossi, L.; Ruggieri, F. Geographical discrimination of red garlic (*Allium sativum* L.) produced in Italy by means of multivariate statistical analysis of ICP-OES data. *Food Chem.* **2019**, *275*, 333–338. [CrossRef] [PubMed]
4. Kubec, R.; Velíšek, J.; Doležal, M.; Kubelka, V. Sulfur-containing volatiles arising by thermal degradation of alliin and deoxyalliin. *J. Agric. Food Chem.* **1997**, *45*, 3580–3585. [CrossRef]
5. Shang, A.; Cao, S.-Y.; Xu, X.-Y.; Gan, R.-Y.; Tang, G.-Y.; Corke, H.; Mavumengwana, V.; Li, H.-B. Bioactive Compounds and Biological Functions of Garlic (Allium Sativum, L.). *Foods Basel Switz.* **2019**, *8*, 246. [CrossRef]
6. Fratianni, F.; Ombra, M.N.; Cozzolino, A.; Riccardi, R.; Spigno, P.; Tremonte, P.; Coppola, R.; Nazzaro, F. Phenolic constituents, antioxidant, antimicrobial and anti-proliferative activities of different endemic Italian varieties of garlic (*Allium sativum* L.). *J. Funct. Foods* **2016**, *21*, 240–248. [CrossRef]
7. Gorinstein, S.; Leontowicz, H.; Leontowicz, M.; Namiesnik, J.; Najman, K.; Drzewiecki, J.; Cvikrová, M.; Martincová, O.; Katrich, E.; Trakhtenberg, A. Comparison of the main bioactive compounds and antioxidant activities in garlic and white and red onions after treatment protocols. *J. Agric. Food Chem.* **2008**, *56*, 4418–4426. [CrossRef]
8. Pedisić, S.; Zorić, Z.; Miljanović, A.; Šimić, D.; Repajić, M.; Dragović-Uzelac, V. Retention of bioactive compounds during domestic processing of croatian domestic garlic (*Allium sativum* L.). *Food Technol. Biotech.* **2018**, *56*, 590–596. [CrossRef]
9. Negri, V.; Maxted, N.; Veteläinen, M. European landrace conservation: An introduction. In *European landraces: On-farm conservation, management and use*; Veteläinen, V., Negri, N., Maxted, N., Eds.; Biodiversity Technical Bulletin No. 15; Biodiversity International: Rome, Italy, 2009; pp. 1–22.
10. Conversa, G.; Bonasia, A.; Elia, A. Collecting landrace vegetable crops in Puglia region: Case study of Foggia Province. *Acta Hortic.* **2018**, *1215*, 453–458. [CrossRef]
11. International plant genetic resources institute (Itália). *Descriptors of Allium Allium Spp.*; International plant genetic resources institute: Rome, Italy, 2001.
12. Mccleary, B.V.; Gibson, T.S.; Mugford, D.C.; Lukow, O.; Jackson, D.S.; Rabe, E.; Patel, N.; Williams, P.C.; Gelroth, J.; Camire, M.E.; et al. Measurement of Total Starch in Cereal Products by Amyloglucosidase-α-Amylase Method: Collaborative Study. *J. AOAC Int.* **1997**, *80*, 571–579. [CrossRef]
13. Rohrer, J. Analysis of Carbohydrates by High-Performance Anion-Exchange Chromatography with Pulsed Amperometric Detection (HPAE-PAD). *Thermo Fish. Sci.* **2012**, *Techincal note 20*, 12.
14. Calvo-Gómez, O.; Morales-López, J.; López, M.G. Solid-Phase Microextraction-Gas Chromatographic-Mass Spectrometric Analysis of Garlic Oil Obtained by Hydrodistillation. *J. Chromatogr. A* **2004**, *1036*, 91–93. [CrossRef] [PubMed]
15. Gambacorta, G.; Trani, A.; Fasciano, C.; Paradiso, V.M.; Faccia, M. Effects of pre-fermentative cold soak on poly-phenols and volatiles of Aglianico, Primitivo and Nero di Troia red wines. *Food Sci. Nutr.* **2019**, *7*, 483–491. [CrossRef] [PubMed]
16. Trani, A.; Verrastro, V.; Punzi, R.; Faccia, M.; Gambacorta, G. Phenols, Volatiles and Sensory Properties of Primitivo Wines from the "Gioia Del Colle" PDO Area. *South Afr. J. Enol. Vitic.* **2016**, *37*, 139–148. [CrossRef]
17. Pasqualone, A.; Punzi, R.; Trani, A.; Summo, C.; Paradiso, V.M.; Caponio, F.; Gambacorta, G. Enrichment of Fresh Pasta with Antioxidant Extracts Obtained from Artichoke Canning By-Products by Ultrasound-Assisted Technology and Quality Characterisation of the End Product. *Int. J. Food Sci. Technol.* **2017**, *52*, 2078–2087. [CrossRef]
18. Hammer, Ø.; Harper, D.; Ryan, P.D. PAST: Paleontological statistics software package for education and data analysis. *Palaeontol. Electron.* **2001**, *4*, 9–18.
19. Figliuolo, G.; Candido, V.; Logozzo, G.; Miccolis, V.; Spagnoletti Zeuli, P.L. Genetic evaluation of cultivated garlic germplasm (*Allium sativum* L. and *A. ampeloprasum* L.). *Euphytica* **2001**, *121*, 325–334. [CrossRef]

20. Mohammadi, B.; Khodadadi, M.; Karami, E.; Shaaf, S. Variation in Agro-Morphological Characters in Iranian Garlic Landraces. *Int. J. Veg. Sci.* **2014**, *20*, 202–215. [CrossRef]
21. Fanaei, H.; Narouirad, M.; Farzanjo, M.; Ghasemi, M. Evaluation of yield and some agronomical traits in garlic genotypes (*Allium sativum* L.). *Annu. Res. Rev. Biol.* **2014**, *4*, 3386–3391. [CrossRef]
22. Lisciani, S.; Gambelli, L.; Durazzo, A.; Marconi, S.; Camilli, E.; Rossetti, C.; Gabrielli, P.; Aguzzi, A.; Temperini, O.; Marletta, L. Carbohydrates Components of Some Italian Local Landraces: Garlic (*Allium sativum* L.). *Sustainability* **2017**, *9*, 1922. [CrossRef]
23. Vadalà, R.; Mottese, A.F.; Bua, G.D.; Salvo, A.; Mallamace, D.; Corsaro, C.; Vasi, S.; Giofrè, S.V.; Alfa, M.; Cicero, N.; et al. Statistical analysis of mineral concentration for the geographic identification of garlic samples from Sicily (Italy), Tunisia and Spain. *Foods* **2016**, *5*, 20. [CrossRef] [PubMed]
24. Petropoulos, S.A.; Fernandes, Â.; Ntatsi, G.; Petrotos, K.; Barros, L.; Ferreira, I.C.F.R. Nutritional value, chemical characterization and bulb morphology of greek garlic landraces. *Molecules* **2018**, *23*, 319. [CrossRef] [PubMed]
25. Pardo, J.E.; Escribano, J.; Gómez, R.; Alvarruiz, A. Physical-chemical and sensory quality evaluation of garlic cultivars. *J. Food Qual.* **2007**, *30*, 609–622. [CrossRef]
26. Rao, P.G.P.; Rao, L.J.; Raghavan, B. Chemical composition of essential oils of garlic (*Allium sativum* L.). *J. Spices Aromatic Crop.* **1999**, *8*, 41–47.
27. Magwaza, L.S.; Opara, U.L. Analytical methods for determination of sugars and sweetness of horticultural products-A review. *Sci. Hortic.* **2015**, *184*, 179–192. [CrossRef]
28. Ritota, M.; Casciani, L.; Han, B.Z.; Cozzolino, S.; Leita, L.; Sequi, P.; Valentini, M. Traceability of Italian garlic (*Allium sativum* L.) by means of HRMAS-NMR spectroscopy and multivariate data analysis. *Food Chem.* **2012**, *135*, 684–693. [CrossRef] [PubMed]
29. Mashayekhi, K.; Chiane, S.M.; Mianabadi, M.; Ghaderifar, F.; Mousavizadeh, S.J. Change in carbohydrate and enzymes from harvest to sprouting in garlic. *Food Sci. Nutr.* **2015**, *4*, 370–376. [CrossRef]
30. Priecina, L.; Karklina, D. Composition of major organic acids in vegetables and spices. Proceedings of CBU International Conference, Prague, Czech Republic, 25–27 March 2015; p. 3.
31. Ramirez, D.A.; Locatelli, D.A.; González, R.E.; Cavagnaro, P.F.; Camargo, A.B. Analytical methods for bioactive sulfur compounds in *Allium*: An integrated review and future directions. *J. Food Compos. Anal.* **2017**, *61*, 4–19. [CrossRef]
32. Abe, K.; Hori, Y.; Myoda, T. Volatile Compounds of Fresh and Processed Garlic (Review). *Exp. Ther. Med.* **2020**, *19*, 1585–1593. [CrossRef]
33. Jabbes, N.; Arnault, I.; Auger, J.; Dridi, BAl-M.; Hannachi, C. Agro-morphological markers and organo-sulphur compounds to assess diversity in Tunisian garlic landraces. *Sci. Hortic.* **2012**, *148*, 47–54. [CrossRef]
34. Singh, C.V.; Gupta, P.; Kasana, B.S. Response of Garlic to Sulphur and Boron Application in Terms of Biochemical Parameters. *Int. J. Curr. Microbiol. Appl. Sci.* **2018**, *7*, 2677–2687. [CrossRef]
35. Huchette, O.; Arnault, I.; Auger, J.; Bellamy, C.; Trueman, L.; Thomas, B.; Ochatt, S.J.; Kahane, R. Genotype, Nitrogen Fertility and Sulphur Availability Interact to Affect Flavour in Garlic (Allium Sativum L.). *J. Hortic. Sci. Biotechnol.* **2007**, *82*, 79–88. [CrossRef]
36. Lanzotti, V. The Analysis of Onion and Garlic. *J. Chromatogr. A* **2006**, *1112*, 3–22. [CrossRef] [PubMed]
37. Shan, C.; Wang, C.; Liu, J.; Wu, P. The analysis of volatile flavor components of Jin Xiang garlic and Tai'an garlic. *Agric. Sci.* **2013**, *4*, 744–748. [CrossRef]
38. Horníčková, J.; Kubec, R.; Cejpek, K.; Velíšek, J.; Ovesná, J.; Stavelíková, H. Profiles of S-alk(en)ylcysteine sulfoxides in various garlic genotypes. *Czech, J. Food Sci.* **2010**, *28*, 298–308. [CrossRef]
39. Yu, T.H.; Wu, C.M.; Liou, Y.C. Volatile compounds from garlic. *J. Agric. Food Chem.* **1989**, *37*, 730–734. [CrossRef]
40. Locatelli, D.A.; Nazareno, M.A.; Fusari, C.M.; Camargo, A.B. Cooked garlic and antioxidant activity: Correlation with organo-sulfur compound composition. *Food Chem.* **2017**, *220*, 219–224. [CrossRef]
41. Ascrizzi, R.; Flamini, G. Leek or Garlic? A Chemical Evaluation of Elephant Garlic Volatiles. *Molecules* **2020**, *25*, 2082. [CrossRef]
42. Farag, M.A.; Ali, S.E.; Hodaya, R.H.; El-Seedi, H.R.; Sultani, H.N.; Laub, A.; Eissa, T.F.; Abou-Zaid, F.O.F.; Wessjohann, L.A. Phytochemical Profiles and Antimicrobial Activities of Allium Cepa Red Cv. and A. Sativum Subjected to Different Drying Methods: A Comparative MS-Based Metabolomics. *Molecules* **2017**, *22*, 761. [CrossRef]
43. Abe, K.; Hori, Y.; Myoda, T. Characterization of Key Aroma Compounds in Aged Garlic Extract. *Food Chem.* **2020**, *312*, 126081. [CrossRef]

44. Di Ferdinando, M.; Brunetti, C.; Fini, A.; Tattini, M. Flavonoids as Antioxidants in Plants Under Abiotic Stresses. In *Abiotic Stress Responses in Plants*; Ahmad, P., Prasad, M., Eds.; Springer: New York, NY, USA, 2012; pp. 159–179.
45. Akula, R.; Ravishankar, G.A. Influence of Abiotic Stress Signals on Secondary Metabolites in Plants. *Plant Signal. Behav.* **2011**, *6*, 1720–1731. [CrossRef]
46. Drozd, M.; Thomas, M.; Nowak, R. Determination of phenolic acids in raw garlic (*Allium sativum* L.) and onion (*Allium cepa* L.) bulbs. *Annales UMCS* **2011**, *24*, 121–127.
47. Kim, S.R.; Jung, Y.R.; An, H.J.; Kim, D.H.; Jang, E.J.; Choi, Y.J.; Moon, K.M.; Park, M.H.; Park, C.H.; Chung, K.W.; et al. Anti-Wrinkle and Anti-Inflammatory Effects of Active Garlic Components and the Inhibition of MMPs via NF-KB Signaling. *PLoS ONE* **2013**, *8*, e73877. [CrossRef] [PubMed]
48. Pan, Y.; Zheng, Y.M.; Ho, W.S. Effect of quercetin glucosides from *Allium* extracts on HepG2, PC-3 and HT-29 cancer cell lines. *Oncol. Lett.* **2018**, *15*, 4657–4661. [CrossRef] [PubMed]
49. Alarcón-Flores, M.I.; Romero-González, R.; Martínez Vidal, J.L.; Garrido Frenich, A. Determination of Phenolic Compounds in Artichoke, Garlic and Spinach by Ultra-High-Performance Liquid Chromatography Coupled to Tandem Mass Spectrometry. *Food Anal. Methods* **2014**, *7*, 2095–2106. [CrossRef]
50. Beato, V.M.; Orgaz, F.; Mansilla, F.; Montaño, A. Changes in Phenolic Compounds in Garlic (Allium Sativum L.) Owing to the Cultivar and Location of Growth. *Plant Foods Hum. Nutr.* **2011**, *66*, 218–223. [CrossRef]
51. Bennett, R.N.; Wallsgrove, R.M. Secondary Metabolites in Plant Defence Mechanisms. *New Phytol.* **1994**, *127*, 617–633. [CrossRef]
52. Leiss, K.A.; Choi, Y.H.; Abdel-Farid, I.B.; Verpoorte, R.; Klinkhamer, P.G.L. NMR Metabolomics of Thrips (*Frankliniella Occidentalis*) Resistance in Senecio Hybrids. *J. Chem. Ecol.* **2009**, *35*, 219–229. [CrossRef]
53. Wink, M. Plant Breeding: Importance of Plant Secondary Metabolites for Protection against Pathogens and Herbivores. *Theor. Appl. Genet.* **1988**, *75*, 225–233. [CrossRef]
54. Shahidi, F.; Wanasundara, P.K. Phenolic Antioxidants. *Crit. Rev. Food Sci. Nutr.* **1992**, *32*, 67–103. [CrossRef]
55. Kallel, F.; Driss, D.; Chaari, F.; Belghith, L.; Bouaziz, F.; Ghorbel, R.; Chaabouni, S.E. Garlic (Allium Sativum L.) Husk Waste as a Potential Source of Phenolic Compounds: Influence of Extracting Solvents on Its Antimicrobial and Antioxidant Properties. *Ind. Crops Prod.* **2014**, *62*, 34–41. [CrossRef]
56. Carotenuto, A.; Fattorusso, E.; Lanzotti, V.; Magno, S.; De Feo, V.; Cicala, C. The flavonoids of *Allium neapolitanum*. *Phytochemistry* **1997**, *44*, 949–957. [CrossRef]
57. Lu, X.; Liu, T.; Chen, K.; Xia, Y.; Dai, W.; Xu, S.; Xu, L.; Wang, F.; Wu, L.; Li, J.; et al. Isorhamnetin: A hepatoprotective flavonoid inhibits apoptosis and autophagy via P38/PPAR-α pathway in mice. *Biomedicine Pharmacother.* **2018**, *103*, 800–811. [CrossRef] [PubMed]
58. Baliga, M.S.; Saxena, A.; Kaur, K.; Kalekhan, F.; Chacko, A.; Venkatesh, P.; Fayad, R. Polyphenols in the prevention of ulcerative colitis: Past, present and future. In *Polyphenols in human health and disease*; Watson, R.R., Preedy, V.R., Zibadi, S., Eds.; Academic Press: San Diego, CA, USA, 2014; pp. 655–663.
59. Vlase, L.; Parvu, M.; Parvu, E.A.; Toiu, A. Chemical Constituents of Three Allium Species from Romania. *Molecules* **2013**, *18*, 114–127. [CrossRef] [PubMed]
60. Abdel-Wahhab, M.A.; Said, A.; Huefner, A. NMR and Radical Scavenging Activities of Patuletin from Urtica Urens. Against Aflatoxin B1. *Pharm. Biol.* **2005**, *43*, 515–525. [CrossRef]
61. Alvarado-Sansininea, J.J.; Sánchez-Sánchez, L.; López-Muñoz, H.; Escobar, M.L.; Flores-Guzmán, F.; Tavera-Hernández, R.; Jiménez-Estrada, M. Quercetagetin and Patuletin: Antiproliferative, Necrotic and Apoptotic Activity in Tumor Cell Lines. *Mol. J. Synth. Chem. Nat. Prod. Chem.* **2018**, *23*, 2579. [CrossRef]
62. Edwards, R.L.; Lyon, T.; Litwin, S.E.; Rabovsky, A.; Symons, J.D.; Jalili, T. Quercetin Reduces Blood Pressure in Hypertensive Subjects. *J. Nutr.* **2007**, *137*, 2405–2411. [CrossRef]
63. Liu, Z.; Tao, X.; Zhang, C.; Lu, Y.; Wei, D. Protective Effects of Hyperoside (Quercetin-3-o-Galactoside) to PC12 Cells against Cytotoxicity Induced by Hydrogen Peroxide and Tert-Butyl Hydroperoxide. *Biomed. Pharmacother. Biomedecine Pharmacother.* **2005**, *59*, 481–490. [CrossRef] [PubMed]
64. Ku, S.-K.; Zhou, W.; Lee, W.; Han, M.-S.; Na, M.; Bae, J.-S. Anti-Inflammatory Effects of Hyperoside in Human Endothelial Cells and in Mice. *Inflammation* **2015**, *38*, 784–799. [CrossRef] [PubMed]

foods

MDPI

Article

Castanea spp. Agrobiodiversity Conservation: Genotype Influence on Chemical and Sensorial Traits of Cultivars Grown on the Same Clonal Rootstock

Gabriele L. Beccaro [1,2], Dario Donno [1,2,*], Guglielmo Gianni Lione [1,2], Marta De Biaggi [1], Giovanni Gamba [1], Sabrina Rapalino [1], Isidoro Riondato [1], Paolo Gonthier [1,2] and Maria Gabriella Mellano [1,2]

1 Dipartimento di Scienze Agrarie, Forestali e Alimentari, Università degli Studi di Torino, 10124 Torino, Italy; gabriele.beccaro@unito.it (G.L.B.); guglielmo.lione@unito.it (G.G.L.); m.debiaggi@hotmail.com (M.D.B.); giovanni.gamba@unito.it (G.G.); rapalino-sabrina@hotmail.it (S.R.); isidoro.riondato@unito.it (I.R.); paolo.gonthier@unito.it (P.G.); gabriella.mellano@unito.it (M.G.M.)

2 Chestnut R&D Center—Piemonte, 12013 Chiusa di Pesio, Italy

* Correspondence: dario.donno@unito.it; Tel.: +39-011-670-8751

Received: 10 July 2020; Accepted: 31 July 2020; Published: 5 August 2020

Abstract: A large species diversity characterises the wide distribution of chestnuts in Asia, North America, and Europe, hence reflecting not only the adaptation of the genus *Castanea* to diverse environmental conditions, but also to different management strategies encompassing orchards. The characterisation and description of chestnut populations and cultivars are crucial to develop effective conservation strategies of one of the most important Italian and European fruit and wood species. Chestnut cultivars grown in the same pedoclimatic conditions and on the same clonal rootstock were characterised with sensory, spectrophotometric, and chromatographic analysis to determine the phytochemical composition and nutraceutical properties. A multivariate approach, including principal component analysis and conditional inference tree models, was also performed. The ease of peeling, seed colour, and intensity of sweetness were the sensory descriptors that allowed us to differentiate *C. sativa* cultivars. Antioxidant capacity ranged from 9.30 ± 0.39 mmol Fe^{+2} kg^{-1} DW ('Bouche de Bètizac') to 19.96 ± 1.89 mmol Fe^{+2} kg^{-1} DW ('Garrone Rosso'). Monoterpenes represented the main component, reaching 88% for hybrids, followed by polyphenols (10–25% for hybrids and chestnuts, respectively). A multivariate approach showed that phenolic acids and tannins were the bioactive classes with the highest discriminating power among different genotypes, and that genotype is a significant variable ($p < 0.05$). In addition, most of the analysed chestnut cultivars showed a content of bioactive compounds similar to or higher than the main hazelnut, walnut, and almond varieties. Chestnut agrobiodiversity could be intended as strictly associated to the genotype effect and underlines the large variability within the genus *Castanea*, and therefore, the importance of in farm and ex situ conservation of local germplasm is part of a global strategy aimed at increasing the levels of agrobiodiversity.

Keywords: chestnut; characterisation tool; bioactive compounds; sensory analysis; multivariate approach

1. Introduction

Landscape modification, habitat loss, and fragmentation stand among the main agrobiodiversity conservation issues that affect the ecosystem structure and functioning with negative effects on plants populations and communities [1]. The consequences of these pressures is particularly true for species with a relevant environmental and historical economic role in agroforestry systems, such as chestnut (*Castanea* spp.) [2].

A large species diversity characterises the wide distribution of chestnut in Asia, North America, and Europe, hence reflecting not only the adaptation of the genus *Castanea* to diverse environmental conditions, but also to different management strategies encompassing orchards for fruit production, coppices for timber production, and naturalised populations providing several ecosystem services [3]. Indeed, the high levels of chestnut diversity are largely acknowledged as the result of the co-existence between *Castanea* spp. and human populations [4].

Natural and planted forests of sweet chestnut cover the species' ecological limits [5], spreading from the Caucasus to Portugal, reaching the southern United Kingdom, Canary Islands, and the Azores archipelago. It is also locally present in Lebanon and Syria [6]. Sweet chestnut is one of the oldest domesticated species, widespread throughout the Roman Empire and commonly cultivated during the Medieval period, becoming so indispensable for the survival of mountain populations that these cultures were identified as "chestnut civilizations" [7]. Therefore, sweet chestnut represents an important resource in Europe for its great ecological (large ecosystem biodiversity and landscape value), economic (fruit, wood, honey, and tannin production), and cultural relevance. Noteworthily, sweet chestnut is considered one of the most important trees in Italy [8], underlined by the presence of ancient forests and orchards with monumental trees [2,9].

However, since the beginning of the 20th century, the growing areas of sweet chestnut have dramatically decreased because of several social, cultural, and environmental changes. Such challenges include the progressive depopulation of mountain areas, diet changes [10,11], climate change [12], and the establishment and spread of diseases and pests. The latter encompass ink disease caused by the oomycetes *Phytophthora cambivora* (Petri) Buisman and *P. cinnamomi* Rands, chestnut blight associated with the ascomycete *Cryphonectria parasitica* (Murrill) M.E. Barr, the emerging nut rot due to *Gnomoniopsis castaneae* G. Tamietti, and the infestation and control of the Asian gall wasp *Dryocosmus kuriphilus* Yasumatsu [13–16]. The preservation of chestnut intra- and interspecific diversity together with a management hinging on regular agronomic treatments seems to be the key factors for a correct management of the plantations [17]. The above diversity can also contribute to the selection of varieties more tolerant or resistant to diseases and pests, and not surprisingly, breeding programs and preservation strategies for the existing varieties are sought, implemented, and supported worldwide [18]. Moreover, genetic and ecophysiological investigations on *C. sativa* showed the wide plasticity of the species to cope with different environments (water and nutrient uptake efficiency) [19].

For these reasons, the characterisation and description of chestnut populations and cultivars [20] are crucial to develop effective conservation strategies of one of the most important Italian and European fruit and wood species [2,17]. A genetic landscape study on the European sweet chestnut was recently performed to evaluate the geographical patterns of its diversity and to identify high-priority areas characterised by high allelic richness [20]. The results showed that the most interesting areas in terms of conservation priority are located in Italy, Georgia, and eastern Turkey, thus pointing out the crucial role played by the Italian peninsula in global biodiversity conservation strategies. Different approaches are under development for the conservation of *Castanea* spp., and on farm and ex situ conservation combined with other initiatives represent the first step to prevent the loss of biodiversity [3,4].

Moreover, an increased economic interest for sweet chestnut in the food industry increased the demand for selected varieties, which is also driven by several research studies on the potential positive health benefits that can be derived from the consumption of fresh and processed chestnut products [21,22]. In addition, the recognition of quality certification, such as EU Protected Designation of Origin (PDO) and Protected Geographical Indication (PGI), represent strategic marketing tools that are widely applied in the Italian chestnut market, with particular reference to the "Marrone-type" (MT) cultivars, which stand among the most appreciated varieties by the processing industry worldwide. Consequently, the increased market demand and consumers awareness impose the development of reliable methods for the description and identification of the cultivars [23], and for the characterisation of the fruit traits for the selection of high-quality products (good sensorial and qualitative properties, high bioactive compound content) [24]. For these reasons, several techniques are applied for the characterisation of chestnut cultivars, from the implementation of molecular typing such as SSR markers [25], to morpho-biological, sensory, phenological and chemical analysis. Furthermore, chemical and sensory analyses are on the rise because they allow us to define the most proper technological uses for each cultivar by the study of the nut properties and merceological traits [26]. Moreover, characterisation and conservation of chestnut germplasm represent crucial aspects to ensure adaptability and productivity of the crop, also in relation to the recent climate changes [27].

In this study, a combination of different approaches was applied to describe and characterise 18 chestnut cultivars in order to assess their diversity in bioactive compounds and sensory attributes. It is worth noting that the attempt of silencing the exogenous sources of variability on the phenotypical expression maximises the likelihood of detecting the genotype influence on chestnut chemical and sensorial properties, hence providing an element of novelty and added value to the current state-of-the-art method in the subject of chestnut cultivars characterisation. The diversity that was detected in bioactive compound profiles and sensory traits among the chestnut cultivars is likely related to the effect of the genotype. Thus, chemical and sensorial analyses such as the ones that were applied in this study could be, in the future, an important mean of characterisation, while also allowing the evaluation and the selection of new cultivars [26].

2. Materials and Methods

2.1. Plant Material and Sampling Site

Samples of *Castanea* spp. fruits were collected in October 2018 from 18 accessions grown in the collection field (about 5 ha) of the Chestnut R&D Center—Piemonte in Chiusa Pesio, Cuneo Province (North-Western Italy) (lat. 44°18′27.5″; long. 7°40′57.3″; elevation 575 m above sea level). These cultivars were grown under the same pedoclimatic conditions, agronomic management practices and, for the first time, on the same clonal rootstock. The origin and identification codes of the sampled material are listed in Table 1.

The experimental design allowed the assessment of the genotype effect on the chestnut composition while controlling for the potential confounding effect potentially exerted by the above environmental and agronomic variables. This approach resulted in the implementation of an effective tool to distinguish the different cultivars and, in particular, the "Marrone-type" (MT) cultivars from the other categories as sweet chestnut (SC) and Euro-Japanese hybrids (EH). The germplasm repository includes the main local and Italian chestnut cultivars, several European varieties, and accessions from China, Japan, and USA [27]. The area is characterised by a temperate climate and is located in the phytoclimatic transition zone between "cold *Castanetum*" and "hot *Fagetum*", following Mayr-Pavari's classification [4]. The annual mean temperature and precipitations are 13.3 °C and 993 mm, respectively. All climatic data were extracted from the meteorological station placed in the collection field. Figure 1 shows climate data from April to December, the period from vegetative awakening to winter dormancy. Soils are composed of fluvial deposits, with a high concentration of sand, and the soil depth is limited (between 30 and 60 cm) by the presence of coarse gravel [4].

Table 1. Origin and identification codes (number of the tree in the orchard) of the sampled raw material.

Category	Species	Cultivar	Nut	ID	Country
Sweet chestnut (SC)	*C. sativa*	Canepina		78	Italy

Table 1. *Cont.*

Category	Species	Cultivar	Nut	ID	Country
	C. sativa	Brunette		9	Italy
	C. sativa	Mansa		390	Spain
	C. sativa	Madonna		1	Italy
	C. sativa	Gentile		186	Italy
	C. sativa	Garrone Rosso		39	Italy
	C. sativa	Bouche Rouge		286	France
	C. sativa	Marrubia		215	Italy
	C. sativa	Contessa		30	Italy
	C. sativa	Gabiana		45	Italy
	C. sativa	Neirana della Val di Susa		218	Italy

Table 1. *Cont.*

Category	Species	Cultivar	Nut	ID	Country
	C. sativa	Tarvisò		25	Italy
Marrone-type chestnut (MT)	*C. sativa*	Marrone di Castel del Rio		130	Italy
	C. sativa	Marrone di Marradi IGP		198	Italy
	C. sativa	Marrone della Val di Susa		47	Italy
Euro-Japanese hybrid (EH)	*C. sativa* x *C. crenata*	Precoce Migoule		258	France
	C. sativa x *C. crenata*	Marsol		5	France
	C. sativa x *C. crenata*	Bouche de Bétizac		159	France

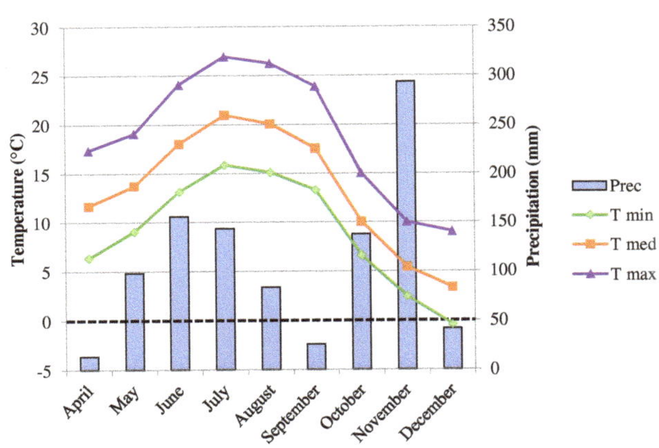

Figure 1. Monthly (maximum, mean, and minimum) temperature and rainfall values available from April to December 2018.

2.2. Sample Preparation

Chestnut fruits (1 kg for each plant of each cultivar—three plants for each cultivar) from 18 accessions were randomly collected, hand selected to remove damaged fruits, and divided into two lots for sensory and chemical analysis. The nuts for the sensory analysis were boiled (100 °C for 45 min) and given to the trained panellists for the tasting sessions (see Sensory analysis section).

Raw nuts used for the chemical analysis were hand-pealed, fragmented in small pieces, dried in an oven (WIPA, Stadtlohn, Germany) at 40 °C for 2 days, and then ground to a fine powder and subdivided into small portions (500 g for each one) sealed in plastic bags.

2.3. Sensory Analysis

Sensory evaluation was used to identify and quantify the product organoleptic traits [28]. Currently, the technique is widely applied to a large range of food [29], and it is very important for the selection of cultivars to be used by agri-food industry.

Sensory analysis was performed in a specific sensory laboratory, by 12 selected panellists (gender ratio: 50:50; age range: 20–50-year-old) from ONAFrut (National Organization of Fruit Tasters). Three training sessions were carried out with 12 panellists to ensure a common lexicon of terms for flavour and aroma. During the three training sessions, the panellists worked in a group, but they individually evaluated the samples. After each panel training session, a discussion was held to decide the appropriate set of descriptors to use [30]. Quantitative Descriptive Analysis (QDA) was carried out as analytical-descriptive method [31]. Selected descriptors were: ease of peeling, seed colour, intensity of flavour, intensity of sweetness, intensity of saltiness, intensity of bitterness, flouryness, and chestnut aroma. Descriptive terms, definitions, and associated reference standards used in the sensory analysis of chestnuts were reported in Supplementary Table S1. Each descriptor was evaluated on a continuous scale partially structured into 10 segments as reported in literature [31–34]. Finally, the same scale was used to evaluate the descriptor of personal judgement of each panellist, based on a subjective approval rating.

2.4. Extraction Protocols

All the chemicals/reagents are reported in Supplementary Materials. Polyphenolic compounds were extracted with a mixture of methanol: water: 37% HCl (95:4.5:0.5, *v*/*v*/*v*). Methanolic extracts were filtered through a membrane microfilter (polytetrafluoroethylene membrane, PTFE; pore size 0.45 μm), and then were stored for a few days at normal atmosphere (NA), at 4 °C and 95% RH.

Monoterpenes, sugars, and organic acids were extracted with 95% ethanol solution. Samples were then stored until analysis in NA, at 4 °C and 95% RH.

Ascorbic acid and dehydroascorbic acid were extracted by an extraction solution (0.1 mol·L^{-1} citric acid, 2 mmol·L^{-1} ethylenediaminetetraacetic acid (EDTA) disodium salt, and 4 mmol·L^{-1} sodium fluoride in methanol-water, 5:95, *v*/*v*). *o*-Phenylenediamine (OPDA) solution (18.8 mmol·L^{-1}) was added to 750 μL of extracted samples for dehydroascorbic acid (DHAA) derivatisation to a fluorophore, 3-(1,2-dihydroxyethyl)furo(3,4-b) quinoxaline-1-one (DFQ).

2.5. Spectrophotometric Analysis

Antioxidant capacity in the chestnut fruits was assessed by a ferric reducing antioxidant power (FRAP) assay [35], and results were expressed as millimoles of Fe^{2+} equivalents per kilogram (solid food) of dried weight (DW).

The total polyphenol content (TPC) was evaluated following the Folin–Ciocalteu colourimetric method [36], and the results were expressed as grams of gallic acid equivalents (GAE) per kilogram of DW.

Absorbance at 595 nm (for antioxidant capacity) and 765 nm (for TPC) with a UV/Vis spectrophotometer (1600-PC, VWR International) was recorded.

2.6. Chromatographic Analysis

An Agilent 1200 High Performance Liquid Chromatograph, equipped with a G1311A quaternary pump, a manual injection valve, and a 20 μL sample loop, coupled to an Agilent GI315D UV-Vis diode array detector (Agilent Technologies, Santa Clara, CA, USA), was used for the analysis.

Six different chromatographic methods were used to analyse the samples. Chromatographic separations were carried out on a Kinetex—C18 column (4.6 × 150 mm, 5 μm, Phenomenex, Torrance, CA, USA), and a SphereClone—NH_2 column (4.6 × 250 mm, 5 μm, Phenomenex, Torrance, CA, USA). Different chromatographic conditions were used to analyse the samples according to the methods described by other studies [37,38], with some modifications, and previously were validated for fresh and dried fruits, herbal medicines, and other food products. Identification and detection were performed with an UV—Vis Diode Array Detector by scanning from 190 to 600 nm. The chromatographic conditions of each method are reported in Table 2. The external standard method was used for quantitative determinations. All results were expressed as g·kg^{-1} of DW.

Table 2. Chromatographic conditions of the used methods.

Method	Compounds of Interest	Stationary Phase	Mobile Phase	Flow (mL min^{-1})	Wavelength (nm)
A	cinnamic acids, flavonols	KINETEX—C18 column (4.6 × 150 mm, 5 μm)	A: 10 mM KH_2PO_4/H_3PO_4, pH = 2.8 B: CH_3CN	1.5	330
B	benzoic acids, catechins, Tannins	KINETEX—C18 column (4.6 × 150 mm, 5 μm)	A: $H_2O/CH_3OH/HCOOH$ (5:95:0.1 *v/v/v*), pH = 2.5 B: $CH_3OH/HCOOH$ (100:0.1 *v/v*)	0.6	280
C	monoterpenes	KINETEX—C18 column (4.6 × 150 mm, 5 μm)	A: H_2O B: CH_3CN	1.0	250
D	organic acids	KINETEX—C18 column (4.6 × 150 mm, 5 μm)	A: 10 mM KH_2PO_4/H_3PO_4, pH = 2.8 B: CH_3CN	0.6	214
E	vitamins	KINETEX—C18 column (4.6 × 150 mm, 5 μm)	A: 5 mM $C_{16}H_{33}N(CH_3)_3Br$/50 mM KH_2PO_4, pH = 2.5 B: CH_3OH	0.9	261, 348
F	sugars	SphereClone—NH_2 column (4.6 × 250 mm, 5 μm)	A: H_2O B: CH_3CN	0.5	286

Method A—gradient analysis: 5% B to 21% B in 17 min + 21% B in 3 min (2 min conditioning time). Method B—gradient analysis: 3% B to 85% B in 22 min + 85% B in 1 min (2 min conditioning time). Method C—gradient analysis: 30% B to 56% B in 15 min + 56% B in 2 min (3 min conditioning). Method D—gradient analysis: 5% B to 14% B in 10 min + 14% B in 3 min (2 min conditioning time). Method E—isocratic analysis: ratio of phase A and B: 95:5 in 10 min (5 min conditioning time). Method F—isocratic analysis: ratio of phase A and B: 5:85 in 12 min (3 min conditioning time).

2.7. Data Analysis

Sensory and nutraceutical data of 18 chestnut cultivars were subjected to one-way analysis of variance (ANOVA), and the averages were compared with the Tukey's HSD post-hoc comparison test (n = 3) [39]. Correlation between sensory and phytochemical data was evaluated with Pearson's coefficient (r) [39]. A principal component analysis (PCA) [39,40] was carried out on the data matrix including 54 rows (3 repetitions for 18 samples) and 11 fields, each one representing a variable obtained from the chemical analyses. Such variables included the content of nine chemical classes CA (cinnamic acids), FL (flavonols), BE (benzoic acids), CAT (catechins), TA (tannins), MO (monoterpenes),

OA (organic acids), VC (vitamin C), SU (sugars), the TPC (total polyphenol content), and AA (antioxidant activity). The Bartlett's test of sphericity was carried out and the Kaiser–Meyer–Olkin (KMO) index was calculated from the data matrix [39,40]. The data matrix was subsequently centred and scaled columnwise and the corresponding cell values were, thus, transformed into Z-scores [41]. Based on the outcomes of the Bartlett's test of sphericity and of the KMO index, a principal component analysis (PCA) was performed on the transformed data matrix. Varimax rotation of the principal axes was applied [39,40]. The minimum number of principal components (PCs) accounting for at least the 50% of the total variance was retained. The association between the chemical variables and the retained PCs was assessed from the plots displaying the loadings of each chemical variable in the PCs plane [39,40]. Points coordinates in the PCs plane were analysed as reported in Lione et al. (2015) [42] and Lione and Gonthier (2016) [43]. The spatial distribution pattern of all the points plotted in the PCs plane was analysed with the Clark-Evans test [44]. The spatial distribution pattern of the points associated with the MT group in the PC plane was analysed with the Mean Distance Randomisation Test Left Tailed ($MDRT_{LT}$), performed on a subset of 106 permutations [43]. Similarly, the spatial distribution of cases associated with SC and EH groups was assessed by the Mean Distance Randomisation Test Right Tailed ($MDRT_{RT}$), carried out by setting the same permutation number.

The effect of the genotype on the chemical fingerprint was tested by fitting a conditional inference tree model [45,46] on CA, FL, BE, CAT, TA, MO, OA, VC, SU, TPC, and AA as response variables and on the genotype as predictor. The unbiased recursive partitioning algorithm described in Hothorn et al. (2006) [45] and Hothorn and Zeileis (2015) [46] was used for model fitting. The algorithm was run by setting the Bonferroni p-value correction for multiple comparisons and the 95% criterion to implement the model splits [39,45,46].

ANOVA and PCA were performed with statistical software package IBM SPSS Statistics 22.0 (IBM, Armonk, NY, USA), the Mean Distance Randomisation Tests were run with the MDT software (https://apsjournals.apsnet.org/doi/suppl/10.1094/PHYTO-05-15-0112-R) [43], and the conditional inference tree model fitting was performed in R [47] with the package partykit [46]. For all statistical tests, the significance threshold was set at 5%.

3. Results and Discussion

3.1. Sensory Analysis

Eight sensory attributes were used to characterise, qualitatively and quantitatively, the 18 chestnut cultivars analysed in this study, including three textural and visual (ease of peeling, seed colour, and flouryness) and five flavour descriptors (intensity of flavour, intensity of sweetness, intensity of saltiness, intensity of bitterness, and chestnut aroma). An overall subjective estimation (subjective judgement) was also expressed in order to describe the panellist personal rating. A 0 to 10 linear scale was used to evaluate the intensity of each attribute.

Although different cultivars showed the same sensory attributes, they differed in terms of intensity [48]. Within the SC and MT groups, cultivars of *C. sativa* showed significant differences ($p < 0.05$) for all the descriptors, except for intensity of saltiness. The EH group was more homogeneous and samples differed ($p < 0.05$) only for ease of peeling, seed colour, intensity of flavour, and intensity of bitterness.

SC cultivars achieved high values in terms of seed colour and intensity of sweetness, according to Kunsch et al. [49], but they were the bitterest ones, with low values for intensity of flavour, a descriptor very appreciated by consumers (Figure 2).

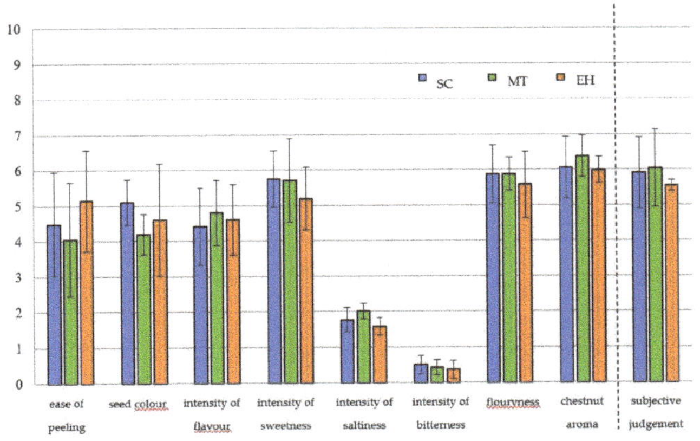

Figure 2. Sensory profiles for analysed chestnut groups (Sweet chestnut—SC, Marrone-type—MT, Euro-Japanese hybrid—EH). *Y*-axis represents the intensity value of sensory descriptors in a continuous scale partially structured into 10 segments.

Significant differences ($p < 0.05$) were observed between chestnut samples for ease of peeling and seed colour, two important attributes. 'Marrubia' (SC) presented the highest value of ease of peeling (7.00 ± 0.65), while 'Marrone di Castel del Rio' and 'Marrone di Marradi IGP' showed a good ease of peeling in the MT group (5.21 ± 2.32 and 5.67 ± 1.94, respectively) as shown in other studies [48]. 'Marsol' presented the lowest ease of peeling value (3.50 ± 0.71) for the EH group, in particular if compared to 'Bouche de Betizac' (5.86 ± 2.14), which was in agreement with the values reported in other studies [49]. A sensory profile of all the chestnut samples is reported in Table 3.

MT cultivars are commonly appreciated for fresh and processing consumption thanks to their positive traits (kernel easily separable from episperm, ease of seed peeling, reddish colour epicarp, good sweet flavour). Sensory analysis on the three cultivars from this group partially confirmed the results published in previous studies [48,50]. They showed the highest ratings for intensity of saltiness and chestnut aroma. Intensity of sweetness level was also considerably high, and it was comparable with the other SC chestnuts (Figure 3).

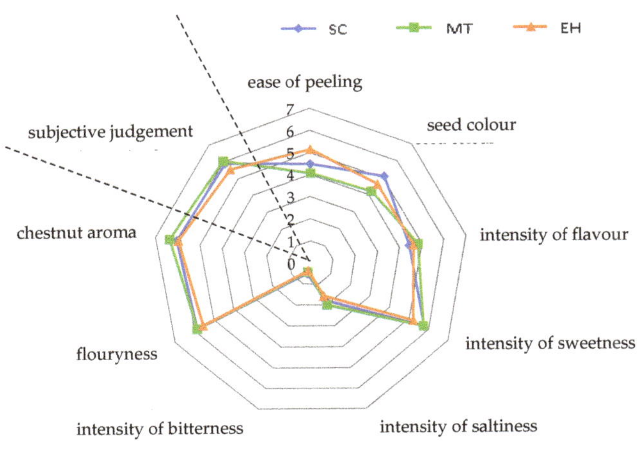

Figure 3. Radar chart for sensory analysis of considered chestnut groups (Sweet chestnut—SC, Marrone-type chestnut—MT, Euro-Japanese hybrid—EH). Subjective judgement is not part of radar chart for sensory analysis described by Quantitative Descriptive Analysis (QDA), but it was added together to other descriptors as complementary information.

Table 3. Sensory profiles of the analysed chestnuts.

Cultivar	Cv Type	ID	Ease of Peeling	Seed Colour	Intensity of Flavour	Intensity of Sweetness	Intensity of Saltiness	Intensity of Bitterness	Flouryness	Chestnut Aroma	Subjective Judgement
Bouche de Betizac ***	EH	159	5.86 ± 2.14 ab	2.64 ± 1.97 b	3.71 ± 1.60 b	6.07 ± 1.30 a	1.43 ± 0.98 a	0.29 ± 0.49 ab	6.86 ±0.80 a	6.00 ± 1.55 a	5.43 ± 2.24 a
Brunette *	SC	9	4.21 ± 2.12 abcd	5.21 ± 0.99 ab	3.93 ± 2.01 ab	6.14 ± 0.85 abc	1.43 ± 1.27 a	0.21 ± 0.39 ab	5.36 ± 1.93 ab	5.50 ± 1.12 ab	5.79 ± 1.07 abcd
Canepina *	SC	78	3.43 ± 1.30 bcd	4.93 ± 0.84 ab	3.00 ± 1.15 b	4.36 ± 1.18 b	1.07 ± 1.02 a	0.43 ± 0.79 ab	6.71 ± 1.15 a	5.14 ± 0.90 b	4.86 ± 1.35 cd
Contessa *	SC	30	4.65 ± 1.73 abcd	5.90 ± 1.78 ab	5.65 ± 1.93 ab	6.85 ± 1.70 abc	1.97 ± 1.13 a	0.00 ± 0.00 b	6.35 ± 1.06 ab	7.35 ± 1.60 ab	7.10 ± 1.30 abcd
Gabiana *	SC	45	5.33 ± 2.06 abcd	4.44 ± 1.51 ab	3.39 ± 1.32 ab	5.50 ± 1.70 abc	1.78 ± 1.30 a	1.22 ± 0.79 a	3.94 ± 1.24 b	4.72 ± 1.95 b	4.56 ± 2.07 d
Garrone Rosso *	SC	39	5.43 ± 2.15 ab	5.21 ± 1.35 ab	5.21 ± 2.48 ab	5.86 ± 1.70 abc	1.71 ± 0.49 a	0.29 ± 0.76 ab	6.86 ± 1.35 a	6.79 ± 1.47 ab	7.07 ± 1.02 abcd
Gentile *	SC	186	5.11 ± 1.27 abcd	3.56 ± 0.98 b	4.33 ± 1.22 ab	4.72 ± 1.20 bc	1.94 ± 0.73 a	0.78 ± 0.79 ab	6.06 ± 1.67 ab	5.78 ± 1.99 ab	6.00 ±1.58 abcd
Mansa *	SC	390	3.13 ± 1.64 bcd	5.13 ± 1.46 ab	3.13 ± 0.79 b	6.00 ± 1.60 abc	1.94 ± 1.21 a	1.31 ± 1.03 a	6.13 ± 2.10 ab	6.19 ± 1.13 ab	5.17 ± 1.12 abcd
Marrone di Castel del Rio **	MT	130	5.21 ± 2.32 abcd	4.36 ± 1.86 ab	5.79 ± 2.14 ab	7.07 ± 1.59 a	1.86 ± 1.49 a	0.14 ± 0.24 ab	5.21 ± 2.10 ab	7.14 ± 1.55 a	7.43 ± 1.24 a
Marrone di Marradi IGP **	MT	198	5.67 ± 1.94 abc	4.78 ± 1.30 ab	5.22 ± 1.72 ab	6.22 ± 1.12 abc	1.83 ± 1.12 a	0.44 ± 0.73 ab	6.17 ± 2.12 ab	6.33 ± 1.20 ab	6.00 ± 1.32 abcd
Marrone della Val di Susa **	MT	47	2.64 ± 1.25 cd	4.21 ± 1.78 b	4.57 ± 2.23 ab	5.21 ± 1.78 abc	2.07 ± 1.84 a	0.57 ± 0.79 ab	6.21 ± 2.06 ab	6.33 ± 1.60 ab	5.92 ± 2.08 abcd
Marrubia *	SC	215	7.00 ± 0.65 a	5.07 ± 1.64 ab	6.21 ± 1.38 a	7.00 ± 0.82 ab	1.71 ± 1.38 a	0.00 ± 0.00 b	6.14 ± 1.11 ab	7.57 ± 0.98 a	7.71 ± 0.76 a
Marsol ***	EH	5	3.50 ± 0.71 b	4.00 ± 1.70 ab	6.00 ± 0 a	4.20 ± 1.96 a	1.90 ± 1.75 a	0.90 ± 0.74 a	5.10 ± 1.29 a	5.80 ± 1.44 a	5.60 ± 1.64 a
Neirana della Val di Susa *	SC	218	5.90 ± 1.65 ab	5.56 ± 2.01 ab	5.06 ± 1.83 ab	6.00 ± 1.98 abc	2.11 ± 1.41 a	0.44 ± 0.71 ab	5.89 ± 1.70 ab	5.72 ± 1.16 ab	5.56 ± 1.20 abcd
Precoce Migoule ***	EH	258	4.50 ± 1.53 ab	5.93 ± 1.37 a	4.14 ± 1.97 ab	4.71 ± 1.73 a	1.64 ± 1.11 a	0.29 ± 0.57 ab	5.64 ± 2.43 a	6.50 ± 0.87 a	5.43 ± 1.37 a
Tarvisò *	SC	25	3.21 ± 1.22 bcd	5.43 ± 1.99 ab	3.57 ± 1.75 ab	5.14 ± 2.25 abc	1.50 ± 1.38 a	0.14 ± 0.38 ab	5.14 ± 2.73 ab	5.86 ± 1.68 ab	5.93 ± 1.59 abcd

Mean value and standard deviation (SD) of each sample is given ($n = 3$). Different letters (a,b,c,d) for each descriptor indicate the significant differences at $p \leq 0.01$. * Sweet chestnut (SC). ** Marrone-type chestnut (MT). *** Euro-Japanese hybrid (EH).

In particular, the descriptor "intensity of sweetness" varied significantly ($p < 0.05$) among the SC and MT cultivars: 'Canepina' presented the lowest value (4.36 ± 1.18) and 'Marrone di Castel del Rio' the highest one (7.07 ± 1.59), while there were no significant differences ($p > 0.05$) among the EH group. Data showed a high correlation level, evaluated by Pearson's coefficient (r), between intensity of sweetness and sugar content in the analysed cultivars ($r = 0.71$ for SC, $r = 0.89$ for MT, and $r = 0.81$ for EH). 'Marrubia' showed no intensity of bitterness, recording the lowest value for this attribute, while 'Marsol' was the most bitter Euro-Japanese hybrid (0.90 ± 0.74). A high correlation coefficient was also found between intensity of bitterness and tannin content ($r = 0.73$ for SC, $r = 0.68$ for MT, and $r = 0.85$ for EH).

Significant differences ($p < 0.05$) were also observed in intensity of flavour and chestnut aroma (flavour descriptors). The first one is usually measured during the seed breakup and refers to the smell sense, while the other is the expression of chestnut aroma measurable by multiple senses as described in other studies [48,51]. Results highlighted a high level of intensity of flavour in the 'Marrubia' cultivar (6.21 ± 1.38), significantly higher ($p < 0.05$) than all the other sweet chestnuts. In the case of chestnut aroma, both 'Marrubia' and 'Marrone di Castel del Rio' significantly differed ($p < 0.05$) from the other cultivars that showed higher ratings. Euro-Japanese hybrids did not differ significantly ($p > 0.05$) in terms of chestnut aroma, while significant differences ($p < 0.05$) were observed in intensity of flavour, which was higher in 'Marsol' (6.00 ± 0.23).

Panellists were also asked to give a personal preference to each cultivar or hybrid, although not planned in the Quantitative Descriptive Analysis (QDA), in order to assess a subjective general rating. 'Marrubia' (7.71 ± 0.76) and 'Contessa' (7.10 ± 1.30) for the SC chestnuts, and 'Marrone di Castel del Rio' (7.43 ± 1.24) for the MT group, displayed significantly higher ($p < 0.05$) values than all the other cultivars, while EH chestnuts showed lower but not significant values ($p > 0.05$) in the group. Data pointed out a good correlation between judgment and sugar content ($r = 0.57$ for SC, $r = 0.89$ for MT, and $r = 0.73$ for EH). These values seemed to be lower than the Pearson's coefficients between intensity of sweetness and sugar content due to the influence of other sensory traits on panellist judgment. Indeed, a significant correlation was also observed between judgment and chestnut aroma, with values similar to the correlation between judgment and sugar content ($r = 0.62$ for SC, $r = 0.80$ for MT, and $r = 0.78$ for EH). These results showed that judgment was equally influenced by chestnut aroma and sugar content. Moreover, a significant correlation between intensity of sweetness and chestnut aroma was observed ($r = 0.52$ for SC, $r = 0.72$ for MT, and $r = 0.54$ for EH). In any case, further statistical assessments are necessary to confirm this hypothesis. Moreover, a good correlation was found between intensity of sweetness and judgement ($r = 0.56$ for SC, $r = 0.873$ for MT, and $r = 0.89$ for EH group), suggesting the influence of this sensory parameter on panellist personal preference. Pearson's correlation data were reported in Supplementary Table S2. Even if MT cultivars were very floury and with dark seed colour, they were the most appreciated chestnuts according to the personal panellist judgment, confirming the findings of Mellano et al., 2007 [52]. Euro-Japanese hybrids, widely spread and cultivated because of their resistance to diseases and their high kernel quality [53], scored as the easiest to peel and the lowest in terms of flouryness, intensity of bitterness, and intensity of saltiness. Nevertheless, hybrids showed values lower than other cultivars for intensity of sweetness and chestnut aroma, as remarked by the low score assigned by the panellists.

3.2. Phytochemical Composition, Antioxidant Capacity, and Nutritional Properties

The phytochemical and nutritional profile (contents of polyphenols, monoterpenes, vitamin C, organic acids, and sugars), complemented by the measurement of the antioxidant capacity, of the 18 cultivars and hybrids of chestnuts were defined by chemical analysis [54].

Mean TPC and antioxidant capacity values are reported in Figures 4 and 5. TPC values (Figure 4) ranged from 0.55 ± 0.02 g_{GAE} kg^{-1} DW for the French cultivar 'Bouche Rouge' to 1.40 ± 0.05 g_{GAE} kg^{-1} DW for the Italian cultivar 'Marrubia,' in agreement with other studies [55,56]. Chestnut TPC values were similar or higher than values detected in other tree nuts [57]. The highest phenolic content was observed for the Piemonte Region cultivars. Significantly different antioxidant capacity values ($p < 0.05$), expressed as a FRAP assay, were observed among the cultivars with a trend similar to the one observed for TPC levels. Antioxidant capacity ranged from 9.30 ± 0.39 mmol Fe^{+2} kg^{-1} DW ('Bouche de Bètizac') to 19.96 ± 1.89 mmol Fe^{+2} kg^{-1} DW ('Garrone Rosso'), as shown in Figure 5, in agreement with previous studies [51,58]. Chestnuts are a potential source of bioactive molecules, with a good antioxidant capacity, as highlighted by similar studies [51,59]. However, establishing the contribution of each single bioactive compound to the total antioxidant activity may be difficult because of the synergistic combination and interaction between the different substances (phytocomplex). Each antioxidant compound may improve the effectiveness of the others, and this action could influence the overall response (total antioxidant capacity) [60]. This additive effect may explain the significant differences between the antioxidant activities of the different analysed samples; for this reason, samples with the highest values of TPC and vitamin C did not always show the highest antioxidant capacity.

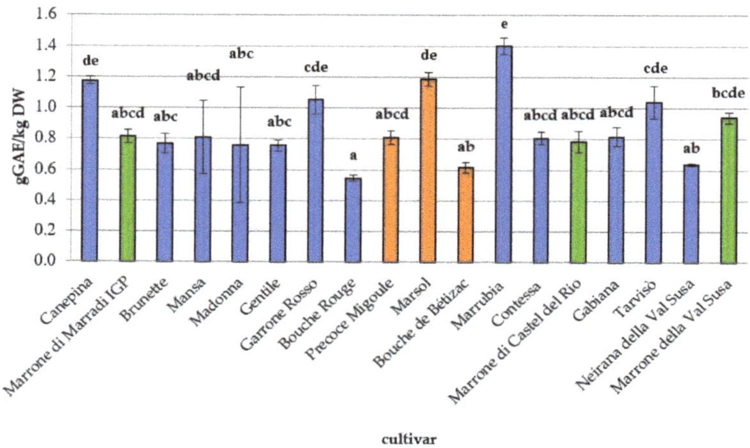

Figure 4. Total polyphenol content of the 18 chestnut cultivars. Different letters for each cultivar indicate the significant differences at $p < 0.05$. Blue colour: Sweet chestnut—SC; orange colour: Euro-Japanese hybrid—EH; green colour: Marrone-type chestnut—MT.

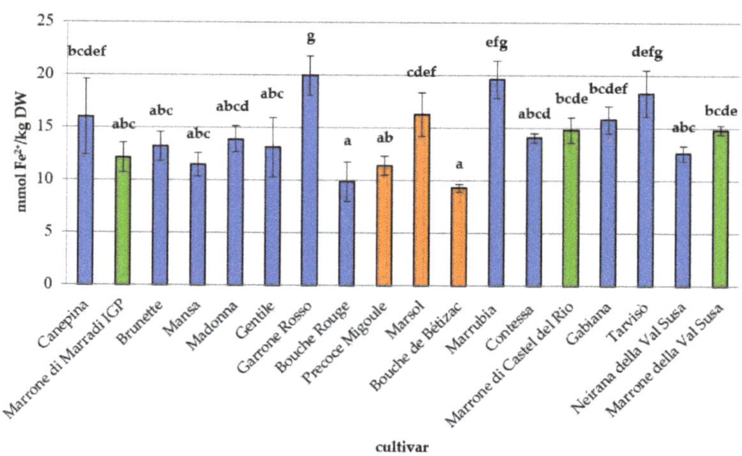

Figure 5. Antioxidant activity of the 18 chestnut cultivars. Different letters for each cultivar indicate the significant differences at $p < 0.05$. Blue colour: Sweet chestnut—SC; orange colour: Euro-Japanese hybrid—EH; green colour: Marrone-type chestnut—MT.

The phytochemical composition of the analysed cultivars identified 21 biomarkers by HPLC-DAD. To evaluate the contribution of each class to the total phytocomplex composition, the bioactive compounds were grouped in the following classes: polyphenols (as the sum of cinnamic acids, flavonols, benzoic acids, catechins, and tannins), monoterpenes, and vitamin C (mean values were considered) (Figure 6).

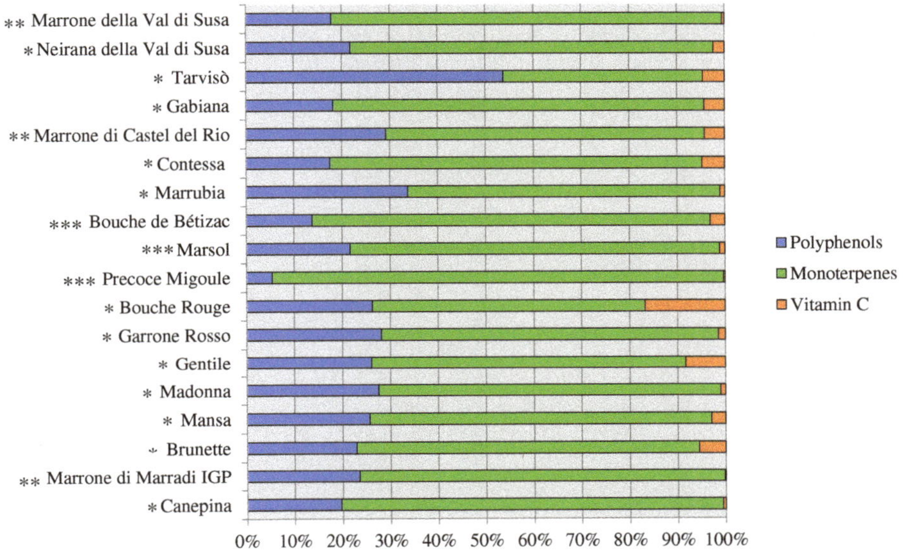

Figure 6. Phytocomplex representation of the analysed chestnut cultivars. The colours identify the different classes: blue corresponds to polyphenols, green to monoterpenes, and orange to vitamin C. * Sweet chestnut (SC). ** Marrone-type chestnut (MT). *** Euro-Japanese hybrid (EH).

Monoterpenes, recognised for their anti-tumour and anti-inflammatory properties [61], represented the main component of the phytocomplex, reaching the 88% of EH cultivars, followed by polyphenols, characterised by antioxidant, anti-bacterial, and anti-tumour properties [62], and is between 10 and 25% for EH and SC, respectively, and has vitamin C in low quantities (2%). However, the cultivar 'Tarvisò' showed a higher proportion of polyphenols than monoterpenes. The highest levels of polyphenols and monoterpenes were detected in 'Canepina' (0.19 g·kg^{-1} DW and 0.76 g·kg^{-1} DW, respectively), a cultivar from central Italy, while the highest values of vitamin C were observed in the French cultivar 'Bouche Rouge' (0.19 g·kg^{-1} DW) as shown in Table 4.

Within the polyphenolic group, differences were observed among chestnut genotypes, but most of the cultivars showed phenolic levels similar to hazelnut ones [63,64], and higher than walnut ones [65]. Cinnamic acids and flavonols were the most important classes among phenolics (20–40% and 15–20%, respectively), followed by tannins (27%, 22%, and 43% for MT, EH, and SC chestnuts, respectively) as shown in Figure 7. Flavonols, catechins, and benzoic acids were detected in very low quantities (<10%).

Table 4. Chestnut phytochemical class profiles.

		Cinnamic Acids		Flavonols		Benzoic Acids		Catechins		Tannins		Monoterpenes		Vitamin C	
		Mean Value	SD	Mean Value	SD	Mean Value	SD	Mean Value	SD	Mean Value	SD	Mean Value	SD	Mean Value	SD
CV	ID	$g \cdot kg^{-1}$ DW		$g \cdot kg^{-1}$ DW		$g \cdot kg^{-1}$ DW		$g \cdot kg^{-1}$ DW		$g \cdot kg^{-1}$ DW		$g \cdot kg^{-1}$ DW		$g \cdot kg^{-1}$ DW	
Bouche de Bétizac ***	159	0.204	0.001	0.089	0.007	0.014	0.001	0.079	0.007	0.173	0.022	3.365	0.118	0.124	0.018
Bouche Rouge *	286	0.151	0.080	0.011	0.000	0.025	0.000	0.049	0.001	0.060	0.008	0.643	0.158	0.189	0.029
Brunette *	9	0.158	0.080	0.085	0.001	0.036	0.001	0.065	0.002	0.174	0.008	1.614	0.095	0.126	0.008
Canepina *	78	0.208	0.001	0.244	0.034	0.064	0.007	0.159	0.034	1.201	0.348	7.590	1.788	0.055	0.023
Contessa *	30	0.199	0.005	0.145	0.024	0.027	0.000	0.019	0.002	0.196	0.004	2.595	0.576	0.160	0.023
Gabiana *	45	0.137	0.010	0.173	0.011	0.066	0.001	0.063	0.013	0.170	0.014	2.589	0.051	0.142	0.065
Garrone Rosso *	39	0.066	0.001	0.083	0.008	0.018	0.001	0.100	0.005	0.148	0.011	1.039	0.172	0.022	0.008
Gentile *	186	0.066	0.000	0.112	0.009	0.019	0.006	0.078	0.007	0.205	0.013	1.207	0.073	0.154	0.011
Madonna *	1	0.117	0.079	0.247	0.043	0.023	0.000	0.076	0.009	0.224	0.028	1.783	0.051	0.025	0.008
Mansa *	390	0.100	0.079	0.005	0.001	0.014	0.001	0.019	0.002	0.026	0.002	0.457	0.045	0.019	0.002
Marrone di Castel del Rio **	130	0.198	0.014	0.163	0.008	0.021	0.001	0.078	0.017	0.232	0.016	1.581	0.054	0.100	0.016
Marrone di Marradi IGP **	198	0.206	0.001	0.030	0.005	0.021	0.001	0.077	0.002	0.036	0.003	1.199	0.193	0.002	0.002
Marrone della Val di Susa **	47	0.210	0.003	0.118	0.010	0.022	0.001	0.075	0.003	0.192	0.020	2.812	0.454	0.019	0.002
Marrubia *	215	0.200	0.004	0.171	0.008	0.026	0.001	0.080	0.003	0.105	0.020	1.124	0.083	0.017	0.010
Marsol ***	5	0.199	0.004	0.072	0.005	0.024	0.001	0.100	0.008	0.117	0.009	1.822	0.061	0.027	0.006
Neirana della Val di Susa *	218	0.155	0.001	0.039	0.001	0.012	0.009	0.023	0.012	0.211	0.017	1.526	0.068	0.047	0.017
Precoce Migoule ***	258	0.209	0.001	0.044	0.002	0.025	0.000	0.106	0.005	0.036	0.006	7.371	0.478	0.025	0.005
Tarvisò *	25	0.202	0.001	0.256	0.031	0.015	0.001	0.109	0.006	0.652	0.041	0.958	0.343	0.105	0.015

Mean value and standard deviation (SD) of each sample is given ($n = 3$). * Sweet chestnut (SC). ** Marrone-type chestnut (MT). *** Euro-Japanese hybrid (EH).

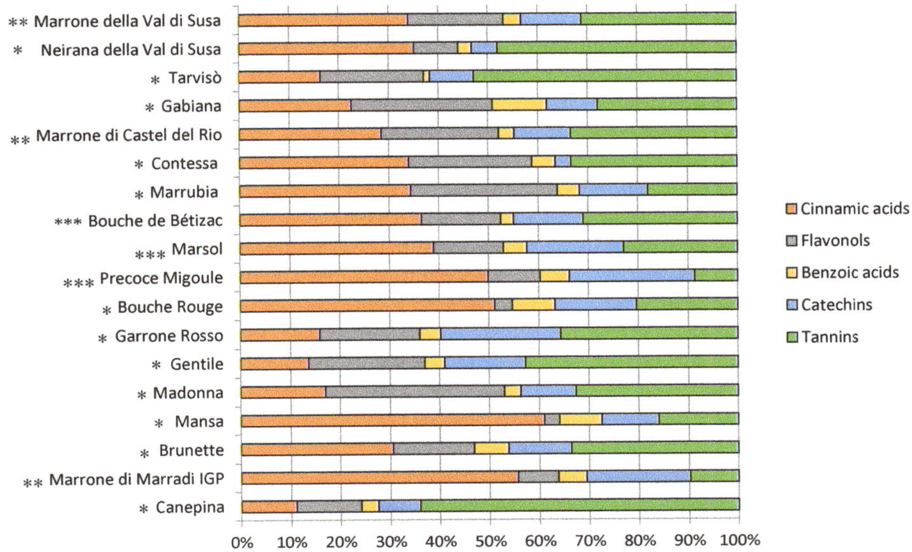

Figure 7. Polyphenol profile of the analysed chestnut cultivars. * Sweet chestnut (SC). ** Marrone-type chestnut (MT). *** Euro-Japanese hybrid (EH).

Data on each bioactive and nutritional compound content are reported in Supplementary Table S3. Tannins were the main polyphenolic class in the analysed chestnuts, followed by cinnamic acids and flavonols, the latter with a great variability in the results. The 'Canepina' cultivar displayed high tannin levels (1.20 ± 0.35 g·kg^{-1} DW). This value was higher than the average of the other samples, as confirmed by Tukey's test ($p < 0.05$), which included this sample in a separate group. The same holds true for the catechin class, which was represented mainly by epicatechin (0.15 ± 0.04 g·kg^{-1} DW). Regarding catechins, the MT varieties were included within the same group, together with 'Gentile' and 'Madonna' cultivars, that showed relevant quantities of these compounds as shown by De Biaggi et al. (2018) [51]. The identification of catechin and epicatechin is an important result as they are involved in the inhibition of lipid peroxidation, and inhibition of human cancer cell line proliferation and cyclooxygenase enzymes [66]. For tannins, except for 'Canepina' and 'Tarvisò' (the highest values) and 'Mansa' and 'Marrone di Marradi IGP' (the lowest values), all the samples contained between 0.10 and 0.25 g·kg^{-1} DW, while French cultivar 'Bouche Rouge', with about 0.06 g·kg^{-1} DW, was classified by the post-hoc test in a separate group ($p < 0.05$). The presence of tannins in adequate amounts increases the nutraceutical properties of chestnuts since these compounds are free radical quenchers [67].

Caffeic and coumaric acids were quantified in all the samples, although in low quantities (<0.01 g·kg^{-1} DW). Ferulic acid was not detected in all the analysed chestnut cultivars, except in 'Canepina' and 'Madonna' (about 0.01 g·kg^{-1} DW). In the flavonol class, no traces of quercetin were detected in any of the analysed samples, and quercitrin and rutin occurred only at low concentrations (always below 0.01 g·kg^{-1} DW), except in 'Marrubia', which showed a quercitrin value of 0.120 ± 0.001 g·kg^{-1} DW. The most representative flavonol was isoquercetin, which is detected in large quantities, especially in 'Tarvisò' (0.26 ± 0.03 g·kg^{-1} DW), 'Canepina' (0.23 ± 0.03 g·kg^{-1} DW), and 'Madonna' (0.22 ± 0.04 g·kg^{-1} DW). 'Tarvisò' cultivar displayed the highest flavonol content (0.26 ± 0.03 g·kg^{-1} DW), followed by 'Madonna' (0.25 ± 0.04 g·kg^{-1} DW), and 'Canepina' (0.24 ± 0.03 g·kg^{-1} DW), while 'Mansa', 'Bouche Rouge', and 'Marrone di Marradi IGP' showed lower values. Flavonols quench active oxygen species and inhibit in vitro oxidation of low-density lipoproteins [68].

Ellagic acid was the most abundant benzoic acid in 'Gabiana' and 'Canepina' (about 0.05 g·kg^{-1} DW), while gallic acid showed high variability among the samples, except for 'Marrone della Val

Pellice,' which did not contain gallic acid and, in general, was characterised by the lowest content of benzoic acids. These molecules are very important in human nutrition and are related to many biological properties, including anticancer, anti-atherosclerotic, anti-inflammatory, antihepatotoxic, and anti-HIV replication activities [69].

As well as polyphenols, the analysed samples also showed different monoterpenes and discrete contents of vitamin C, as reported in Supplementary Table S3. Monoterpenes are a large class of naturally bioactive molecules used extensively for their aromatic qualities combined with their antioxidant capacity and anti-inflammatory properties [70]. Many of these molecules have antibacterial and antitumor activity [71]. EH cultivars showed high contents of monoterpenes, in particular 'Precoce Migoule,' but 'Canepina' (SC) was characterised by the highest content of these compounds (0.76 ± 0.18 g·kg^{-1} DW). Limonene was predominant and reached quantities of 6.35 ± 1.77 g·kg^{-1} DW in the 'Canepina' cultivar. High limonene amounts were also found in the EH group. γ-terpinene was detected in high quantities, in particular in 'Precoce Migoule' (0.38 ± 0.70 g·kg^{-1} DW), 'Marrone della Val Pellice' (0.21 ± 0.13 g·kg^{-1} DW), and 'Gabiana' (0.16 ± 0.05 g·kg^{-1} DW), similar to other studies [62,72]. Several studies reported their chemopreventive activity against rodent mammary, skin, liver, lung, and forestomach cancers [73]. Terpinolene, sabinene, and phellandrene were also identified even if at trace levels.

Vitamin C was evaluated as the sum of ascorbic and dehydroascorbic acids due to their biological activity in human organisms as reported in other studies [60,74]. The maximum vitamin C value was detected in the 'Bouche Rouge' cultivar (0.19 ± 0.03 g·kg^{-1} DW), followed by 'Marrone della Val Pellice' (0.18 ± 0.09 g·kg^{-1} DW), while the minimum amount was detected in 'Marrone di Marradi IGP'. Among EH chestnuts, 'Bouche de Bétizac' provided the largest amount of vitamin C (0.12 ± 0.02 g· kg^{-1} DW), which was comparable to the values reported in De Biaggi et al. (2018) [51]. The majority of the chestnut cultivars showed vitamin C content similar to walnut and almond ones [57,75], higher than hazelnut varieties [76].

Large and significant differences ($p < 0.05$) in organic acid and sugar content values were detected among cultivars. Mean values for SC, MT, and EH chestnuts are reported in Table 5.

Table 5. Nutritional properties of analysed chestnut cultivars.

		Organic Acids		Sugars	
		Mean Value	SD	Mean Value	SD
CV	ID	g·kg^{-1} DW		g·kg^{-1} DW	
Bouche de Bétizac ***	159	4.08	0.32	32.61	4.10
Bouche Rouge *	286	3.34	0.48	10.64	2.33
Brunette *	9	7.97	0.25	48.79	9.68
Canepina *	78	4.41	0.37	114.80	10.63
Contessa *	30	6.00	0.15	18.36	2.92
Gabiana *	45	3.88	0.43	38.32	3.86
Garrone Rosso *	39	1.20	0.21	14.90	2.07
Gentile *	186	2.63	0.19	48.97	2.19
Madonna *	1	4.34	0.80	62.47	3.38
Mansa *	390	6.58	0.71	273.38	21.16
Marrone di Castel del Rio **	130	2.49	0.39	26.80	3.91
Marrone di Marradi IGP **	198	3.18	0.31	29.22	2.03
Marrone della Val di Susa **	47	1.30	0.15	14.26	1.76
Marrubia *	215	1.79	0.31	25.29	4.01
Marsol ***	5	5.01	0.45	29.07	9.82
Neirana della Val Susa *	218	3.22	0.21	20.00	2.67
Precoce Migoule ***	258	7.43	0.09	38.97	1.40
Tarvisò *	25	4.75	0.35	25.75	1.34

Mean value and standard deviation (SD) of each sample is given ($n = 3$). * Sweet chestnut (SC). ** Marrone-type chestnut (MT). *** Euro-Japanese hybrid (EH).

High levels of organic acids were observed in hybrids, in particular for 'Precoce Migoule' (7.43 ± 0.09 g·kg^{-1} DW), while the chestnut cultivar 'Garrone Rosso' showed the lowest values (1.20 ± 0.21 g· kg^{-1} DW). Citric acid was the most abundant organic acid in the analysed chestnuts, with high levels in 'Mansa' and 'Precoce Migoule' cultivars (5.31 ± 0.25 g·kg^{-1} DW and 3.28 ± 0.25 g·kg^{-1} DW, respectively), followed by quinic acid (maximum value of 3.55 ± 0.28 g·kg^{-1} DW in 'Brunette cultivar'), as reported by similar studies [77]. Quinic acid was detected in all the samples, except in 'Marrubia,' which only contained citric acid. Malic acid was not detected in the analysed chestnuts, due to the intrinsic characteristics of the considered varieties and the effect of the drying treatment applied during the sample preparation [49]. Tukey's test highlighted significant differences ($p < 0.05$) in the organic acid composition among different cultivars of the SC, MT, and EH groups, leading to the identification of different groups composed by one or a few compounds. This result could be due to the differences associated with the different genotypes but, since organic acids are volatile molecules, other factors could have slightly influenced the results, such as the extraction technique, sample storage, and applied drying technique [78].

The highest quantity of sugars was observed for the cultivar 'Mansa' (273.38 ± 21.16 g·kg^{-1} DW), while the average values ranged from 10.64 ± 2.33 g·kg^{-1} DW ('Bouche Rouge') to 114.80 ± 10.63 g·kg^{-1} DW ('Canepina'); chestnut cultivars showed higher values if compared to the sugar levels of other tree nuts such as walnut and almond [79]. MT chestnuts showed sugar levels (15–30 g·kg^{-1} DW) similar to other chestnut cultivars, which was in agreement with other studies [51,80,81]. Although sucrose was the most abundant sugar in many analysed chestnuts, some samples, including 'Marrone di Marradi IGP,' 'Marsol,' 'Precoce Migoule,' 'Brunette,' 'Garrone Rosso,' 'Marrone della Val Pellice,' and 'Marrubia,' showed higher contents of glucose than sucrose. Other cultivars (e.g., 'Mansa,' 'Marrone della Val di Susa,' and 'Neirana della Val di Susa') showed higher contents of fructose than glucose. The higher fructose amount compared to the glucose one may be important to define chestnuts as a potential functional food for consumers suffering from type 2 diabetes, as fructose shows a lower glycemic index than glucose and, consequently, the postprandial glycemic peak due to fructose is lower than the glycemic peak due to glucose, as well as the insulin response [82]. As evidenced by the Tukey's test, the 'Mansa' cultivar significantly differed ($p < 0.05$) from the other samples, followed by 'Canepina', as mentioned above, and 'Brunette' and 'Gentile', which reported quantities close to 50 g·kg^{-1} DW.

3.3. Multivariate Analysis

Rather than hinging on the action of a single compound, therapeutic effects obtainable from the consumption of fresh fruit and derived-products are the result of the synergistic or additive interaction of several phytochemicals that jointly contribute to disease prevention [83]. For this reason, compounds belonging to the same chemical class were combined in bioactive classes for multivariate data analysis. The outcome of the Bartlett's test of sphericity ($p < 0.05$) showed a significant collinearity among variables. The KMO index attained a value of 0.74. The PCA resulted in two principal components accounting for 50.71% of the total variance, with 32.47% explained by PC1, and 18.24% by PC2. The location in the PCs plane of the 18 samples (mean values of three repetitions for each cultivar) in relation to phytochemical composition, nutritional properties, and nutraceutical traits is shown in the score plot (Figure 8).

PCA showed that cultivars belonging to the MT group, which are highlighted in Figure 8, presented similar traits according to the chemical results. PCA loadings plot showed an association between polyphenolic classes, vitamin C, monoterpenes and PC1, and a correlation between TPC, antioxidant activity, organic acids, sugars, and PC2 (Figure 9).

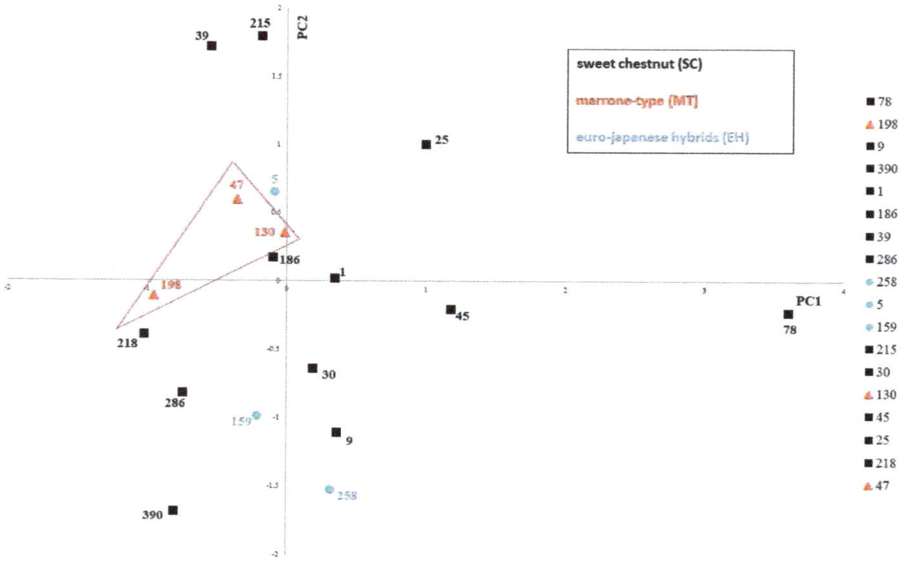

Figure 8. Principal component analysis (PCA) score plot of analysed chestnut cultivars. Mean values ($n = 3$) were considered for each cultivar. Cultivar name (ID): Bouche de Bétizac (159); Bouche Rouge (286); Brunette (9); Canepina (78); Contessa (30); Gabiana (45); Garrone Rosso (39); Gentile (186); Madonna (1); Mansa (390); Marrone di Castel del Rio (130); Marrone di Marradi IGP (198); Marrone della Val di Susa (47); Marrubia (215); Marsol (5); Neirana della Val di Susa (218); Precoce Migoule (258); Tarvisò (25).

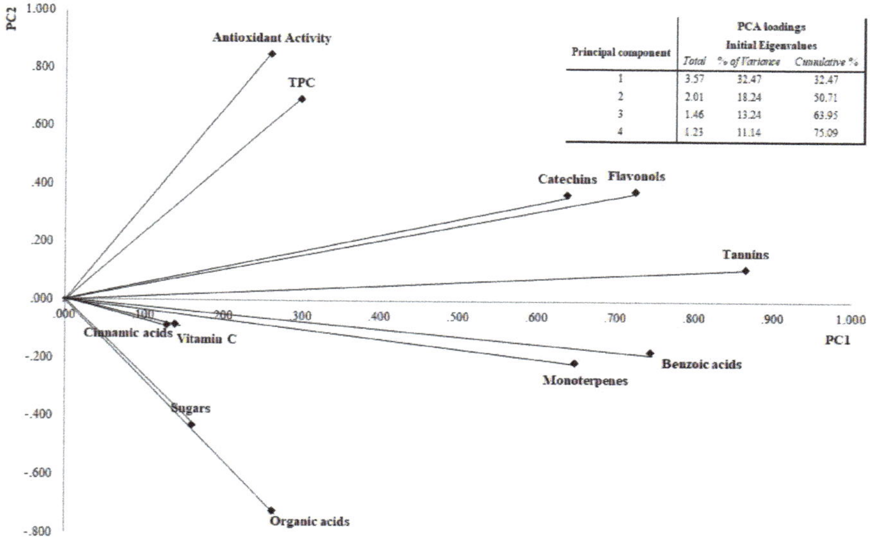

Principal component	PCA loadings Initial Eigenvalues		
	Total	*% of Variance*	*Cumulative %*
1	3.57	32.47	32.47
2	2.01	18.24	50.71
3	1.46	13.24	63.95
4	1.23	11.14	75.09

Figure 9. PCA loading plot of considered variables.

Phenolic acids and tannins, associated with PC1, were identified as bioactive classes with the most discriminating power among different genotypes; these phytochemical classes included compounds displaying significant differences ($p < 0.05$) in their bioactive content among the different cultivars. Moreover, monoterpenes showed also a good discriminating power among chestnut samples. For this reason, all these molecules could represent the most important markers in order to build a discriminant model between chestnut genotypes, but further studies are necessary to confirm this hypothesis.

In this study, a multivariate analysis as PCA allowed for the visualisation of the information included in the fingerprints. The results showed that PCA classification characterised the samples according to the different chemical composition, providing information on the bioactive classes and chemical markers that most influence the phytocomplex. A chemometric method was applied with the HPLC fingerprint technique for a better recognition of the analysed extracts as reported by Cirlini et al. [84]. Different marker compounds were detected as the variables most relevant for the discrimination of chestnut cultivars, which could be applied to accurate composition control of a chestnut flour derived from a specific cultivar. In this study, PCA results showed that MT genotypes formed a single group within a larger group of SC and EH cultivars: the HPLC fingerprint combined with chemometrics could be considered as a tool of traceability in order distinguish different genotypes by their phytochemical composition and antioxidant properties, as is reported in other research [85,86].

The tests about the spatial distribution pattern further supported the interpretation of the PCA results. The Clark-Evans test showed that the points in the PCs plane were significantly clustered ($r = 0.643$, $p < 0.05$). The $MDRT_{LT}$ test showed a significant clustering of points associated with MT group ($p < 0.05$) in the PC plan, while the $MDRT_{RT}$ test showed that the points associated with SC + EH groups are characterised by a significantly dispersed spatial distribution ($p < 0.05$) in the same PCs plane. These results suggest that the MT cultivars represented a homogenous group with less variable traits than the SC + EH groups (Figure 10).

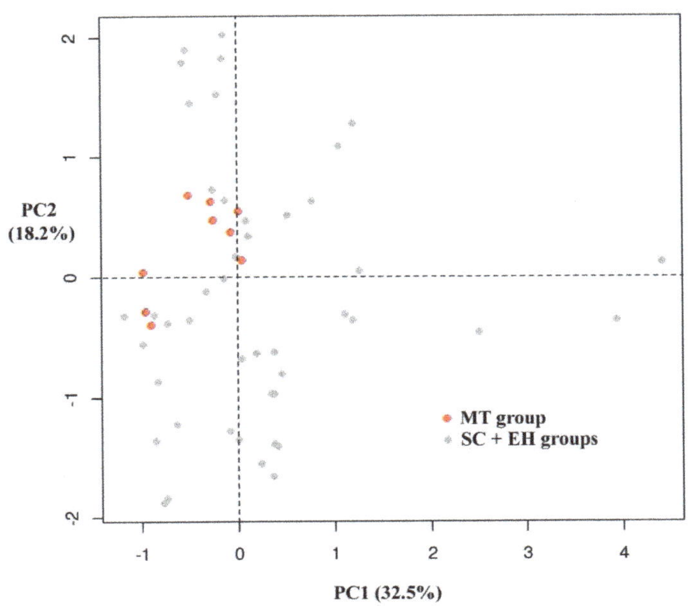

Figure 10. Distribution of the points associated with "Marrone" and "Chestnut" groups in the principal components (PCs) plane. A significant clustering of points associated with MT group ($p < 0.05$) is observed in the PCs plane, while the points associated with SC + EH groups are characterised by a significantly dispersed spatial distribution ($p < 0.05$) in the same PCs plane. Sweet chestnut = SC. Marrone-type chestnut = MT. Euro-Japanese hybrid = EH.

In the PCs plane, spatial proximity between points is interpretable in terms of similarity of the underlying features [42]; hence, MT genotypes, closely located in the PCA plane, analysed in this study are characterised by a similar profile of chemical composition and of the associated properties beneficial to human health. The reduced variability in terms of chemical composition of MT cultivars that was pointed out by the geostatistical tests suggests that MT fruits are more suitable for specific uses than SC and EH ones. For instance, quality tracking and certifications are likely easier to be obtained by fruits showing constant chemical and organoleptic features than by fruit whose characteristics may undergo large fluctuations. Moreover, strict dietary requirements might be more likely respected by the consumers if the food they eat is endowed with a stable composition. This holds true not only for direct consumption, but also for the industrial transformation, in general aiming at producing standardised products with constant nutritional and chemical properties.

The outcomes of the conditional inference tree model pointed out that the genotype plays a significant ($p < 0.05$) role and may account for most of the differences detected among the chemical fingerprints of the chestnut fruits. In fact, the tree model (Supplementary Figure S1) displayed four highly significant ($p < 0.001$) splits resulting in two intermediate and four terminal nodes.

Each terminal node clustered the genotypes whose overall chemical fingerprint was homogeneous ($p > 0.05$), while genotypes included in separate nodes were characterised by significantly different chemical fingerprints ($p < 0.05$). It is worth noting that while the PCA clearly pointed out the role played by different chemical compounds in profiling, the samples analysed, the conditional inference tree model accounted for the role of the genotype, a categorical variable that cannot be included directly within a PCA, since this can only handle continuous covariates [39,40]. In combination, both analyses strongly support our hypothesis about the genotype influence on the chemical composition of chestnut cultivars grown on the same clonal rootstock and in the same agri-environmental conditions. Although replicates within genotypes were not particularly abundant, the algorithm run to fit the conditional inference tree model [45,46] has been specifically designed to be robust and reliable. Even when the number of covariates is high, the sample size is low or the data are unbalanced, as confirmed by a recent study on chestnut [87].

The combination of chromatographic fingerprint and chemometric evaluation could be a potential tool for chestnut product traceability and quality control, in order to select the best raw material based on the desired traits and properties. In addition, the above tools could be used to avoid potential voluntary or involuntary adulterations and contaminations. These hyphenated techniques could also contribute to the analysis of several processed products. Chemical fingerprint coupled to chemometrics could also be a useful tool to obtain label certifications for the valorisation of specific local genotypes. Moreover, the approach used in this study could also contribute to the selection of new varieties more tolerant or resistant to pests and diseases, in particular considering the new issues related to the climate change, and to support and improve breeding programs and preservation strategies for the existing cultivars.

4. Conclusions

Chromatographic and spectrophotometric data confirmed the high variability within the genus *Castanea*, and multivariate analysis allowed to explain such variability in terms of phytochemical and nutritional composition, characterising the different genotypes. Monoterpenes, important for their anti-inflammatory and anti-tumour activities, represented the main component of the chestnut composition (in particular, 88% for EH cultivars); followed by polyphenols, which are characterised by antioxidant and anti-bacterial properties (10–25% for EH and SC); and vitamin C in trace (about 2%). Tannins were the main polyphenolic compounds detected in the analysed chestnuts, followed by phenolic acids and flavonols. In particular, 'Canepina' presented higher phenolic amounts than almost all the analysed cultivars. Moreover, the majority of the analysed chestnut cultivars showed a content of bioactive compounds, as phenolics and vitamin C, whose levels were similar to, or higher than those reported for the main hazelnut, walnut, and almond varieties.

As genetic and phytochemical diversity represent fundamental aspects to ensure the productivity and adaptability of chestnut orchards, different approaches need to be developed to ensure the correct characterisation strategy. Hence, different techniques were combined in this study to define a suitable strategy for the characterisation of the chestnut cultivars as a key prerequisite to allow the conservation of *Castanea* germplasm.

The analysed cultivars were selected as part of a core collection that maximises the chestnut agro-biodiversity. The diversity observed among the analysed cultivars could be strictly associated to the genotype effect and underlines the large variability of the genus *Castanea*, and therefore, the importance of in farm and ex situ conservation of local germplasm as part of a global strategy, and also in relation to an active utilisation of agrobiodiversity.

Supplementary Materials: The following are available online at http://www.mdpi.com/2304-8158/9/8/1062/s1.

Author Contributions: Conceptualization, G.L.B., D.D. and M.G.M.; methodology, D.D. and M.G.M.; software, D.D. and G.G.L.; validation, D.D. and G.G.L.; formal analysis, D.D., M.G.M. and S.R.; investigation, S.R.; resources, G.G., M.G.M. and I.R.; data curation, M.D.B., D.D., G.G., G.G.L., M.G.M. and S.R.; writing—original draft preparation, M.D.B., D.D., G.G., G.G.L. and M.G.M.; writing—review and editing, G.L.B., D.D. and G.G.L.; supervision, G.L.B. and P.G.; project administration, G.L.B.; funding acquisition, G.L.B. All authors have read and agreed to the published version of the manuscript.

Funding: This research received no external funding.

Conflicts of Interest: The authors declare no conflict of interest.

References

1. Fischer, J.; Lindenmayer, D.B. Landscape modification and habitat fragmentation: A synthesis. *Glob. Ecol. Biogeogr.* **2007**, *16*, 265–280. [CrossRef]
2. Beccaro, G.; Alma, A.; Bounous, G.; Gomes-Laranjo, J. *The Chestnut Handbook: Crop & Forest Management*; CRC Press: Boca Raton, FL, USA, 2019.
3. Martín, M.A.; Mattioni, C.; Cherubini, M.; Villani, F.; Martín, L.M. A comparative study of European chestnut varieties in relation to adaptive markers. *Agrofor. Syst.* **2017**, *91*, 97–109. [CrossRef]
4. Mellano, M.; Beccaro, G.; Donno, D.; Marinoni, D.T.; Boccacci, P.; Canterino, S.; Cerutti, A.; Bounous, G. *Castanea* spp. biodiversity conservation: Collection and characterization of the genetic diversity of an endangered species. *Genet. Resour. Crop Evol.* **2012**, *59*, 1727–1741. [CrossRef]
5. Pitte, J.-R. *Terres de Castanide: Hommes et Paysages du Châtaignier de L'Antiquité à nos Jours*; Fayard: Paris, France, 1986.
6. Bounous, G. *Il Castagno: Coltura, Ambiente ed Utilizzazioni in Italia e nel Mondo*; Edagricole: Bologna, Italy, 2002.
7. Conedera, M.; Krebs, P.; Tinner, W.; Pradella, M.; Torriani, D. The cultivation of *Castanea sativa* (Mill.) in Europe, from its origin to its diffusion on a continental scale. *Veg. Hist. Archaeobotany* **2004**, *13*, 161–179. [CrossRef]
8. Agnoletti, M. The degradation of traditional landscape in a mountain area of Tuscany during the 19th and 20th centuries: Implications for biodiversity and sustainable management. *For. Ecol. Manag.* **2007**, *249*, 5–17. [CrossRef]
9. Bounous, G. The chestnut: A multipurpose resource for the new millennium. In Proceedings of the III International Chestnut Congress, Chaves, Portugal, 20–23 October 2004; pp. 33–40.
10. Conedera, M.; Manetti, M.; Giudici, F.; Amorini, E. Distribution and economic potential of the Sweet chestnut (*Castanea sativa* Mill.) in Europe. *Ecol. Mediterr.* **2004**, *30*, 179–193.
11. Gullino, P.; Larcher, F.; Devecchi, M. The importance of the chestnut cultivation and its evolution in the piedmont landscape (North-West Italy). In Proceedings of the I European Congress on Chestnut-Castanea, Cuneo, Italy, 13–16 October 2009; pp. 37–42.
12. Spathelf, P.; Van Der Maaten, E.; Van Der Maaten-Theunissen, M.; Campioli, M.; Dobrowolska, D. Climate change impacts in European forests: The expert views of local observers. *Ann. For. Sci.* **2014**, *71*, 131–137. [CrossRef]

13. Lione, G.; Danti, R.; Fernandez-Conradi, P.; Ferreira-Cardoso, J.; Lefort, F.; Marques, G.; Meyer, J.; Prospero, S.; Radócz, L.; Robin, C. The emerging pathogen of chestnut Gnomoniopsis castaneae: The challenge posed by a versatile fungus. *Eur. J. Plant Pathol.* **2019**, *153*, 671–685. [CrossRef]
14. Rigling, D.; Prospero, S. Cryphonectria parasitica, the causal agent of chestnut blight: Invasion history, population biology and disease control. *Mol. Plant Pathol.* **2018**, *19*, 7–20. [CrossRef]
15. Sartor, C.; Dini, F.; Marinoni, D.T.; Mellano, M.; Beccaro, G.; Alma, A.; Quacchia, A.; Botta, R. Impact of the Asian wasp Dryocosmus kuriphilus (Yasumatsu) on cultivated chestnut: Yield loss and cultivar susceptibility. *Sci. Hortic.* **2015**, *197*, 454–460. [CrossRef]
16. Vettraino, A.; Morel, O.; Perlerou, C.; Robin, C.; Diamandis, S.; Vannini, A. Occurrence and distribution of Phytophthora species in European chestnut stands, and their association with Ink Disease and crown decline. *Eur. J. Plant Pathol.* **2005**, *111*, 169. [CrossRef]
17. Turchetti, T.; Maresi, G. Biological control of chestnut diseases in Italy: Effectiveness of blight and ink disease management. In Proceedings of the International Workshop on Chestnut Management in Mediterranean Countries-Problems and Prospects, Bursa, Turkey, 23–25 October 2007; pp. 253–260.
18. Anagnostakis, S.L. Chestnut breeding in the United States for disease and insect resistance. *Plant Dis.* **2012**, *96*, 1392–1403. [CrossRef] [PubMed]
19. Lauteri, M.; Monteverdi, M.; Scarascia-Mugnozza, G. Preservation of chestnut (*Castanea sativa* Mill.) genetic resources and Adaptive potential in relation to environmental changes. In Proceedings of the I European Congress on Chestnut-Castanea, Cuneo, Italy, 13–16 October 2009; pp. 677–682.
20. Mattioni, C.; Martin, M.A.; Chiocchini, F.; Cherubini, M.; Gaudet, M.; Pollegioni, P.; Velichkov, I.; Jarman, R.; Chambers, F.M.; Paule, L. Landscape genetics structure of European sweet chestnut (*Castanea sativa* Mill): Indications for conservation priorities. *Tree Genet. Genomes* **2017**, *2*, 1–14. [CrossRef]
21. De Vasconcelos, M.C.; Bennett, R.N.; Rosa, E.A.; Ferreira-Cardoso, J.V. Composition of European chestnut (*Castanea sativa* Mill.) and association with health effects: Fresh and processed products. *J. Sci. Food Agric.* **2010**, *90*, 1578–1589. [CrossRef]
22. Donno, D.; Beccaro, G.L.; Mellano, M.G.; Bonvegna, L.; Bounous, G. Castanea spp. buds as a phytochemical source for herbal preparations: Botanical fingerprint for nutraceutical identification and functional food standardisation. *J. Sci. Food Agric.* **2014**, *94*, 2863–2873. [CrossRef]
23. Botta, R.; Marinoni, D.; Beccaro, G.; Akkak, A.; Bounous, G. Development of a DNA typing technique for the genetic certification of chestnut cultivars. *For. Snow Landsc. Res.* **2001**, *76*, 425–428.
24. Donno, D.; Cerutti, A.; Prgomet, I.; Mellano, M.; Beccaro, G. Foodomics for mulberry fruit (*Morus* spp.): Analytical fingerprint as antioxidants' and health properties' determination tool. *Food Res. Int.* **2015**, *69*, 179–188. [CrossRef]
25. Marinoni, D.; Akkak, A.; Bounous, G.; Edwards, K.J.; Botta, R. Development and characterization of microsatellite markers in *Castanea sativa* (Mill.). *Mol. Breed.* **2003**, *11*, 127–136. [CrossRef]
26. Botta, R.; Guaraldo, P.; Bounous, G.; Akkak, A.; Mellano, M. DNA Typing and Quality Evaluation of Chestnut ("*Castanea sativa*" Mill.) Cultivars. *Adv. Hortic. Sci.* **2006**, *20*, 96–100.
27. Boccacci, P.; Canterino, S.; Torello Marinoni, D.; Mellano, M.; Beccaro, G.; Bounous, G. Genetic characterization of chestnut accessions from 'Centro Regionale di Castanicoltura' (Cuneo, Italy). In Proceedings of the I European Congress on Chestnut-Castanea, Cuneo, Italy, 13–16 October 2009; pp. 97–101.
28. Piggott, J.R.; Simpson, S.J.; Williams, S.A. Sensory analysis. *Int. J. Food Sci. Technol.* **1998**, *33*, 7–12. [CrossRef]
29. Donno, D.; Beccaro, G.; Mellano, M.; Torello Marinoni, D.; Cerutti, A.; Canterino, S.; Bounous, G. Application of sensory, nutraceutical and genetic techniques to create a quality profile of ancient apple cultivars. *J. Food Qual.* **2012**, *35*, 169–181. [CrossRef]
30. Mellano, M.G.; Rapalino, S.; Donno, D. Sensory profiles of *Castanea sativa* cultivars and eurojapanese hybrids. *Castanea* **2017**, *2*, 8–9.
31. Mellano, M.; Bounous, G.; Botta, G. Valutazione mediante analisi sensoriale di cultivar piemontesi di castagno. *Italus Hortus* **2005**, *12*, 64.
32. Zeppa, G.; Rolle, L.; Gerbi, V. Studio per la caratterizzazione dei prodotti tradizionali regionali. *Il Marrone della Valle di Susa. Quad. della Reg. Piemonte* **2003**, *38*, 35–39.
33. Cartier, R.; Rytz, A.; Lecomte, A.; Poblete, F.; Krystlik, J.; Belin, E.; Martin, N. Sorting procedure as an alternative to quantitative descriptive analysis to obtain a product sensory map. *Food Qual. Prefer.* **2006**, *17*, 562–571. [CrossRef]

34. Lima, V.; Melo, E.A.; Maciel, M.I.S.; Prazeres, F.G.; Musser, R.S.; Lima, D.E.S. Total phenolic and carotenoid contents in acerola genotypes harvested at three ripening stages. *Food Chem.* **2005**, *90*, 565–568. [CrossRef]
35. Benzie, I.F.; Strain, J. Ferric reducing/antioxidant power assay: Direct measure of total antioxidant activity of biological fluids and modified version for simultaneous measurement of total antioxidant power and ascorbic acid concentration. *Methods Enzymol.* **1999**, *299*, 15–27.
36. Slinkard, K.; Singleton, V.L. Total Phenol Analysis: Automation and Comparison with Manual Methods. *Am. J. Enol. Vitic.* **1977**, *28*, 49–55.
37. Donno, D.; Mellano, M.G.; Prgomet, Z.; Beccaro, G.L. Advances in Ribes x nidigrolaria Rud. Bauer and A. Bauer fruits as potential source of natural molecules: A preliminary study on physico-chemical traits of an underutilized berry. *Sci. Hortic.* **2018**, *237*, 20–27. [CrossRef]
38. Soifoini, T.; Donno, D.; Jeannoda, V.; Rakotoniaina, E.; Hamidou, S.; Achmet, S.M.; Solo, N.R.; Afraitane, K.; Giacoma, C.; Beccaro, G.L. Bioactive compounds, nutritional traits, and antioxidant properties of *Artocarpus altilis* (Parkinson) fruits: Exploiting a potential functional food for food security on the Comoros Islands. *J. Food Qual.* **2018**, in press. [CrossRef]
39. Crawley, M. *The R book* , 2nd ed.; John Wiley & Sons. Ltd.: Chichester, UK, 2013.
40. Field, A. *Discovering Statistics Using IBM SPSS Statistics*; SAGE: Thousand Oaks, CA, USA, 2013.
41. Dinç, İ.; Sigdel, M.; Dinç, S.; Sigdel, M.S.; Pusey, M.L.; Aygün, R.S. Evaluation of normalization and pca on the performance of classifiers for protein crystallization images. In Proceedings of the IEEE Southeastcon, Lexington, KY, USA, 10 November 2014; pp. 1–6.
42. Lione, G.; Giordano, L.; Sillo, F.; Gonthier, P. Testing and modelling the effects of climate on the incidence of the emergent nut rot agent of chestnut Gnomoniopsis castanea. *Plant Pathol.* **2015**, *64*, 852–863. [CrossRef]
43. Lione, G.; Gonthier, P. A permutation-randomization approach to test the spatial distribution of plant diseases. *Phytopathology* **2016**, *106*, 19–28. [CrossRef] [PubMed]
44. Clark, P.J.; Evans, F.C. Distance to nearest neighbor as a measure of spatial relationships in populations. *Ecology* **1954**, *35*, 445–453. [CrossRef]
45. Hothorn, T.; Hornik, K.; Zeileis, A. Unbiased recursive partitioning: A conditional inference framework. *J. Comput. Graph. Stat.* **2006**, *15*, 651–674. [CrossRef]
46. Hothorn, T.; Zeileis, A. partykit: A modular toolkit for recursive partytioning in R. *J. Mach. Learn. Res.* **2015**, *16*, 3905–3909.
47. Team, R.C. *R: A Language and Environment Statistical Computing*; R Foundation for Statistical Computing: Vienna, Austria, 2019.
48. Warmund, M.R.; Elmore, J.R.; Adhikari, K.; McGraw, S. Descriptive sensory analysis and free sugar contents of chestnut cultivars grown in North America. *J. Sci. Food Agric.* **2011**, *91*, 1940–1945. [CrossRef]
49. Künsch, U.; Schärer, H.; Patrian, B.; Höhn, E.; Conedera, M.; Sassella, A.; Jermini, M.; Jelmini, G. Effects of roasting on chemical composition and quality of different chestnut (*Castanea sativa* Mill) varieties. *J. Sci. Food Agric.* **2001**, *81*, 1106–1112. [CrossRef]
50. Künsch, U.; Schärer, H.; Patrian, B.; Hurter, J.; Conedera, M.; Sassella, A.; Jermini, M.; Jelmini, G. Quality assessment of chestnut fruits. In Proceedings of the II International Symposium on Chestnut, Bordeaux, France, 19–23 October 1998; pp. 119–128.
51. De Biaggi, M.; Rapalino, S.; Donno, D.; Mellano, M.G.; Beccaro, G.L. Genotype influence on chemical composition and sensory traits of chestnut in 18 cultivars grown on the same rootstock and at the same agronomic conditions. In Proceedings of the VI International Chestnut Symposium, Samsun, Turkey, 9–13 October 2017; pp. 215–220.
52. Mellano, M.; Beccaro, G.; Bounous, G.; Trasino, C.; Barrel, A. Morpho-biological and sensorial quality evaluation of chestnut cultivars in Aosta Valley (Italy). In Proceedings of the International Workshop on Chestnut Management in Mediterranean Countries-Problems and Prospects, Bursa, Turkey, 23–25 October 2007; pp. 125–132.
53. Rodríguez, L.; Cuenca, B.; Pato, B.; Cámara, M.; Ocaña, L. Cost and efficiency of propagating Castanea sativa hybrids in vitro and by cutting in a commercial nursery. In Proceedings of the III International Chestnut Congress, Chaves, Portugal, 20–23 October 2004; pp. 305–312.
54. Isabelle, M.; Lee, B.L.; Lim, M.T.; Koh, W.P.; Huang, D.J.; Ong, C.N. Antioxidant activity and profiles of common fruits in Singapore. *Food Chem.* **2010**, *123*, 77–84. [CrossRef]

55. Korel, F.; Balaban, M.Ö. Chemical Composition and Health Aspects of Chestnut (*Castanea* spp.). *Tree Nuts: Compos. Phytochem. Health Eff.* **2008**. [CrossRef]
56. Zhu, F. Properties and Food Uses of Chestnut Flour and Starch. *Food Bioprocess Technol.* **2017**, *10*, 1173–1191. [CrossRef]
57. Alasalvar, C.; Shahidi, F. *Tree Nuts: Composition, Phytochemicals, and Health Effects*; CRC Press: Boca Raton, FL, USA, 2008.
58. Esposito, T.; Celano, R.; Pane, C.; Piccinelli, A.L.; Sansone, F.; Picerno, P.; Zaccardelli, M.; Aquino, R.P.; Mencherini, T. Chestnut (*Castanea sativa* Miller.) Burs Extracts and Functional Compounds: UHPLC-UV-HRMS Profiling, Antioxidant Activity, and Inhibitory Effects on Phytopathogenic Fungi. *Molecules* **2019**, *24*, 302. [CrossRef] [PubMed]
59. De Vasconcelos, M.D.C.B.M.; Bennett, R.N.; Rosa, E.A.S.; Cardoso, J.V.F. Primary and Secondary Metabolite Composition of Kernels from Three Cultivars of Portuguese Chestnut (*Castanea sativa* Mill.) at Different Stages of Industrial Transformation. *J. Agric. Food Chem.* **2007**, *55*, 3508–3516. [CrossRef]
60. Donno, D.; Mellano, M.G.; De Biaggi, M.; Riondato, I.; Rakotoniaina, E.N.; Beccaro, G.L. New Findings in Prunus padus L. Fruits as a Source of Natural Compounds: Characterization of Metabolite Profiles and Preliminary Evaluation of Antioxidant Activity. *Molecules (Basel, Switz.)* **2018**, *23*, 725. [CrossRef]
61. de Cassia da Silveira e Sa, R.; Andrade, L.N.; de Sousa, D.P. A review on anti-inflammatory activity of monoterpenes. *Molecules* **2013**, *18*, 1227–1254. [CrossRef]
62. de Vasconcelos, M.; Bennett, R.N.; Quideau, S.; Jacquet, R.; Rosa, E.A.S.; Ferreira-Cardoso, J.V. Evaluating the potential of chestnut (*Castanea sativa* Mill.) fruit pericarp and integument as a source of tocopherols, pigments and polyphenols. *Ind. Crop. Prod.* **2010**, *31*, 301–311. [CrossRef]
63. Ghirardello, D.; Bertolino, M.; Belviso, S.; Dal Bello, B.; Giordano, M.; Rolle, L.; Gerbi, V.; Antonucci, M.; Spigolon, N.; Zeppa, G. Phenolic composition, antioxidant capacity and hexanal content of hazelnuts (*Corylus avellana* L.) as affected by different storage conditions. *Postharvest Biol. Technol.* **2016**, *112*, 95–104. [CrossRef]
64. Slatnar, A.; Mikulic-Petkovsek, M.; Stampar, F.; Veberic, R.; Solar, A. HPLC-MSn identification and quantification of phenolic compounds in hazelnut kernels, oil and bagasse pellets. *Food Res. Int.* **2014**, *64*, 783–789. [CrossRef]
65. Pereira, J.A.; Oliveira, I.; Sousa, A.; Ferreira, I.C.F.R.; Bento, A.; Estevinho, L. Bioactive properties and chemical composition of six walnut (*Juglans regia* L.) cultivars. *Food Chem. Toxicol.* **2008**, *46*, 2103–2111. [CrossRef]
66. Seeram, N.P. Berry fruits: Compositional elements, biochemical activities, and the impact of their intake on human health, performance, and disease. *J. Agric. Food Chem.* **2008**, *56*, 627–629. [CrossRef]
67. Ammar, I.; Ennouri, M.; Bouaziz, M.; Ben Amira, A.; Attia, H. Phenolic Profiles, Phytchemicals and Mineral Content of Decoction and Infusion of Opuntia ficus-indica Flowers. *Plant Food Hum. Nutr.* **2015**, *70*, 388–394. [CrossRef]
68. Del Rio, D.; Rodriguez-Mateos, A.; Spencer, J.P.; Tognolini, M.; Borges, G.; Crozier, A. Dietary (poly) phenolics in human health: Structures, bioavailability, and evidence of protective effects against chronic diseases. *Antioxid. Redox Signal.* **2013**, *18*, 1818–1892. [CrossRef] [PubMed]
69. Landete, J.M. Ellagitannins, ellagic acid and their derived metabolites: A review about source, metabolism, functions and health. *Food Res. Int.* **2011**, *44*, 1150–1160. [CrossRef]
70. Papaefthimiou, D.; Papanikolaou, A.; Falara, V.; Givanoudi, S.; Kostas, S.; Kanellis, A.K. Genus Cistus: A model for exploring labdane-type diterpenes' biosynthesis and a natural source of high value products with biological, aromatic, and pharmacological properties. *Front. Chem.* **2014**, *2*. [CrossRef] [PubMed]
71. Trombetta, D.; Castelli, F.; Sarpietro, M.G.; Venuti, V.; Cristani, M.; Daniele, C.; Saija, A.; Mazzanti, G.; Bisignano, G. Mechanisms of antibacterial action of three monoterpenes. *Antimicrob. Agents Chemother.* **2005**, *49*, 2474–2478. [CrossRef]
72. Neri, L.; Dimitri, G.; Sacchetti, G. Chemical composition and antioxidant activity of cured chestnuts from three sweet chestnut (*Castanea sativa* Mill.) ecotypes from Italy. *J. Food Compos. Anal.* **2010**, *23*, 23–29. [CrossRef]
73. Crowell, P.L. Prevention and therapy of cancer by dietary monoterpenes. *J. Nutr.* **1999**, *129*, 775S–778S. [CrossRef]

74. De Biaggi, M.; Donno, D.; Mellano, M.G.; Riondato, I.; Rakotoniaina, E.N.; Beccaro, G.L. *Cornus mas* (L.) Fruit as a Potential Source of Natural Health-Promoting Compounds: Physico-Chemical Characterisation of Bioactive Components. *Plant Food Hum. Nutr.* **2018**, *73*, 89–94. [CrossRef]
75. Alasalvar, C.; Bolling, B.W. Review of nut phytochemicals, fat-soluble bioactives, antioxidant components and health effects. *Br. J. Nutr.* **2015**, *113*, S68–S78. [CrossRef]
76. Köksal, A.İ.; Artik, N.; Şimşek, A.; Güneş, N. Nutrient composition of hazelnut (*Corylus avellana* L.) varieties cultivated in Turkey. *Food Chem.* **2006**, *99*, 509–515. [CrossRef]
77. Ertürk, Ü.; Mert, C.; Soylu, A. Chemical composition of fruits of some important chestnut cultivars. *Braz. Arch. Biol. Technol.* **2006**, *49*, 183–188. [CrossRef]
78. Delgado, T.; Pereira, J.A.; Ramalhosa, E.; Casal, S. Comparison of different drying methods on the chemical and sensory properties of chestnut (*Castanea sativa* M.) slices. *Eur Food Res Technol* **2017**, *243*, 1957–1971. [CrossRef]
79. Moodley, R.; Kindness, A.; Jonnalagadda, S.B. Elemental composition and chemical characteristics of five edible nuts (almond, Brazil, pecan, macadamia and walnut) consumed in Southern Africa. *J. Environ. Sci. Health Part B* **2007**, *42*, 585–591. [CrossRef] [PubMed]
80. Bernardez, M.M.; Miguelez, J.D.M.; Queijeiro, J.G. HPLC determination of sugars in varieties of chestnut fruits from Galicia (Spain). *J. Food Compos. Anal.* **2004**, *17*, 63–67. [CrossRef]
81. Beccaro, G.; Bounous, G.; De Biaggi, M.; Donno, D.; Marinoni, D.T.; Zou, F.; Mellano, M.G. Botany, Anatomy, and Nut Composition. In *The Chestnut Handbook*; CRC Press: Boca Raton, FL, USA, 2019; pp. 5–34.
82. Lamothe, L.M.; Lê, K.-A.; Samra, R.A.; Roger, O.; Green, H.; Macé, K. The scientific basis for healthful carbohydrate profile. *Crit. Rev. food sci. Nutr.* **2019**, *59*, 1058–1070. [CrossRef] [PubMed]
83. Nicoletti, M. Nutraceuticals and botanicals: Overview and perspectives. *Int. J. Food Sci. Nutr.* **2012**, *63* (Suppl. 1), 2–6. [CrossRef] [PubMed]
84. Cirlini, M.; Dall'Asta, C.; Silvanini, A.; Beghe, D.; Fabbri, A.; Galaverna, G.; Ganino, T. Volatile fingerprinting of chestnut flours from traditional Emilia Romagna (Italy) cultivars. *Food Chem.* **2012**, *134*, 662–668. [CrossRef]
85. Donno, D.; Beccaro, G.L.; Carlen, C.; Ancay, A.; Cerutti, A.K.; Mellano, M.G.; Bounous, G. Analytical fingerprint and chemometrics as phytochemical composition control tools in food supplement analysis: Characterization of raspberry bud-preparations of different cultivars. *J. Sci. Food Agric.* **2016**, *96*, 3157–3168. [CrossRef]
86. Karabagias, I.K.; Koutsoumpou, M.; Liakou, V.; Kontakos, S.; Kontominas, M.G. Characterization and geographical discrimination of saffron from Greece, Spain, Iran, and Morocco based on volatile and bioactivity markers, using chemometrics. *Eur. Food Res. Technol.* **2017**, *243*, 1577–1591. [CrossRef]
87. Lione, G.; Giordano, L.; Turina, M.; Gonthier, P. Hail-induced infections of the chestnut blight pathogen Cryphonectria parasitica depend on wound size and may lead to severe diebacks. *Phytopathology* **2020**. [CrossRef]

foods

MDPI

Article

Screening and Characterization of Phenolic Compounds and Their Antioxidant Capacity in Different Fruit Peels

Hafiz A. R. Suleria [1,2,*], Colin J. Barrow [2] and Frank R. Dunshea [1,3]

1 School of Agriculture and Food, Faculty of Veterinary and Agricultural Sciences, The University of Melbourne, Parkville, VIC 3010, Australia; fdunshea@unimelb.edu.au
2 Centre for Chemistry and Biotechnology, School of Life and Environmental Sciences, Deakin University, Waurn Ponds, VIC 3217, Australia; colin.barrow@deakin.edu.au
3 Faculty of Biological Sciences, The University of Leeds, Leeds LS2 9JT, UK
* Correspondence: hafiz.suleria@unimelb.edu.au; Tel.: +61-4-7043-9670

Received: 6 August 2020; Accepted: 28 August 2020; Published: 1 September 2020

check for updates

Abstract: Fruit peels have a diverse range of phytochemicals including carotenoids, vitamins, dietary fibres, and phenolic compounds, some with remarkable antioxidant properties. Nevertheless, the comprehensive screening and characterization of the complex array of phenolic compounds in different fruit peels is limited. This study aimed to determine the polyphenol content and their antioxidant potential in twenty different fruit peel samples in an ethanolic extraction, including their comprehensive characterization and quantification using the LC-MS/MS and HPLC. The obtained results showed that the mango peel exhibited the highest phenolic content for TPC (27.51 ± 0.63 mg GAE/g) and TFC (1.75 ± 0.08 mg QE/g), while the TTC (9.01 ± 0.20 mg CE/g) was slightly higher in the avocado peel than mango peel (8.99 ± 0.13 mg CE/g). In terms of antioxidant potential, the grapefruit peel had the highest radical scavenging capacities for the DPPH (9.17 ± 0.19 mg AAE/g), ABTS (10.79 ± 0.56 mg AAE/g), ferric reducing capacity in FRAP (9.22 ± 0.25 mg AA/g), and total antioxidant capacity, TAC (8.77 ± 0.34 mg AAE/g) compared to other fruit peel samples. The application of LC-ESI-QTOF-MS/MS tentatively identified and characterized a total of 176 phenolics, including phenolic acids (49), flavonoids (86), lignans (11), stilbene (5) and other polyphenols (25) in all twenty peel samples. From HPLC-PDA quantification, the mango peel sample showed significantly higher phenolic content, particularly for phenolic acids (gallic acid, 14.5 ± 0.4 mg/g) and flavonoids (quercetin, 11.9 ± 0.4 mg/g), as compared to other fruit peel samples. These results highlight the importance of fruit peels as a potential source of polyphenols. This study provides supportive information for the utilization of different phenolic rich fruit peels as ingredients in food, feed, and nutraceutical products.

Keywords: fruit peels; polyphenols; phenolic acids; flavonoids; flavan-3-ols; hydrolysable and condensed tannins; antioxidant activities; LC-MS and HPLC

1. Introduction

Food processing industries discard huge amounts of fruit wastes, particularly peels, seeds, and some other fruit residues [1]. These fruit wastes have different challenges for many countries, including Australia. Inappropriate landfill management results in emissions of gases including methane and carbon dioxide, while incomplete incineration involves the subsequent formation of secondary wastes such as dioxins, furans, acid gases, and releases of other dangerous pollutants that can cause serious environmental and health issues [2]. For these reasons, there is an urgent need to find uses for these food wastes, including fruit peel wastes. Some fruit peels have been recycled into products ranging from agricultural compost, biofuel, and citric acid [3]. However, fruit peels also provide an excellent source of carbohydrates, fibre, proteins, and phytochemicals, particularly phenolic compounds with high antioxidant capacities [4]. These components are not generally recovered from peels and so provide a future source of valuable antioxidant ingredients. Polyphenols are a large group of secondary metabolites commonly present in fruits and vegetables, which play a prominent role in human health and nutrition [5]. Phenolic compounds consist of aromatic rings with hydroxyl groups, organic acids, and acylated sugars. These phenolic moieties have high antioxidant activity which prevents the formation of free radicals [6]. The most abundant polyphenols in different fruit peels include flavan-3-ols, flavonols, phenolic acids, anthocyanins, and hydroquinones [7].

The fruit juice industries generate substantial quantities of peel residues during juice processing [8]. The major phenolic compounds present in different fruit peels (apple, pomegranates, mango, pineapple, and citrus peels) include hydroxybenzoic and hydroxycinnamic acids (caffeic acid, gallic acid, protocatechuic acid, and chlorogenic acid), hydrolysable tannins (pedunculagin, punicalin, punicalagin, and ellagic and gallic acids) and flavonoids including anthocyanins [9]. The phenolic compounds identified in avocado and custard apple peels include high contents of condensed tannins and flavonoids including procyanidins [10], whereas those in banana peels are mainly gallocatechin, catechin, and epicatechin [11]. *Prunus* cultivars such as nectarine, peaches, and apricot peels are rich in hydroxycinnamates and flavan-3-ols that have potential antioxidant activities [12]. The antioxidant potential of polyphenols can be estimated with different in vitro spectrophotometric-based methods, including (i) the determination of total phenolics, (ii) free radical scavenging methods, (iii) non-radical redox potential-based methods, and (iv) metal-chelating methods [13]. In addition, polyphenols can also inactivate the Fenton reaction by reacting with different metal ions [14]. For this reason, a set of different in vitro spectrophotometric-based assays with different mechanisms, including total phenolic content (TPC), total flavonoid content (TFC), total tannin content (TTC), 2, 2′-diphenyl-1-picrylhydrazyl (DPPH) free radical scavenging assay, ferric reducing assay (FRAP), 2, 2′-azinobis-(3-ethylbenzothiazoline-6-sulfonic acid) (ABTS) assay and total antioxidant capacity (TAC), were used to estimate overall phenolic contents and map their antioxidant potential [15].

In recent years, there is increasing interest in the extraction of phenolic compounds from different plant materials. Extraction, identification, and characterization of novel phenolics from different plant-based materials are challenging due to their chemical and structural diversity and complexity. The liquid chromatography coupled with electrospray-ionization and quadrupole time-of-flight mass spectrometry (LC-ESI-QTOF-MS/MS) is an innovative tool with high sensitivity and is the most effective method for the characterization of both low and high molecular weight phenolic and non-phenolic compounds [16]. Also, high-performance liquid chromatography (HPLC) coupled with the photodiode array detector (PDA) is a useful tool for quantifying targeted polyphenols. Although several studies have quantified selected phenolic compounds from a range of different fruit by-products using conventional HPLC-UV-based techniques, there is limited literature available on the relative abundance and distribution of numerous phenolic compounds in Australia's grown fruit peels, particularly using advanced LC-MS/MS characterization methods. As far we know, only some selected phenolic compounds have been characterized in fruit peels using LC-MS/MS [17]. Therefore, extraction, identification, and characterization of phenolics from different fruit peels using advanced analytical techniques including the LC-ESI-QTOF-MS/MS will provide further information in developing innovative functional foods, nutraceuticals, and pharmaceuticals on a commercial scale from these food wastes.

The objective of this study is to determine the phenolic content including TPC, TFC, TTC in twenty (20) different fruit peel samples and assess their antioxidant potential by determining DPPH, TAC, FRAP, and ABTS. Moreover, the identification and characterization of untargeted phenolic compounds were achieved through the LC-ESI-QTOF-MS/MS followed by the quantification of twenty targeted phenolics through HPLC-PDA. This study provides supportive information for the use of different phenolic rich fruit peels as ingredients in food, feed, and nutraceutical products.

2. Materials and Methods

2.1. Chemicals

In this study, most of the chemicals, reagents, and standards were analytical grade and purchased from Sigma-Aldrich (Castle Hill, NSW, Australia). Gallic acid, L-ascorbic acid, vanillin, hexahydrate aluminium chloride, Folin-Ciocalteu's phenol reagent, sodium phosphate, iron(III) chloride hexahydrate ($Fe[III]Cl_3.6H_2O$), hydrated sodium acetate, hydrochloric acid, sodium carbonate anhydrous, ammonium molybdate, quercetin, catechin, 2,2′-diphenyl-1-picrylhy-drazyl (DPPH), 2,4,6tripyridyl-s-triazine (TPTZ), and 2,2′-azinobis-(3-ethylbenzothiazoline-6-sulfonic acid) (ABTS) were purchased from the Sigma-Aldrich (Castle Hill, NSW, Australia) for the estimation of polyphenols and antioxidant potential. Sulfuric acid (H_2SO_4) with 98% purity was purchased from RCI Labscan (Rongmuang, Thailand). HPLC standards including gallic acid, *p*-hydroxybenzoic acid, caftaric acid, caffeic acid, protocatechuic acid, sinapinic acid, chlorogenic acid, syringic acid, ferulic acid, coumaric acid, catechin, quercetin, quercetin-3-galactoside, diosmin, quercetin-3-glucuronide, epicatechin gallate, quercetin-3-glucoside, kaempferol and kaempferol-3-glucoside were produced by Sigma-Aldrich (Castle Hill, NSW, Australia) for quantification proposes. HPLC and LC-MS grade reagents including methanol, ethanol, acetonitrile, formic acid, and glacial acetic acid were purchased from Thermo Fisher Scientific Inc. (Scoresby, VIC, Australia). To perform various in vitro bioactivities and antioxidant assays, 96 well-plates were bought from the Thermo Fisher Scientific (VIC, Australia). Additionally, HPLC vials (1 mL) were procured from the Agilent technologies (VIC, Australia).

2.2. Sample Preparation

Twenty different Australian grown fresh and mature fruits varieties (2–3 kg) including apple (Royal gala), apricot (Mystery), avocado (Hass), banana (Cavendish), custard apple (African Pride), dragon fruit (Red-fleshed), grapefruit (Thompson), kiwifruit (Hayward), mango (Kensington Pride), lime (Tahitian), melon (Rock melons), nectarine (Fantasia), orange (Navels), papaya (Sunrise Solo), passionfruit (Misty Gem), peach (Florda gold), pear (Packham's Triumph), pineapple (Aussie Rough), plum (Angeleno), and pomegranate (Griffith) were purchased from a local produce market in Melbourne, Australia. The fruits were manually cleaned, and peels were removed and freeze-dried according to the method of Peng, et al. [18], described in the supplementary material. Figure 1 represents the graphical and schematic layout of our study.

2.3. Extraction of Phenolic Compounds

To extract the phenolic compounds, 2.0 ± 0.5 g of each fruit peel powder was mixed with 20 mL 70% ethanol by modifying the method of Gu, et al. [19], explained in the supplementary material.

2.4. Estimation of Phenolics and Antioxidant Potential

For the phenolic estimation in selected fruit peel samples, TPC, TFC, and TTC assays were performed, while for measuring their antioxidant capacities, four different types of antioxidant assays including FRAP, DPPH, ABTS, and TAC were performed by adopting our previously published methods of Tang, et al. [20], explained in the supplementary material. The data was determined using a Multiskan® Go microplate photometer (Thermo Fisher Scientific, Waltham, MA, USA).

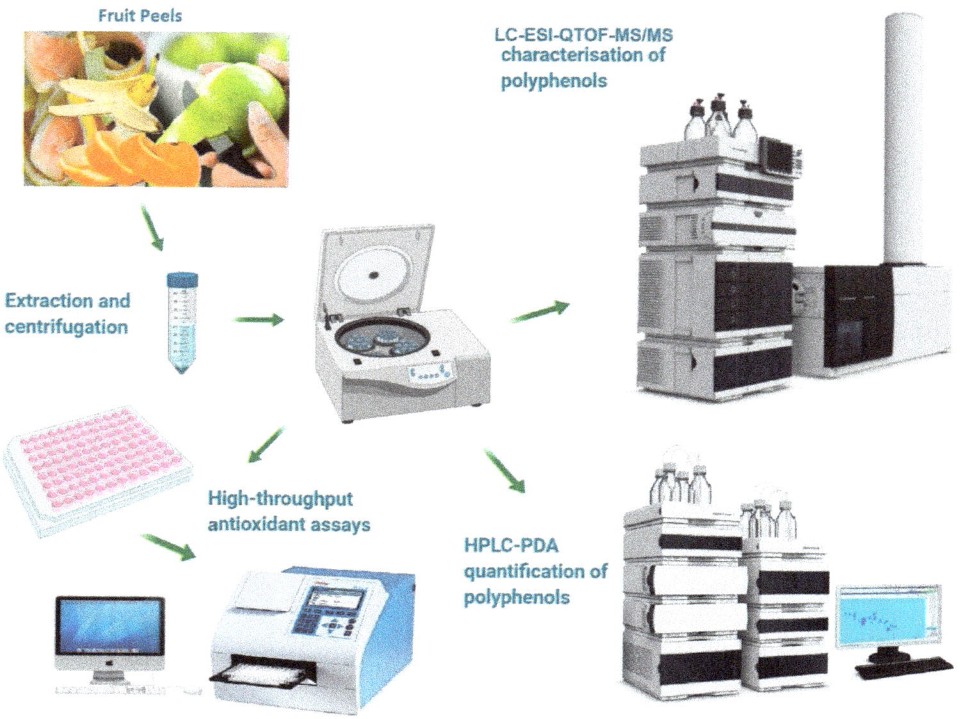

Figure 1. Graphical overview and schematic layout of the proposed research study.

2.5. Characterization and Quantification of Phenolics Using LC-ESI-QTOF-MS/MS and HPLC-PDA

The phenolic compound characterization was performed on an Agilent 1200 HPLC with 6520 Accurate-Mass Q-TOF-MS (Agilent Technologies, Santa Clara, CA, USA). The separation and characterization of phenolics were conducted by adopting our previously published method of Zhong, et al. [21], elaborated in the supplementary material. However, for the quantification of targeted phenolics present in different fruit peel samples was achieved with an Agilent 1200 HPLC coupled with a photodiode array (PDA) detector by following the protocol of Ma, et al. [22], explained in the supplementary material.

2.6. Statistical Analysis

All analyses were performed in triplicate, and the results are presented as mean ± standard deviation (n = 3). The mean differences between different samples were analyzed by one-way analysis of variance (ANOVA) and Tukey's honestly significant differences (HSD) multiple rank test at $p \leq 0.05$. ANOVA was carried out via Minitab 19.0 (Minitab, LLC, State College, PA, USA) and GraphPad Prism 7.05 Software for Windows (GraphPad 7.05 Software, San Diego, CA, USA, www.graphpad.com). For correlations between polyphenol content and antioxidant activities, Pearson's correlation coefficient at $p \leq 0.05$ and multivariate statistical analysis including a principal component analysis (PCA), XLSTAT-2019.1.3 were used by Addinsoft Inc. New York, NY, USA.

3. Results and Discussion

This study involved the screening and characterization of phenolic compounds with antioxidant potential from twenty different fruit peel samples. An untargeted polyphenol identification and characterization were achieved by the LC-ESI-QTOF-MS/MS, an advanced analytical technique which can provide comprehensive phytochemical screening and MS/MS characterization. For the quantification of phenolic compounds, the twenty most abundant phenolic compounds including (10) phenolic acids and (10) flavonoids present in different fruit peels were targeted and quantified by the HPLC-PDA. A strong correlation between phenolic compound levels and antioxidant activities was observed in all selected fruit peel samples.

3.1. Phenolic Estimation (TPC, TFC and TTC)

Fruit peels contain high concentrations of phenolic compounds including flavonoids, phenolic acids, and tannins. The phenolic contents in different fruit peel samples were determined with TPC, TFC, and TTC assays.

Table 1 summarizes the polyphenol concentrations and antioxidant potentials of twenty selected fruit peel samples. The TPC values of these fruit peel samples varied widely, with mango, grapefruit, and lime peel samples exhibiting the highest TPC values (27.51 ± 0.63, 27.22 ± 1.00 and 23.32 ± 2.07 mg GAE/g, respectively), followed by orange and avocado peel samples. The lowest phenolic contents were detected in dragon fruit, nectarine, and passion fruit peels. Comparing all the peel samples, the mango peel sample had significantly higher phenolic contents ($p < 0.05$) than any other fruit peels. Previously, Nguyen, et al. [23] reported significantly higher phenolic contents in mango peels as compared to other tropical fruits, including passion fruit and dragon fruit, which is consistent with our results. In our study, total phenolic content was measured using the Folin-Ciocalteu reagent that has the ability to react with both phenolics and non-phenolic compounds such as ascorbic acid and other reducing substances [24]. Grape and lime peel were reported to be rich in ascorbic acid, which may be one of the contributors to their high total polyphenol content [25]. However, grapefruit and lime peel were previously found to be abundant in polymethoxylated flavones, phenolic acids, and flavanones including naringin and neohesperidin [26]. Previously, similar trends but with higher TPC values were detected in different fruit juices, including grapefruit (657.65 ± 69.20 mg GAE/g), lime (579.41 ± 91.14 mg GAE/g) and orange (523.44 ± 87.20 mg GAE/g) [27]. Nurliyana, et al. [28] also found that dragon peel has a high phenolic content, most likely due to the abundance of betacyanins (pigments) rather than polyphenols, which increased the TPC values [29]. Considering these facts, polyphenol characterization through advanced analytical techniques including LC-MS/MS can provide more reliable and useful information for their applications in different food, feed, nutraceutical, and pharmaceutical industries.

Table 1. The polyphenol concentrations and antioxidant potentials of twenty different selected fruit peels.

Sample	TPC (mg GAE/g)	TFC (mg QE/g)	TTC (mg CE/g)	DPPH (mg AAE/g)	ABTS (mg AAE/g)	FRAP (mg AAE/g)	TAC (mg AAE/g)
Apple peel	10.82 ± 0.51 ^{e}	1.22 ± 0.10 $^{b–e}$	2.25 ± 0.12 ^{e}	5.20 ± 0.25 ^{b}	4.96 ± 0.17 ^{d}	3.20 ± 0.04 ^{d}	2.97 ± 0.16 ^{d}
Apricot peel	5.60 ± 0.27 ^{f,g}	1.22 ± 0.09 $^{b–e}$	1.07 ± 0.05 ^{f}	3.73 ± 0.55 ^{c}	3.21 ± 0.04 ^{e}	2.27 ± 0.11 ^{e}	2.28 ± 0.04 ^{e,f}
Avocado peel	18.79 ± 1.46 ^{c}	1.24 ± 0.11 $^{b–d}$	9.01 ± 0.20 ^{a}	8.67 ± 0.44 ^{a}	7.19 ± 0.72 ^{c}	3.65 ± 0.07 ^{c}	4.50 ± 0.16 ^{c}
Banana peel	6.13 ± 0.25 ^{f,g}	1.32 ± 0.12 ^{b,c}	1.22 ± 0.08 ^{f}	1.20 ± 0.12 ^{e}	1.31 ± 0.03 ^{g,h}	0.81 ± 0.03 ^{i,j}	2.36 ± 0.22 ^{e,f}
Custard apple peel	15.72 ± 0.74 ^{d}	1.21 ± 0.08 $^{b–e}$	8.32 ± 0.56 $^{a–c}$	2.52 ± 0.52 ^{d}	4.00 ± 0.44 ^{e}	1.51 ± 0.02 ^{f}	2.58 ± 0.04 ^{d,e}
Dragon fruit peel	0.45 ± 0.12 ^{k}	0.03 ± 0.01 ^{h}	0.03 ± 0.01 ^{h}	1.03 ± 0.16 ^{e}	0.56 ± 0.08 ^{h}	0.06 ± 0.01 ^{l}	0.19 ± 0.02 ^{i}
Grapefruit peel	27.22 ± 1.00 ^{a}	0.82 ± 0.14 ^{f}	7.60 ± 0.35 ^{c}	9.17 ± 0.19 ^{a}	10.79 ± 0.56 ^{a}	9.22 ± 0.25 ^{a}	8.77 ± 0.34 ^{a}
kiwi fruit peel	5.30 ± 0.40 ^{g,h}	0.45 ± 0.06 ^{g}	3.51 ± 0.33 ^{d}	5.03 ± 0.39 ^{b}	8.95 ± 0.18 ^{b}	1.13 ± 0.10 $^{g–i}$	0.79 ± 0.05 ^{h}
Lime peel	23.32 ± 2.07 ^{b}	1.14 ± 0.17 $^{c–e}$	8.42 ± 0.63 ^{a,b}	2.73 ± 0.34 ^{d}	1.46 ± 0.14 ^{g}	0.92 ± 0.07 $^{h–j}$	2.27 ± 0.08 ^{e,f}
Mango peel	27.51 ± 0.63 ^{a}	1.75 ± 0.08 ^{a}	8.99 ± 0.13 ^{a}	8.67 ± 0.49 ^{a}	9.32 ± 0.24 ^{b}	6.19 ± 0.26 ^{b}	6.19 ± 0.23 ^{b}
Melon peel	2.39 ± 0.02 $^{i–k}$	0.03 ± 0.01 ^{h}	0.02 ± 0.01 ^{h}	0.48 ± 0.28 ^{e}	1.16 ± 0.20 ^{g,h}	0.08 ± 0.01 ^{l}	0.93 ± 0.23 $^{g–h}$
Nectarine peel	1.53 ± 0.04 ^{j,k}	0.09 ± 0.01 ^{h}	0.23 ± 0.18 ^{h}	1.29 ± 0.09 ^{e}	1.25 ± 0.13 ^{g,h}	0.91 ± 0.07 $^{h–j}$	0.97 ± 0.05 ^{g,h}
Orange peel	21.31 ± 1.37 ^{b}	1.08 ± 0.06 $^{c–f}$	8.12 ± 0.26 ^{b,c}	4.79 ± 0.31 ^{b}	3.36 ± 0.16 ^{e}	2.44 ± 0.12 ^{e}	2.55 ± 0.08 ^{d,e}
Papaya peel	3.13 ± 0.15 $^{h–j}$	1.06 ± 0.07 $^{c–f}$	1.09 ± 0.04 ^{f}	1.13 ± 0.11 ^{e}	3.30 ± 0.17 ^{e}	0.91 ± 0.07 ^{i,j}	1.12 ± 0.13 ^{g,h}
Passion fruit peel	1.55 ± 0.21 ^{j,k}	0.04 ± 0.01 ^{h}	0.19 ± 0.02 ^{h}	0.72 ± 0.13 ^{e}	1.04 ± 0.07 ^{g,h}	0.42 ± 0.04 ^{k}	1.32 ± 0.05 ^{g}
Peach peel	5.84 ± 0.33 ^{f,g}	1.02 ± 0.08 $^{d–f}$	0.16 ± 0.05 ^{h}	1.33 ± 0.11 ^{e}	1.03 ± 0.06 ^{g,h}	0.89 ± 0.07 ^{i,j}	1.13 ± 0.07 ^{g,h}
Pear peel	4.30 ± 0.29 $^{g–i}$	1.07 ± 0.12 $^{c–f}$	0.10 ± 0.03 ^{h}	0.84 ± 0.12 ^{e}	1.21 ± 0.06 ^{g,h}	0.65 ± 0.08 ^{j,k}	1.18 ± 0.03 ^{g,h}
Pineapple peel	7.83 ± 0.35 ^{f}	1.47 ± 0.07 ^{b}	1.23 ± 0.05 ^{f}	1.30 ± 0.07 ^{e}	2.36 ± 0.06 ^{f}	1.30 ± 0.16 ^{f,g}	2.00 ± 0.14 ^{f}
Plum peel	4.81 ± 0.30 ^{g,h}	0.96 ± 0.08 ^{e,f}	0.29 ± 0.05 ^{g,h}	1.01 ± 0.10 ^{e}	1.19 ± 0.08 ^{g,h}	0.71 ± 0.04 ^{j,k}	0.87 ± 0.04 ^{h}
Pomegranate peel	3.89 ± 0.21 $^{g–i}$	0.97 ± 0.10 ^{e,f}	0.99 ± 0.02 $^{f–g}$	4.60 ± 0.08 ^{b,c}	3.34 ± 0.09 ^{e}	1.25 ± 0.13 $^{f–h}$	2.40 ± 0.18 ^{e,f}

All values are expressed as mg/g mean ± standard deviation ($n = 3$). Alphabetic letters indicate the significant difference ($p < 0.05$) in a row using a one-way analysis of variance (ANOVA) and Tukey's test. TPC, Total phenolic content; TFC, total flavonoid content; TTC, total tannins content; FRAP, ferric reducing antioxidant power assay; DPPH, 2,2′-diphenyl-1-picrylhydrazyl assay; ABTS, 2,2′-azino-bis-3-ethylbenzothiazoline-6-sulfonic acid assay; TAC, total antioxidant capacity; GAE, gallic acid equivalents; CE, catechin equivalents; QE, quercetin equivalents; AAE, ascorbic acid equivalents.

Flavonoids are the predominant class of phenolic substances found in almost all plants, which was determined via the aluminium chloride colorimetric method in this study. Aluminium chloride reacts with carbonyl group present in flavonoids, forming a stable complex [30]. The highest amount of flavonoid was found in the mango peel (1.75 ± 0.08 mg QE/g), followed by pineapple and banana peels (1.47 ± 0.07 and 1.32 ± 0.12 mg QE/g, respectively). Marina and Noriham [31] also reported higher flavonoid contents in mango peel than other tropical fruit peels such as papaya and guava peels, which is consistent with our study. Previously, Morais, et al. [32] determined the TFC in different parts of the avocado (*Persea americana*), and found that avocado peels had more flavonoids than seeds and pulp. Ayala-Zavala, et al. [33] also reported that the peels of tropical exotic fruits like avocado, pineapple, banana, papaya, passion fruit, and melon contain more phenolic acids and flavonoids than pulp. Overall, TFC values of our twenty different fruit peels were slightly higher than previously reported values, which may be due to the difference in the growing area, climatic conditions, varietal differences, and extraction. Fruits growing under different climatic regions have different flavonoid content in their peels, the peels being the outer part of fruit bodies exposed to more to sunlight as compared to pulp, leading to the synthesis of the abundant and diverse nature of flavonoids. Nogata, et al. [34] reported that the flavonoid contents in the outer layer of citrus fruits are higher than the inner layers and pulps. The flavonoid profile differs among species and cultivars of the same fruits grown in different regions under different climatic conditions, soil characteristics, and cultivation techniques [35]. Moreover, the efficiency of the extraction of flavonoids also varies under different extraction conditions, such as the type of solvents, solvent concentration, extraction time and temperature, solvent-to-solid ratio, etc. [36,37].

Tannins are also one of the important groups of phenolic compounds which can be classified into hydrolysable tannins and condensed tannins. Avocado peels exhibited the highest TTC values of 9.01 ± 0.20 mg CE/g, followed by mango (8.99 ± 0.13 mg CE/g), lime, and custard apple peels (8.42 ± 0.63 and 8.32 ± 0.56 mg CE/g, respectively), while few tannins were detected in dragon fruit, melon, nectarine, passion fruit, peach and pear peels. Overall, most of our TTC values are in accordance with previously published work, while we also had high values of tannins in mango peel as compared to the previously published literature. Previously, the mango fruit peel has already been reported to be a rich source of hydrolysable tannins, while hydrolysable tannins can decrease significantly during the ripening process [38]. One of the possible reasons might be the difference in sample preparations, storage conditions, and extraction techniques. In our study, all the fruit peels were freeze-dried prior to the extraction of polyphenols; it has been reported that freeze-drying facilitates the overall polyphenol extraction. Freeze-drying can also preserve the highest percentage of condensed tannins as compared to other conventional drying methods. Freeze-drying also helps to accelerate the release of bounded phenolic compound [39], deactivating oxidative and hydrolytic enzymes, improving the extraction and protecting the phenolic compounds [40].

3.2. *Antioxidant Potential (DPPH, ABTS, FRAP and TAC)*

To further investigate the antioxidant potential of the twenty different fruit peels, different antioxidant assays based on different mechanisms were applied in this study. Antioxidant assays including DPPH and ABTS were used to measure the radical scavenging ability, while FRAP and TAC assays were used to determine the reducing power of samples. The results shown in Table 1 were reported in mg ascorbic acid equivalents (AAE) per g of samples (mg AAE/g).

The DPPH assay is widely used to determine the free radical scavenging activity, which is mainly attributed to polyphenols [15]. Grapefruit, mango, and avocado peels exhibit higher DPPH radical scavenging ability (9.17 ± 0.19, 8.67 ± 0.49 and 8.67 ± 0.44 mg AAE/g, respectively). Previously, different varieties of mango peel extracts have shown concentration-dependent DPPH free radical scavenging activity [41]. Most of our DDPH values are in accordance with the previously published literature. Moreover, the DPPH assay showed significantly higher levels of antioxidant capacity in freeze-dried fruit peels as compared to fresh fruit peels. The freeze-drying process generates redox-active metabolites that can scavenge and neutralize free radicals [42]. The DPPH assay is one of the non-specific free radical scavenging assays since it measures scavenged free radicals from both phenolic and non-phenolic compounds, including ascorbic acid. Therefore, the antioxidant potential of plant polyphenols cannot be properly assessed only through DPPH assays. For this reason, a set of different in vitro reagent-based assays can be applied to estimate antioxidant potential, while the confirmation of these antioxidant compounds can be achieved through the LC-MS characterization.

The ABTS assay is another widely used method for determining the antiradical scavenging abilities based on the hydrogen atom donating tendency of phenolic compounds. The scavenged ABTS free radicals were measured using a colorimetric assay where antioxidants in samples reduce $ABTS^{+}$ and form a stable free radical [15]. The ABTS assay exhibits high similarity with that of the DPPH assay with the highest ABTS value from the grapefruit peel with 10.79 ± 0.56 mg AAE/g, followed by the mango peel (9.32 ± 0.24 mg AAE/g), kiwi fruit peel (8.95 ± 0.18 mg AAE/g), and avocado peel (7.19 ± 0.72 mg AAE/g) samples. In comparison, banana, dragon fruit, melon, nectarine, passion fruit, peach, pear, and plum peels exhibited relatively low ABTS radical scavenging ability. Previously, a similar $ABTS^{\cdot+}$scavenging tendency was found in white and pink freeze-dried grapefruit peel extracts [42]. Pal, et al. [43] also found the ABTS radical scavenging ability in kiwi fruit at different ripening stages. Tremocoldi, et al. [44] reported slightly higher ABTS radical scavenging activities in different avocado varieties, including Hass and Fuerte peel samples, as compared to our results. Ortega-Arellano, et al. [45] also reported the ABTS antioxidant activity for both Hass and Reed peels, which is consistent with our results.

The FRAP assay evaluates the ability of samples to donate electrons to reduce a Fe^{+3}-TPTZ complex to a blue Fe^{+2}-TPTZ complex. Grapefruit peel exhibited the highest FRAP reducing power with 9.22 ± 0.25 mg AAE/g, followed by mango peel (6.19 ± 0.26 mg AAE/g), avocado peel (3.65 ± 0.07 mg AAE/g) and apple peel samples (3.20 ± 0.04 mg AAE/g), while the FRAP reducing power from dragon fruit, melon, passion fruit, pear, and plum peels were relatively low as compared to other fruit peels. Previously, Oboh and Ademosun [46] also reported high FRAP activity in orange and apple peels that was attributed to their bound phenolics compounds and flavonoids. Furthermore, FRAP activities previously reported in other fruit peels including kiwifruit, lime, pineapple, banana, and mango was also in accordance with our study [47].

The total antioxidant capacity (TAC) assay is based on an electron transfer mechanism. This assay is very similar to FRAP, where molybdenum (VI) will be reduced to molybdenum (V) through antioxidant compounds or phenolic compounds. Similar to the results of FRAP assay, the highest TAC values were reported in the grapefruit peel (8.77 ± 0.34 mg AAE/g), followed by mango, avocado, and apple peels (6.19 ± 0.23, 4.50 ± 0.16 and 2.97 ± 0.16 mg AAE/g, respectively). In comparison, dragon fruit, kiwi fruit, melon, nectarine, papaya, peach, pear, and plum peels had relatively low TAC values. The strong antioxidant activities including DPPH, ABTS, and FRAP of different citrus fruits have already been reported, while grapefruit had the strongest antioxidant potential [48]. Antioxidant assays involved multiple reactions and mechanisms to estimate the antioxidant potential of any plant material, and unfortunately, there is no single method that can accurately reflect the overall antioxidant potential due to the complex nature of phytochemicals. For this reason, the MS/MS characterization is one of the key areas in phytochemical research to used compute overall phenolic compounds and their antioxidant potential.

In general, grapefruit, mango, and avocado peels exhibit distinctive antioxidant activity in four different types of antioxidant assays. Our polyphenolic and antioxidant results indicated that further research is needed to determine the actual contribution of polyphenols toward the antioxidant potential by minimizing other distracting factors of in vitro reagent-based assays, including the contribution of non-phenolic compounds toward the antioxidant potential.

3.3. LC-ESI-QTOF-MS/MS Characterization

LC-MS/MS has been widely used for the identification and characterization of bioactive compounds, including phenolics from different fruits, vegetable, and medicinal plants. An untargeted qualitative analysis of phenolics from twenty different fruit peel samples was achieved via LC-ESI-QTOF-MS/MS analysis in both negative and positive modes of ionization (Table S1, Figures S1 and S2—Supplementary Data). Phenolics present in different fruit peel samples were tentatively identified and characterized from their *m/z* value and MS spectra in both negative and positive modes of ionization ($[M-H]^-/[M+H]^+$) using Agilent LC-MS Qualitative Software and Personal Compound Database and Library (PCDL). Compounds with mass error < ± 5 ppm and PCDL library score more than 80 were selected for further MS/MS identification and *m/z* characterization and verification purposes. In our study, LC-MS/MS enabled the tentative identification and characterization of 176 phenolics in twenty different fruit peel samples, including phenolic acids (49), flavonoids (86), lignans (11), stilbene (5) and other polyphenols (25) listed in Table S1 (Supplementary data).

3.3.1. Phenolic Acids

Phenolic acids are the most abundant bioactive compounds present in different fruits [5]. In our study, a total of 49 phenolic acids were tentatively characterized, including hydroxybenzoic acids (12), hydroxycinnamic acids (31), hydroxyphenylacetic acids (2), and hydroxyphenylpropanoic acids (4).

Hydroxybenzoic acids are widely present in different fruits such as mango, apple, custard apple, citrus, strawberries, and raspberries with significant antioxidant potential. Compound **1** presenting in mango, pear and kiwifruit was proposed as vanillic acid 4-sulfate based on the observed *m/z* at 246.9911 in negative ionization mode and further confirmed by the MS/MS experiment which displayed a characteristic loss of SO_3 (80 Da) at *m/z* 167 [49]. Most of the phenolic acids showed the loss of CO_2 (44 Da) and hexosyl moiety (162 Da) [50]. Compound **3** (*m/z* 169.0146), compound **6** (*m/z* 137.0244) and compound **8** (*m/z* 153.0193) were identified as gallic acid, 2-hydroxybenzoic acid and 2,3-dihydroxybenzoic acid, showing product ions at *m/z* 125, at *m/z* 93 and at *m/z* 109, represented the loss of CO_2 from the precursor ions [50,51]. Previously, Kim, et al. [52] had also tentatively identified gallic acid from white and red dragon fruit peel and pulp samples.

Hydroxycinnamic acids contained collectively a larger number of detected compounds than in any other subclass in this study. In our study, a total of 31 hydroxycinnamic acids were identified with remarkable antioxidant potential. Six caffeic acid derivatives were successfully identified in our work. Compound **15** (*m/z* 341.0861) and compound **26** (*m/z* 355.0686) exhibited a product ion at *m/z* 179 (caffeic acid ion) by losing glucoside (162 Da) and glucuronide (176 Da) in negative mode and identified as caffeoyl glucose and caffeic acid 3-*O*-glucuronide [53].

Ferulic acid (Compound **23**) was also observed in eight different peel samples. In an MS^2 experiment, ferulic acid displayed the product ions at *m/z* 178, *m/z* 149, and *m/z* 134, indicating the loss of CH_3, CO_2, and CH_3 with CO_2 from the precursor, respectively [54]. Compound **25** (RT = 19.319 min) was tentatively identified as *m*-coumaric acid with the precursor $[M-H]^-$ *m/z* at 163.0406 and confirmed by the MS/MS spectra (Figure 2), which exhibited the fragments at *m/z* 119 due to the loss of CO_2 [54]. Compound **47** (dihydroferulic acid 4-*O*-glucuronide, *m/z* at 371.0986) and compound **49** (dihydrocaffeic acid 3-*O*-glucuronide, *m/z* at 357.0811) were both detected only in the negative ionization mode, and the characteristic loss of the glucuronide (176 Da) moiety was observed in both compounds, which produced the fragment ions at *m/z* 195 and at *m/z* 181, respectively [55].

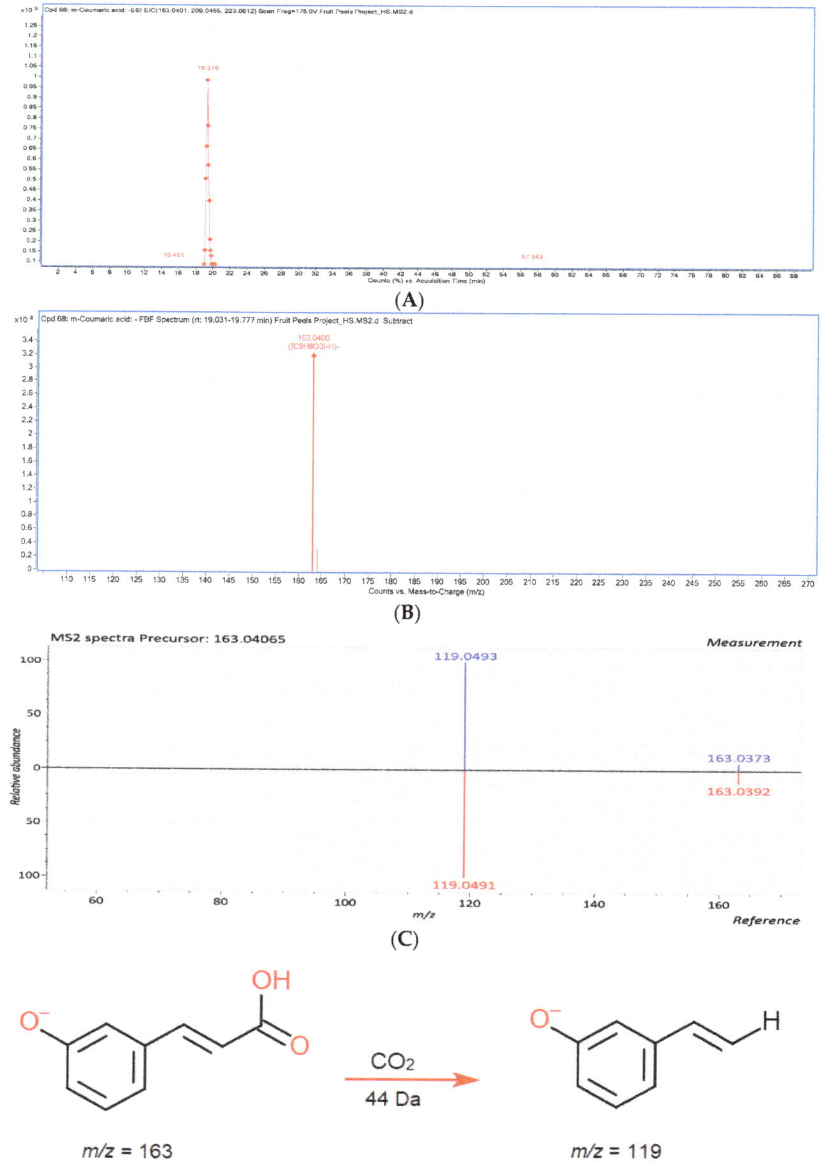

Figure 2. The LC-ESI-QTOF-MS/MS characterization of *m*-coumaric acid. (**A**) A chromatograph of *m*-coumaric acid (Compound 25, Table S1—Supplementary Data) in the negative mode $[M - H]^-$ which was tentatively identified and characterized in fifteen different fruit peel samples; (**B**) a mass spectrum of *m*-coumaric acid with a precursor of *m/z* 163.0406 in the apple peel; (**C**) MS/MS spectrum of *m*-coumaric acid with the product ion of *m/z* 119 (confirmed from online LC-MS library and database); (**D**) a fragmentation pattern of the *m*-coumaric acid in negative mode $[M - H]^-$, with precursor of *m/z* 163 and a product ion of *m/z* 119 due to the loss of CO_2.

3.3.2. Flavonoids

Flavonoids (in total 86) are the most abundant class with antioxidant potential found in the fruit peels. Flavonoids were divided into eight subclasses, including flavanols (11), flavones (12), flavanones (8), flavonols (19), dihydrochalcones (3), dihydroflavonols (2), anthocyanins (12) and Isoflavonoids (19).

A total of eight flavanones was discovered in the peels. Quercetin 3′-*O*-glucuronide (Compound **82**) and myricetin 3-*O*-arabinoside (Compound **83**) were found in both modes and tentatively identified by the precursor ions $[M - H]^-$ *m/z* at 477.067 and $[M - H]^-$ *m/z* at 449.0716. The product ion at *m/z* 301 in the MS^2 spectrum of quercetin 3′-*O*-glucuronide was produced by the loss of glucuronide (176 Da) from the precursor [56], and the peaks at *m/z* 317 (loss of pentose moiety, 132 Da) confirmed the identity of myricetin 3-*O*-arabinoside [57].

3.3.3. Other Polyphenols

A total of 25 other polyphenols were identified from the peels, which were further divided into hydroxycoumarins (5), hydroxybenzaldehydes (2), hydroxybenzoketones (2), hydroxyphenylpropenes (1), curcuminoids (3), furanocoumarins (1), phenolic terpenes (2), tyrosols (5) and other polyphenols (4).

Coumarin (Compound **138**) and scopoletin (Compound **139**) were found in both negative and positive modes and tentatively identified according to the precursors $[M + H]^+$ at *m/z* 147.0448 and $[M - H]^-$ at *m/z* 191.0345. In the MS^2 experiment of 147.0448, peaks at *m/z* 103 $[M + H - CO_2]$ and *m/z* 91 $[M + H - 2CO]$ achieved the identification of coumarin, and in the MS/MS spectra of *m/z* 191.0345, peaks at 176 $[M - H - 15$, loss of $CH_3]$ are characteristic for scopoletin [58,59].

3.3.4. Lignans

A total of eleven lignans were identified in most of the fruit peels. Compounds **161** and **163** presenting in the positive mode were identified as enterolactone and schisandrin C according to the *m/z* 299.1283 and *m/z* 385.1652, respectively. The MS/MS experiment achieved the identification of these lignans. Enterolactone exhibited the fragment ions at *m/z* 281, 187, and 165, representing the loss of H_2O, $C_6H_8O_2$ and $C_9H_8O_2$, respectively [60]. The presence of schisantherin C was verified by the product ions at *m/z* 370 (loss of CH_3, 15 Da), *m/z* 315 (loss of C_5H_{10}, 70 Da) and *m/z* 300 (loss of CH_3 and C_5H_{10}, 85 Da) [61].

3.3.5. Stilbenes

A total of five stilbenes were identified in different fruit peel samples. Resveratrol (Compound **173**, $[M - H]^-$ *m/z* at 227.0709 presenting in custard apple and avocado peels) and resveratrol 5-*O*-glucoside (Compound **174**, $[M - H]^-$ *m/z* at 389.1245 appearing in passion fruit, pomegranate, and kiwi fruit peels) were detected in both ionization modes. In the MS^2 spectra, Resveratrol showed the characteristic *m/z* at 212 (loss of CH_3), 185 (loss of CHCOH), 157 (loss of CHCOH and CO), and 143 (loss of CHCOH and C_2H_2O) [62]. The excepted loss of glucoside (162 Da) was observed in the MS^2 fragmentation of resveratrol 5-*O*-glucoside, which allowed the identification of this compound [63].

The LC-MS/MS characterizations of phenolic compounds presented in different fruit peels have remarkable antioxidant capacities. Most of the hydroxycinnamic hydroxybenzoic acids and their derivatives and flavonoid and their derivatives have strong free radical scavenging ability. The presence of these phenolics in different fruit peel samples indicates that these food wastes could be valuable sources of natural antioxidant compounds. In short, these fruit peels could be utilized in different food, feed, nutraceutical, and pharmaceutical industries.

3.4. Distribution of Phenolic Compounds—Venn Diagram

To further investigate the distribution of phenolic compounds in different fruit peels, Venn diagrams were generated among fruits grown in different climate zones including tropical, sub-tropical, and temperate (Figure 3). Although the aim of this study was not to explore the relationship between growing regions and phenolic contents in different fruit peel samples, we tentatively characterized their phenolic profiling through Venn diagrams. This preliminary analysis indicates that it is worth further exploring the relationship between growing regions and phenolic contents in different fruit peel samples. The comparison showed that there are differences in the phenolic compositions of fruits grown in different climate zones, and so it may be possible to optimize phenolic levels in these fruits and their peels through the targeted selection of the growing location.

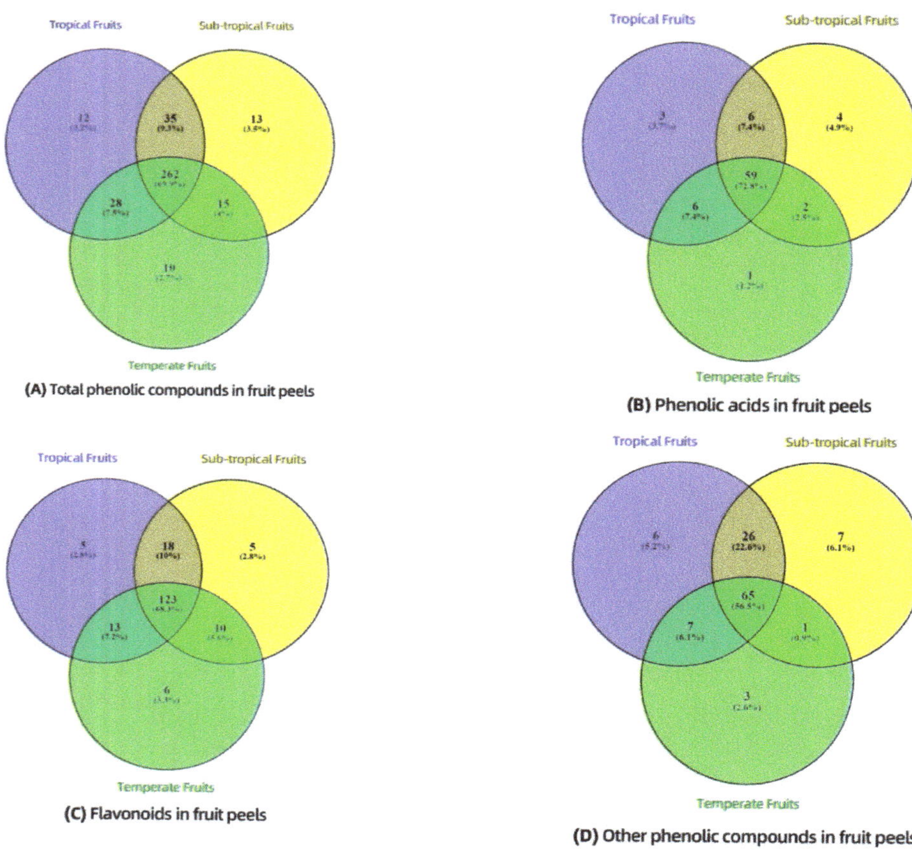

Figure 3. Venn diagram of phenolic compounds presented in different fruit peel samples grown in different regions. (**A**) shows the relations of total phenolic compounds present in different fruit peel samples grown in three different zones. (**B**) shows the relations of phenolic acids present in different fruit peel samples. (**C**) shows the relations of flavonoids present in different fruit peel samples. (**D**) shows the relations of other phenolic compounds present in different fruit peel samples.

Fruit peel samples were divided into three groups according to their growing regions, which were tropical (banana, custard apple, dragon fruit, mango, papaya, and pineapple peels), sub-tropical (pomegranate, passion fruit, orange, grapefruit, avocado, lime peels), and temperate (apple, apricot, kiwi fruit, peach, pear, melon, plum, nectarine peels).

From Figure 3A, a total of 375 phenolics were tentatively identified in all twenty selected fruit peels. Among "*total phenolic compounds*", 69.9% of them were commonly identified in all three zones, including tropical, sub-tropical, and temperate regions. From Figure 3B,C, a total of 72.8% of the "*phenolic acids*" and 68.3% of the "*flavonoids*" were commonly identified in all three zones. The proportions of common phenolic acids and flavonoids shared by all the fruit peels were almost similar to that of total phenolic compounds, which indicated the compositions of these compounds were similar in tropical, sub-tropical, and temperate fruits, despite different growing regions. However, Figure 3D shows that "*other phenolic compounds*" had only 56.5% of commonly identified compounds in the three groups, the proportions of which were much lower than those in the total phenolic compounds. The lower proportion of shared compounds of "*other phenolic compound*" indicated that they might be the main contributors responsible for the differences in overall phenolic concentrations and antioxidant activities of different fruit peels collected from three different climatic zones. Additionally, tropical and sub-tropical fruits were more similar in the compositions of other phenolic compounds, while temperate fruits had a quite different composition. The difference may be explained by a previous study, which indicated that tropical fruits often had richer phenolic contents and stronger antioxidant capabilities than temperate fruits due to the presence of some phenolic compounds in tropical fruits that functioned as lipid peroxidation inhibitors and decreased deleterious effects in plants caused by the strong ultraviolet radiation in tropical regions [64]. For example, stilbenes possess an antioxidant ability that can decrease oxidative stress caused by UV irradiation, as well as for the defense system of plants against fungi and bacteria [65]. Also, other phenolic compounds with anti-insect functions might exist exclusively in tropical fruits, as a warmer climate usually favors pest threats [66,67].

In this work we found that there is a strong relationship between growing regions and phenolic contents in different fruit peel samples, and we elucidate the differences in the compositions of phenolic compounds, particularly "*other phenolic compounds*". Further work is required to explore the impacts of individual phenolics.

3.5. *HPLC-PDA Quantitative Analysis*

HPLC has been widely used as an effective tool for the identification and quantification of phenolic compounds in different fruit and vegetable samples. The twenty most abundant phenolic compounds present in the different fruit peels, including 10 phenolic acids and 10 flavonoids, were selected for quantification. Tables 2 and 3 show the quantified phenolic acids and flavonoids by comparing retention time with reference standards, and results were calculated using standard curves.

3.5.1. Phenolic Acids

In our study, ten targeted phenolic acids were quantified in the twenty fruit peels. Table 2 showed that the mango peel was most abundant in terms of the overall phenolic acids (72.2 ± 4.5 mg/g) and most of the individual phenolic acid, while melon peels significantly had the lowest overall phenolic acid content. Mango peels significantly had the highest phenolic content for seven out of ten targeted phenolic acids, including gallic acid (14.5 ± 0.4 mg/g), chlorogenic acid (13.8 ± 0.9 mg/g), caffeic acid (4.5 ± 0.4 mg/g), *p*-hydroxybenzoic acid (10.5 ± 0.4 mg/g), syringic acid (11.5 ± 0.7 mg/g), ferulic acid (6.3 ± 0.4 mg/g), and coumaric acid (5.1 ± 0.2 mg/g), respectively. Previously, Palafox-Carlos, et al. [68] detected gallic acid, chlorogenic acid, and protocatechuic acids in different mango varieties, including in both pulp and peels. Chlorogenic acid was the most abundant in their study, while gallic acid was the most abundant in our study. However, Kim, et al. [69] reported that gallic acid was the predominant phenolic acid in mango peels, which is in agreement with our results. In another study, Hu, et al. [70], reported a gallic acid concentration of 0.08–0.59 mg/g among mango peel samples, which is much lower than our results. These variations can be explained by the variability of phenolic content with cultivar type and maturity, growing regions, and climatic conditions.

Another study of Marina and Noriham [31] indicated that mango peels had higher phenolic content than papaya peels, which also agrees with our quantification results for the ten targeted phenolic acids. Moreover, Gorinstein, et al. [71] also reported a similar trend that mango had significantly higher phenolic contents than avocado peel samples. However, they also reported a higher phenolic content in kiwifruit than in mango, which is in contrast with our results. The difference might be caused by the difference in varieties and growing conditions as they, used grown fruits from Singapore, while our study was conducted on grown fruits from Australia. Apart from mango, other tropical fruits, including banana, custard apple, dragon fruit, papaya, passion fruit, and pineapple peels, did not show significantly higher phenolic contents than other temperate or subtropical fruits, although some of these fruits, such as the banana, were reported to have high phenolic contents and antioxidant ability [72].

Pomegranate was another fruit other than mango which had a significantly higher content for most of the phenolic acids. Previously, Li, et al. [73] detected 249.4 ± 17.2 mg/g phenolic contents in pomegranate peels, which indicated that this fruit was an excellent source of phenolics. Moreover, the study of Marina and Noriham [31] also indicated that pomegranates possessed high phenolic contents. From our results, similar conclusions can be postulated, as pomegranate has a significantly higher phenolic acid content compared with other fruit peels. Additionally, Pal, et al. [74] reported approximately a three-fold higher phenolic content in the pomegranate peel than in the orange peel, which is consistent with our study. Apart from pomegranate, grapefruit and lime peels were also quantified in our study and significantly showed the highest contents for several phenolic acids. Previously, Sir Elkhatim, et al. [75] compared the phenolic contents between peels of citrus fruits including orange, grapefruit, and lime, and reported that grapefruit peels had the highest phenolic content, followed by lime and orange peels, which showed a similar pattern with our results for the targeted phenolic acids. Li, et al. [76] also reported similar results that the grapefruit peel had the highest phenolic contents compared with other citrus fruit peels.

As for temperate fruits, apple peels had significantly higher contents of protocatechuic acid (7.4 ± 0.4 mg/g) than all other fruits. While the apple peel did not have higher overall phenolic acid contents among all the 20 fruits, it is one of the most widely consumed fruits known for its antioxidant ability [77], and importantly, the peel is often consumed. Previously, Russell, et al. [78] reported a higher content of phenolic acids, including gallic acid, protocatechuic acid, *p*-hydroxybenzoic acid, syringic acid, and sinapinic acid, in apple peels than in the peel of pears of Scottish varieties, which is consistent with our results. The study of Mihailović, et al. [79] indicated that chlorogenic acid was the most dominant phenolic acid presented in the apple peel, which is in agreement with our study, which detected the highest chlorogenic acid content of 11.2 ± 0.1 mg/g for apple peels. Moreover, Veberic, et al. [80] also reported that chlorogenic acid was the most abundant phenolic acid in the apple peel with the content range of 4.1–79.5 mg/100 g. The variation can be attributed to a difference in apple varieties. Previous studies have suggested that most tropical fruits have higher phenolic contents than temperate fruits, as phenolic compounds are essential for inhibiting lipid peroxidation and deleterious effects in plant tissues caused by strong ultraviolet radiation in tropical areas [64]. It can also be concluded from previous studies that, although some temperate fruits were already potential phenolic sources, topical fruits had richer phenolic contents, which makes them better sources of phenolic acids [64]. In our study, the tropical fruit mango showed significantly higher phenolic acid content in the peel than all the sub-tropical and temperate fruits, which is consistent with previous studies.

Table 2. Phenolic acids quantified in different fruit peel samples using HPLC-PDA.

Fruit Peels	Gallic Acid	Protocatechuic Acid	Caftaric Acid	Chlorogenic Acid	*p*-hydroxybenzoic Acid	Caffeic Acid	Syringic Acid	Coumaric Acid	Ferulic Acid	Sinapinic Acid	Sum of Phenolic Acids
APL-P	4.2 ± 0.9^{d}	7.4 ± 0.4^{a}	-	11.2 ± 0.1^{b}	6.5 ± 0.8^{c}	2.1 ± 0.9^{c}	1.1 ± 0.7^{g}	-	1.2 ± 0.3^{h}	-	33.7 ± 1.5^{C}
APR-P	2.1 ± 0.6^{g}	-	2.4 ± 0.4^{e}	5.9 ± 0.2^{e}	3.1 ± 0.3^{f}	3.5 ± 0.1^{b}	-	-	4.5 ± 0.1^{c}	1.3 ± 0.1^{e}	22.8 ± 1.7^{D}
AVO-P	3.2 ± 0.5^{e}	4.2 ± 0.2^{b}	1.2 ± 0.1^{g}	9.5 ± 0.1^{c}	4.5 ± 0.6^{d}	-	3.5 ± 0.3^{d}	4.1 ± 0.1^{b}	-	2.7 ± 0.6^{c}	32.9 ± 2.5^{C}
BNA-P	1.2 ± 0.2^{i}	-	-	4.5 ± 0.5^{f}	2.8 ± 0.1^{g}	-	4.1 ± 0.5^{c}	-	3.7 ± 0.2^{d}	-	16.3 ± 1.9^{G}
CTA-P	1.4 ± 0.2^{h}	-	4.7 ± 0.9^{c}	7.5 ± 0.3^{d}	4.5 ± 0.9^{d}	-	-	1.8 ± 0.4^{f}	1.8 ± 0.3^{g}	3.9 ± 0.8^{b}	25.6 ± 1.6^{D}
DGF-P	1.1 ± 0.5^{i}	-	3.5 ± 0.5^{d}	4.1 ± 0.9^{g}	1.2 ± 0.7^{i}	-	3.1 ± 0.9^{d}	2.8 ± 0.1^{d}	2.7 ± 0.8^{e}	-	18.5 ± 2.1^{F}
GRF-P	5.4 ± 0.9^{c}	3.4 ± 0.4^{c}	7.8 ± 0.5^{a}	-	3.5 ± 0.9^{e}	4.2 ± 0.5^{a}	-	3.1 ± 0.8^{c}	2.1 ± 0.4^{f}	1.7 ± 0.7^{d}	31.2 ± 1.9^{C}
KWF-P	1.1 ± 0.9^{i}	-	4.2 ± 0.4^{c}	3.2 ± 0.5^{i}	2.8 ± 0.1^{g}	-	2.1 ± 0.1^{e}	4.1 ± 0.8^{b}	1.2 ± 0.7^{h}	-	18.7 ± 1.7^{F}
LMN-P	-	1.2 ± 0.8^{e}	2.4 ± 0.5^{e}	-	4.2 ± 0.4^{d}	1.2 ± 0.6^{e}	-	2.1 ± 0.4^{e}	4.5 ± 0.9^{c}	4.9 ± 0.7^{a}	20.5 ± 2.1^{E}
MNG-P	14.5 ± 0.4^{a}	-	2.1 ± 0.1^{f}	13.8 ± 0.9^{a}	10.5 ± 0.4^{a}	4.5 ± 0.4^{a}	11.5 ± 0.7^{a}	5.1 ± 0.2^{a}	6.3 ± 0.4^{a}	3.9 ± 0.9^{b}	72.2 ± 4.5^{A}
MEL-P	-	1.1 ± 0.7^{e}	-	1.6 ± 0.3^{j}	-	-	2.3 ± 0.1^{e}	-	1.2 ± 0.2^{h}	-	6.2 ± 1.2^{M}
NEC-P	1.5 ± 0.7^{h}	1.2 ± 0.2^{e}	-	4.5 ± 0.4^{f}	2.8 ± 0.1^{g}	1.7 ± 0.9^{d}	-	-	1.1 ± 0.1^{h}	-	12.8 ± 1.9^{I}
ORN-P	5.4 ± 0.9^{c}	-	3.1 ± 0.4^{d}	5.6 ± 0.3^{e}	3.6 ± 0.1^{e}	-	1.8 ± 0.2^{f}	-	2.1 ± 0.7^{f}	1.8 ± 0.2^{d}	23.4 ± 2.3^{D}
PAP-P	2.4 ± 0.7^{f}	-	5.6 ± 0.1^{b}	-	2.9 ± 0.2^{g}	-	2.4 ± 0.9^{e}	-	1.8 ± 0.4^{g}	-	15.1 ± 1.1^{H}
PSN-P	-	-	5.2 ± 0.8^{b}	-	2.1 ± 0.4^{h}	-	3.5 ± 0.3^{d}	-	1.1 ± 0.1^{h}	-	11.9 ± 1.9^{J}
PEC-P	1.5 ± 0.4^{h}	1.2 ± 0.1^{e}	-	3.7 ± 0.9^{h}	-	1.8 ± 0.2^{d}	-	-	2.7 ± 0.1^{e}	1.4 ± 0.9^{e}	12.3 ± 1.3^{I}
PER-P	1.1 ± 0.7^{i}	-	2.1 ± 0.8^{f}	-	3.2 ± 0.3^{f}	-	1.5 ± 0.4^{f}	-	1.2 ± 0.1^{h}	-	9.1 ± 1.7^{L}
PIN-P	1.5 ± 0.9^{h}	2.1 ± 0.2^{d}	-	-	1.2 ± 0.1^{i}	1.1 ± 0.5^{e}	-	2.8 ± 0.3^{d}	-	1.9 ± 0.7^{d}	10.6 ± 1.9^{K}
PLM-P	1.4 ± 0.3^{h}	-	-	-	3.8 ± 0.1^{e}	-	1.7 ± 0.7^{f}	-	4.2 ± 0.4^{c}	-	11.1 ± 2.1^{J}
POM-P	6.7 ± 0.1^{b}	7.4 ± 0.6^{a}	-	11.8 ± 0.7^{b}	9.8 ± 0.1^{b}	4.5 ± 0.7^{a}	6.7 ± 0.9^{b}	-	5.8 ± 0.2^{b}	-	52.7 ± 3.9^{B}

All values are expressed as "mg/g", mean ± standard deviation ($n = 3$). Alphabetic letters indicate significant difference ($p < 0.05$) in a row using a one-way analysis of variance (ANOVA) and Tukey's test. Fruit peel samples were mentioned in abbreviations. Apple peel "APL-P", Apricot peel "APR-P", Avocado peel "AVO-P", Banana peel "BNA-P", Custard apple peel "CTA-P", Dragon fruit peel "DGF-P", Grapefruit peel "GRF-P", Kiwifruit peel "KWF-P", Lime peel "LMN-P", Mango peel "MNG-P", Melon peel "MEL-P", Nectarine peel "NEC-P", Orange peel "ORN-P", Papaya peel "PAP-P", Passionfruit peel "PSN-P", Peach peel "PEC-P", Pear peel "PER-P", Pineapple peel "PIN-P", Plum peel "PLM-P", and Pomegranate peel "POM-P".

3.5.2. Flavonoids

Flavonoids are the largest group of phenolics and are present in most of the fruits. Among the fruit peels investigated, the mango peel has the highest content for overall flavonoids (57.1 ± 2.4 mg/g), while passion fruit had the lowest (10.4 ± 1.4 mg/g) listed in Table 3.

Mango peels showed similarly high contents for flavonoids as for phenolic acids, significantly with the highest contents of epicatechin gallate (3.2 ± 0.9 mg/g), quercetin-3-galactoside (10.9 ± 0.1 mg/g), quercetin-3-glucuronide (11.5 ± 0.7 mg/g), quercetin (11.9 ± 0.4 mg/g), and kaempferol (9.8 ± 0.7 mg/g). Previously, catechin and quercetin-3-galactoside were quantified by López-Cobo, et al. [81] in different mango peel samples. Compared with other fruits, Marina and Noriham [31] reported higher catechin and epicatechin contents in mango peels than other tropical fruit peels, such as papaya peel and guava peel, which is consistent with our study. However, a few studies reported lower flavonoids in mango pulp as compared to kiwifruit and avocado pulp, which did not agree with our fruit peel extracts [71]. The contradictory results might be explained by previous literature indicating that peels contained more flavonoids as compared to pulp [68].

Apart from mango peel, dragon fruit peel was also found to be abundant with flavonoids while catechin was dominantly detected in it with a concentration of 7.5 ± 0.9 mg/g. Previously, flavonoids including kaempferol and quercetin derivatives were detected and quantified in dragon fruit peels [82]. The pineapple peel sample had a relatively low flavonoid content among all the twenty fruits which showed a different pattern from mango and dragon fruit peels, but these results agree with the previous study of Silva, et al. [83], who reported significantly higher flavonoid contents in mango, papaya, and passion fruit than in pineapple, in which only a few flavonoids were detected in the pineapple peel sample. The pomegranate peel sample also had higher flavonoids (35.7 ± 4.7 mg/g) similar to phenolic contents. For individual flavonoids, pomegranate peel had the highest epicatechin content (4.1 ± 0.3 mg/g). Previously, Li, Guo, Yang, Wei, Xu and Cheng [73] reported higher flavonoids in pomegranate peel than our results, which may be because we only quantified the ten most abundant flavonoids across the fruit samples observed, and there is a chance that individual fruits may have high concentrations of a flavonoid outside this group. Another study showed that flavonoid contents in pomegranate and mango juices were significantly lower than the phenolic acid contents, which is consistent with our results [68]. Our results indicated that both mango and pomegranate peels are excellent sources of phenolic compounds.

For citrus fruit peels, the kaempferol-3-glucoside content was highest in lime peel (3.7 ± 0.4 mg/g), which is higher than those in orange, grapefruit, and pomegranate. Previously, Singh and Immanuel [84] reported similar results that lime peel had a higher total flavonoid content compared with other citrus species, such as orange. However, a more recent study of Sir Elkhatim, Elagib, and Hassan [75] showed that orange peel contained higher amounts of flavonoids than lime and grapefruit peels, which is in contrast with our results. The variation can be attributed to the difference in fruit varieties and extraction methods. In temperate fruits, quercetin-3-glucoside was the most abundant in apple peel, with a concentration of 4.5 ± 0.9 mg/g. Previously, Schieber, et al. [85] also reported quercetin-3-glucoside was present in apple pomace at a low concentration. Another study of Mihailović, Mihailovic, Kreft, Ciric, Joksović and Djurdjevic [79] reported flavonoids including catechin (0.187 ± 0.007 mg/g) and quercitrin (0.256 ± 0.002 mg/g) from peels of wild apple varieties, which were also detected in our study. In summary, all twenty fruit peel samples have a considerable quantity of phenolic compounds, including both phenolic acids & flavonoids, and these fruit peels are potential commercial sources of these phenolics.

Table 3. Flavonoids quantified in different fruit peel samples using HPLC-PDA.

Fruit Peels	Catechin	Epicatechin	Epicatechin Gallate	Quercetin-3-Galactoside	Quercetin-3-Glucuronide	Quercetin-3-Glucoside	Kaempferol-3-Glucoside	Diosmin	Quercetin	Kaempferol	Sum of Flavonoids
APL-P	3.2 ± 0.8^{b}	-	1.5 ± 0.1^{d}	5.7 ± 0.6^{b}	-	4.5 ± 0.9^{a}	-	2.1 ± 0.4^{b}	9.6 ± 0.9^{b}	-	26.6 ± 1.9^{C}
APR-P	-	2.1 ± 0.8^{b}	2.3 ± 0.1^{b}	-	3.8 ± 0.7^{d}	3.5 ± 0.2^{b}	3.2 ± 0.4^{b}	-	6.9 ± 0.7^{c}	4.5 ± 0.1^{d}	26.3 ± 2.1^{C}
AVO-P	2.1 ± 0.9^{d}	1.8 ± 0.4^{c}	-	4.9 ± 0.7^{c}	1.7 ± 0.8^{f}	2.8 ± 0.1^{c}	1.9 ± 0.5^{e}	1.7 ± 0.1^{c}	3.9 ± 0.9^{g}	2.9 ± 0.4^{f}	23.7 ± 2.7^{D}
BNA-P	1.5 ± 0.7^{e}	-	-	3.8 ± 0.9^{d}	-	1.7 ± 0.1^{e}	-	-	4.8 ± 0.8^{e}	-	11.8 ± 2.1^{H}
CTA-P	2.1 ± 0.4^{d}	-	1.9 ± 0.7^{c}	-	1.4 ± 0.1^{f}	-	1.7 ± 0.5^{e}	-	3.2 ± 0.9^{h}	3.9 ± 0.4^{d}	14.2 ± 1.9^{G}
DGF-P	7.5 ± 0.9^{a}	-	1.5 ± 0.1^{d}	4.5 ± 0.7^{c}	1.7 ± 0.4^{f}	-	2.4 ± 0.7^{d}	-	4.9 ± 0.3^{e}	3.5 ± 0.7^{e}	26.0 ± 1.1^{C}
GRF-P	-	-	2.1 ± 0.7^{b}	2.9 ± 0.1^{e}	-	1.7 ± 0.7^{e}	1.9 ± 0.3^{e}	1.1 ± 0.7^{d}	5.9 ± 0.1^{d}	4.2 ± 0.9^{d}	19.8 ± 1.9^{E}
KWF-P	1.5 ± 0.2^{e}	-	1.1 ± 0.9^{e}	-	4.5 ± 0.3^{c}	-	1.2 ± 0.8^{f}	1.7 ± 0.1^{c}	5.4 ± 0.2^{d}	7.1 ± 0.7^{b}	22.5 ± 2.7^{D}
LMN-P	2.7 ± 0.1^{c}	-	-	4.8 ± 0.7^{c}	8.7 ± 0.2^{b}	-	3.7 ± 0.4^{a}	1.9 ± 0.6^{b}	-	1.2 ± 0.1^{h}	23.0 ± 1.3^{D}
MNG-P	7.1 ± 0.3^{a}	-	3.2 ± 0.9^{a}	10.9 ± 0.1^{a}	11.5 ± 0.7^{a}	-	2.7 ± 0.4^{c}	-	11.9 ± 0.4^{a}	9.8 ± 0.7^{a}	57.1 ± 2.4^{A}
MEL-P	-	1.9 ± 0.3^{c}	-	2.8 ± 0.9^{e}	-	1.7 ± 0.1^{e}	-	-	4.5 ± 0.3^{f}	-	10.9 ± 1.2^{I}
NEC-P	2.1 ± 0.8^{d}	-	1.7 ± 0.1^{d}	1.8 ± 0.7^{g}	-	-	2.9 ± 0.7^{c}	-	2.9 ± 0.9^{i}	3.1 ± 0.1^{e}	14.5 ± 2.1^{G}
ORN-P	3.5 ± 0.1^{b}	2.4 ± 0.4^{b}	-	3.8 ± 0.2^{d}	-	-	3.9 ± 0.1^{a}	-	-	1.9 ± 0.8^{g}	15.5 ± 1.7^{F}
PAP-P	-	1.7 ± 0.3^{c}	1.9 ± 0.9^{c}	-	4.7 ± 0.1^{c}	-	1.7 ± 0.9^{e}	-	4.9 ± 0.4^{e}	2.9 ± 0.2^{f}	17.8 ± 2.1^{F}
PSN-P	1.8 ± 0.7^{e}	-	-	1.7 ± 0.1^{f}	-	2.1 ± 0.9^{d}	-	1.7 ± 0.1^{c}	-	3.1 ± 0.8^{e}	10.4 ± 1.4^{I}
PEC-P	1.2 ± 0.5^{f}	-	1.9 ± 0.3^{c}	2.1 ± 0.4^{f}	-	1.8 ± 0.2^{e}	-	1.1 ± 0.1^{d}	-	4.1 ± 0.7^{d}	12.2 ± 1.7^{H}
PER-P	1.8 ± 0.7^{e}	1.9 ± 0.9^{c}	-	-	1.7 ± 0.4^{f}	-	1.1 ± 0.1^{f}	-	-	7.4 ± 0.7^{b}	13.9 ± 2.1^{G}
PIN-P	-	-	1.5 ± 0.1^{d}	-	-	3.7 ± 0.1^{b}	-	3.1 ± 0.6^{a}	-	6.5 ± 0.7^{c}	14.8 ± 1.9^{G}
PLM-P	3.1 ± 0.1^{b}	-	1.4 ± 0.9^{d}	2.7 ± 0.3^{e}	-	3.1 ± 0.5^{c}	-	1.9 ± 0.7^{b}	-	4.1 ± 0.2^{d}	16.3 ± 2.1^{F}
POM-P	2.1 ± 0.1^{d}	4.1 ± 0.3^{a}	-	5.7 ± 0.1^{b}	2.1 ± 0.7^{e}	-	3.6 ± 0.2^{a}	1.1 ± 0.3^{d}	9.4 ± 0.9^{b}	7.6 ± 0.7^{b}	35.7 ± 4.7^{B}

All values are expressed as "mg/g", mean ± standard deviation ($n = 3$). Alphabetic letters indicate the significant difference ($p < 0.05$) in a row using ANOVA and Tukey's test. Fruit peel samples were mentioned in abbreviations. Apple peel "APL-P", Apricot peel "APR-P", Avocado peel "AVO-P", Banana peel "BNA-P", Custard apple peel "CTA-P", Dragon fruit peel "DGF-P", Grapefruit peel "GRF-P", Kiwifruit peel "KWF-P", Lime peel "LMN-P", Mango peel "MNG-P", Melon peel "MEL-P", Nectarine peel "NEC-P", Orange peel "ORN-P", Papaya peel "PAP-P", Passionfruit peel "PSN-P", Peach peel "PEC-P", Pear peel "PER-P", Pineapple peel "PIN-P", Plum peel "PLM-P" and Pomegranate peel "POM-P".

3.6. Heat Map and Hierarchical Clustering Phenolic Compound Analysis

For further analyzing the hierarchical clustering of targeted phenolic compounds in the twenty selected fruit peels, a heat map was constructed (Figure 4). The distance measure used for determining the similarity between fruits and compounds was the correlation, while the clustering method used for rows and columns was based on average concentration. For tree ordering, the tightest clusters were clustered first.

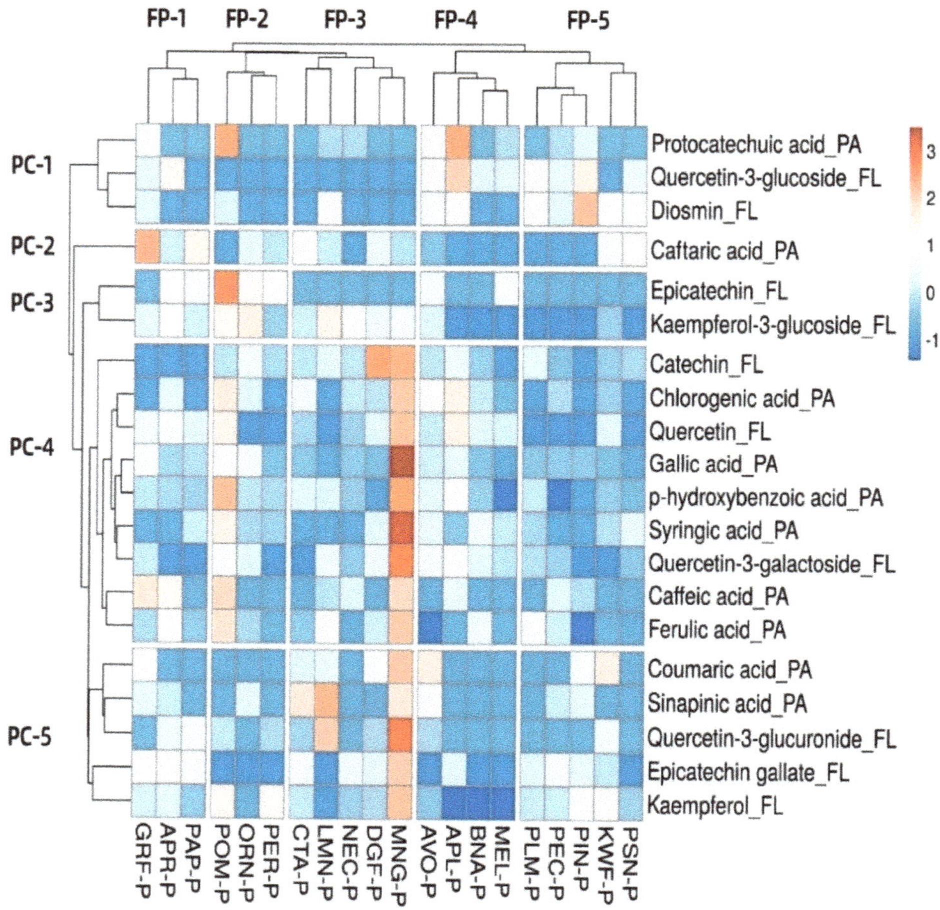

Figure 4. Heatmap showing phenolic compounds' distribution and concentration among twenty fruit peel samples. Red boxes mean concentrations are higher among different fruit peel samples. Blue boxes mean lower concentrations. PA: phenolic acids; FL: flavonoids; FP 1-5: fruit peel clusters 1; PC 1-5: phenolic compound clusters. Fruit peel samples were mentioned in abbreviations. Apple peel "APL-P", Apricot peel "APR-P", Avocado peel "AVO-P", Banana peel "BNA-P", Custard apple peel "CTA-P", Dragon fruit peel "DGF-P", Grapefruit peel "GRF-P", kiwifruit peel "KWF-P", Lime peel "LMN-P", Mango peel "MNG-P", Melon peel "MEL-P", Nectarine peel "NEC-P", Orange peel xORN-P", Papaya peel "PAP-P", Passionfruit peel "PSN-P", Peach peel "PEC-P", Pear peel "PER-P", Pineapple peel "PIN-P", Plum peel "PLM-P" and Pomegranate peel "POM-P".

In the heat map, five clusters in rows and columns were generated and highlighted by the hierarchical clustering; different clusters of samples indicate significant differences in phenolic profiles. The color difference showed the abundance of phenolic acids and flavonoids in different fruit peels. From the results, it can be observed that MGN-P, DGF-P, NEC-P, LMN-P, and CTA-P were clustered together in the group (FP-3), which shared similar patterns of phenolic contents. Within this cluster, MGN-P had red color areas for gallic acid and syringic acid, representing higher contents. Previously, Pereira-Netto [64] reported that tropical fruits shared similarly higher contents of phenolics than temperate fruits, which agrees with the clustering, where tropical fruits MGN-P, DGF-P, LMN-P, and CTA-P were grouped together.

Phenolic compounds were also grouped into five main clusters (PC-1, PC-2, PC-3, PC-4, and PC-5) in the dendrogram and were further grouped into different sub-clusters according to the similarity of their concentration patterns in the twenty fruit peel samples. Overall, PC-1 to PC-5 clusters indicated that several phenolic acids (caffeic acid, ferulic acid, coumaric acid, sinapinic acid) and flavonoids (quercectin-3-glucournoide, epicatechin gallate and kaempferol) had greater similarity in terms of the concentration among different fruit peel samples. However, some phenolic acids (caftaric acid, protocatechuic acid) and flavonoids (epicatechin and kaempferol-3-glucoside) showed variability with respect to other phenolic compound clusters.

3.7. Correlation between Phenolic Compounds, Targeted Phenolics Quantified through HPLC-PDA and Antioxidant Assays

The correlation between phenolic content (TPC, TFC, TTC, phenolic acids and flavonoids—quantified through HPLC-PDA) and antioxidant activities (DPPH, FRAP, ABTS, and TAC) was performed with a Pearson's correlation test (Table 4). In addition, principal components analysis (PCA, Figure 5) was performed to investigate the overall similarities and differences between the phenolic content, targeted phenolic acid, and flavonoids quantified through HPLC in different peels of fruit samples, and the relationship between the various methods used in the evaluation of the antioxidant potential. The targeted (10) phenolic acids and (10) flavonoids were calculated by summarizing the content of the proposed compounds in the HPLC-PDA table to investigate the correlations between overall phenolics and their antioxidant activities.

Table 4. Pearson's correlation coefficients (r) between phenolic content (TPC, TFC, TTC, phenolic acids, and flavonoids) and antioxidant activities (DPPH, FRAP, ABTS, and TAC).

Variables	TPC	TFC	TTC	DPPH	ABTS	FRAP	TAC	Phenolic Acids
TFC	0.488 *							
TTC	0.932 **	0.457 *						
DPPH	0.718 **	0.396	0.720 **					
ABTS	0.591 **	0.270	0.622 **	0.904 **				
FRAP	0.722 **	0.314	0.603 **	0.868 **	0.835 **			
TAC	0.780 **	0.397	0.668 **	0.850 **	0.779 **	0.967 **		
Phenolic acids	0.496 *	0.343	0.515 *	0.761 **	0.628 *	0.614 **	0.640 **	
Flavonoids	0.349	0.232	0.355	0.633 *	0.535 *	0.473 *	0.452 *	0.911 **

* Significant correlation with $p \leq 0.05$; ** Significant correlation with $p \leq 0.01$. Phenolic acids and flavonoids are quantified through HPLC-PDA.

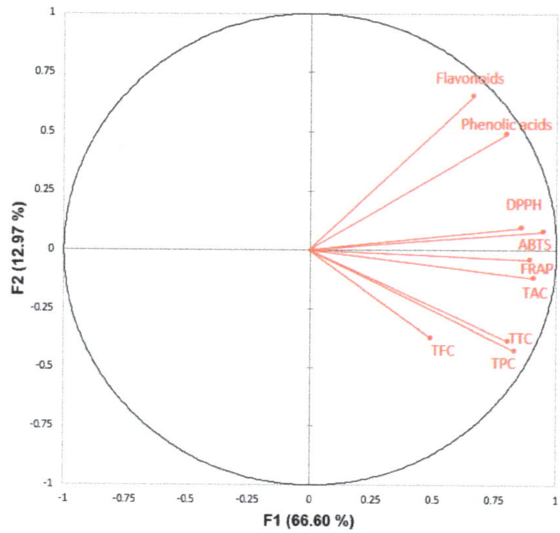

Figure 5. Principal component analysis (PCA) of the phenolic content (TPC, TFC, TTC, phenolic acids and flavonoids—quantified through HPLC-PDA) and antioxidant activities (DPPH, ABTS, FRAP, and TAC) of twenty different fruit peel samples.

A total of 79.57% variability of the initial data can be explained by the first two factors (F1 and F2) in Figure 5. Regarding antioxidant assays, DPPH, FRAP, ABTS, and TAC were strongly correlated with each other ($p \leq 0.01$). This significantly positive correlation was previously reported by Floegel, et al. [86]. They found that both DPPH and ABTS assays evaluate the free radical scavenging ability, and the ABTS assay can better reflect the hydrophilic, lipophilic, and high-pigmented antioxidants in fruits compared to the DPPH assay. The high correlation between DPPH, ABTS, FRAP, and TAC indicated that phenolic compounds present in twenty different fruit peel extracts exhibit the strong scavenging ability of DPPH, ABTS-reducing ability, and ferric ion- and phosphomolybdate ion-reducing abilities, respectively. The significantly positive correlations between FRAP and other antioxidant assays were in agreement with a previous study [87].

The TPC was highly significantly correlated with four antioxidant assays (DPPH, ABTS, FRAP, and TAC), which suggested that phenolic compounds are primary contributors to the antioxidant activities of the twenty different fruit peel samples. These results are in agreement with our previously published studies on phenolic compounds in different fruits and vegetable pulp samples and their antioxidant potential [19]. In addition, TPC were strongly correlated with TTC with $r = 0.932$, $p \leq 0.01$. However, a non-significant correlation between TFC and antioxidant assays was found, indicating that the contribution from flavonoids to the antioxidant potential of some peel samples was limited. The TFC method used in this study only targeted specific flavonoids, because the aluminum chloride selectively reacts with flavonols and the flavone luteolin [88], which may explain the non-significant correlations. In addition, strong correlations between TTC and four antioxidant assays were found, indicating that tannin present in selected fruit peel samples had a significant contribution to the antioxidant activities.

The phenolic acids content detected in HPLC was highly significantly correlated with most of the antioxidant assays (DPPH, ABTS, FRAP, and TAC) with $r = 0.761, 0.628, 0.614, 0.640$, respectively ($p \leq 0.05$), indicating that phenolic acids were one of the significant contributors to the antioxidant activities. Flavonoids detected by HPLC were also significantly correlated with most of the antioxidant assays, which was not consistent with the correlation results between the TFC value and antioxidant assays discussed before. One of the reasons might be that we selected only 10 of the most abundant flavonoids across all the fruit peels for quantification purposes, while TFC assays specifically react with all types of flavonoids. In addition, the overall flavonoids detected by HPLC were not correlated with the TFC value ($r = 0.232$), which might be due to the high proportion of other subclasses of flavonoids rather than our targeted (10) flavonoids. Overall, both phenolic acids and flavonoids were strongly correlated with antioxidant assays, which indicated that both phenolic classes have strong antioxidant activities.

4. Conclusions

In conclusion, most of the selected fruit peels were found to have considerable amounts of phenolic content with very high in vitro antioxidant potential. The TPC, TFC, DPPH, FRAP, TAC and ABTS scavenging activity was higher in mango peel as compared to other fruit peels. The mango peel sample also showed significantly higher phenolic compounds, including gallic acid and quercetin, as compared to other fruit peel samples. The LC-ESI-QTOF-MS/MS technique was successfully applied for characterization of the phenolic compounds in different fruit peels; a total of 176 phenolic compounds were tentatively characterized. Quantification by HPLC-PDA also verified that fruit peels are rich in phenolic compounds. The obtained results supported the idea that fruit peels are a potential food waste source of phenolic compounds, with high antioxidant potential that has potential utility in food, feed, and nutritional supplements. In the future, in vitro digestibility, bioavailability, bioaccessibility, toxicological, and animal studies are required for developing these different fruit peels as commercial ingredients.

Supplementary Materials: The following are available online at http://www.mdpi.com/2304-8158/9/9/1206/s1, Table S1: Characterization of phenolic compounds in different fruit peel samples by LC-ESI-QTOF-MS/MS, Figure S1: Characterization of phenolic compounds in different fruit peels in the negative mode of ionization by LC-ESI-QTOF-MS/MS, Figure S2: Characterization of phenolic compounds in different fruit peels in the positive mode of ionization by LC-ESI-QTOF-MS/MS.

Author Contributions: Conceptualization, methodology, validation and investigation, H.A.R.S., F.R.D. and C.J.B.; resources, H.A.R.S., F.R.D. and C.J.B.; writing—original draft preparation, H.A.R.S.; writing—review and editing, H.A.R.S., F.R.D. and C.J.B.; supervision, F.R.D. and C.J.B.; ideas sharing, H.A.R.S., F.R.D. and C.J.B.; funding acquisition, H.A.R.S., F.R.D. and C.J.B. All authors have read and agreed to the published version of the manuscript.

Funding: This research was funded by the University of Melbourne under the "McKenzie Fellowship Scheme" (Grant No. UoM-18/21) and the "Faculty Research Initiative Funds" funded by the Faculty of Veterinary and Agricultural Sciences, The University of Melbourne, Australia and "the Alfred Deakin Research Fellowship" funded by the Deakin University, Australia.

Acknowledgments: We would like to thank Nicholas Williamson, Shuai Nie, and Michael Leeming from the Mass Spectrometry and Proteomics Facility, Bio21 Molecular Science and Biotechnology Institute, the University of Melbourne, VIC, Australia for providing access and support for the use of HPLC-PDA and LC-ESI-QTOF-MS/MS and data analysis. We would also like to thank Biming Zhong, Zhicong Chen, Zihan Zhu, Zexing Leng, Vigasini Subbiah, Rana Dildar Khan and Akhtar Ali from the School of Agriculture and Food, Faculty of Veterinary and Agricultural Sciences, the University of Melbourne for their incredible support.

Conflicts of Interest: The authors declare no conflict of interest.

References

1. Ain, H.B.U.; Saeed, F.; Barrow, C.J.; Dunshea, F.R.; Suleria, H.A.R. Food processing waste: A potential source for bioactive compounds. In *Bioactive Compounds in Underutilized Fruits and Nuts*; Murthy, H.N., Bapat, V.A., Eds.; Springer International Publishing: Cham, Switzerland, 2020; pp. 625–649.
2. Abu Qdais, H.; Wuensch, C.; Dornack, C.; Nassour, A. The role of solid waste composting in mitigating climate change in jordan. *Waste Manag. Res.* **2019**, *37*, 833–842. [CrossRef] [PubMed]
3. Azizan, A.; Xin, L.A.; Abdul Hamid, N.A.; Maulidiani, M.; Mediani, A.; Abdul Ghafar, S.Z.; Zulaikha Zolkeflee, N.K.; Abas, F. Potentially bioactive metabolites from pineapple waste extracts and their antioxidant and α-glucosidase inhibitory activities by (1)h nmr. *Foods* **2020**, *9*, 173. [CrossRef] [PubMed]
4. Samsuri, S.; Li, T.; Ruslan, M.S.; Amran, N.A. Antioxidant recovery from pomegranate peel waste by integrating maceration and freeze concentration technology. *Int. J. Food Eng.* **2020**. [CrossRef]

5. Abbas, M.; Saeed, F.; Anjum, F.M.; Afzaal, M.; Tufail, T.; Bashir, M.S.; Ishtiaq, A.; Hussain, S.; Suleria, H.A.R. Natural polyphenols: An overview. *Int. J. Food Prop.* **2017**, *20*, 1689–1699. [CrossRef]
6. Borges, G.; Mullen, W.; Crozier, A. Comparison of the polyphenolic composition and antioxidant activity of european commercial fruit juices. *Food Funct.* **2010**, *1*, 73–83. [CrossRef] [PubMed]
7. Brahem, M.; Renard, C.M.; Eder, S.; Loonis, M.; Ouni, R.; Mars, M.; Le Bourvellec, C. Characterization and quantification of fruit phenolic compounds of european and tunisian pear cultivars. *Food Res. Int.* **2017**, *95*, 125–133. [CrossRef]
8. Ozturk, B.; Parkinson, C.; Gonzalez-Miquel, M. Extraction of polyphenolic antioxidants from orange peel waste using deep eutectic solvents. *Sep. Purif. Technol.* **2018**, *206*, 1–13. [CrossRef]
9. Fazio, A.; Iacopetta, D.; La Torre, C.; Ceramella, J.; Muià, N.; Catalano, A.; Carocci, A.; Sinicropi, M.S. Finding solutions for agricultural wastes: Antioxidant and antitumor properties of pomegranate akko peel extracts and β-glucan recovery. *Food Funct.* **2018**, *9*, 6618–6631. [CrossRef]
10. Hurtado-Fernández, E.; Carrasco-Pancorbo, A.; Fernández-Gutiérrez, A. Profiling lc-dad-esi-tof ms method for the determination of phenolic metabolites from avocado (persea americana). *J. Agric. Food Chem.* **2011**, *59*, 2255–2267. [CrossRef]
11. Toh, P.Y.; Leong, F.S.; Chang, S.K.; Khoo, H.E.; Yim, H.S. Optimization of extraction parameters on the antioxidant properties of banana waste. *Acta Sci. Pol. Technol. Aliment.* **2016**, *15*, 65–78. [CrossRef]
12. Zhao, X.; Zhang, W.; Yin, X.; Su, M.; Sun, C.; Li, X.; Chen, K. Phenolic composition and antioxidant properties of different peach [prunus persica (l.) batsch] cultivars in china. *Int. J. Mol. Sci.* **2015**, *16*, 5762–5778. [CrossRef]
13. Suárez, B.; Álvarez, Á.L.; García, Y.D.; del Barrio, G.; Lobo, A.P.; Parra, F. Phenolic profiles, antioxidant activity and in vitro antiviral properties of apple pomace. *Food Chem.* **2010**, *120*, 339–342. [CrossRef]
14. Rodrigo, R.; Gil, D.; Miranda-Merchak, A.; Kalantzidis, G. Antihypertensive role of polyphenols. *Adv. Clin. Chem.* **2012**, *58*, 225.
15. Yang, D.; Dunshea, F.R.; Suleria, H.A.R. LC-ESI-QTOF/MS characterization of Australian herb and spices (garlic, ginger, and onion) and potential antioxidant activity. *J. Food Process Preserv.* **2020**. [CrossRef]
16. Lucci, P.; Saurina, J.; Núñez, O. Trends in lc-ms and lc-hrms analysis and characterization of polyphenols in food. *Trac. Trends Anal. Chem.* **2017**, *88*, 1–24. [CrossRef]
17. Wojdylo, A.; Oszmianski, J.; Laskowski, P. Polyphenolic compounds and antioxidant activity of new and old apple varieties. *J. Agric. Food Chem.* **2008**, *56*, 6520–6530. [CrossRef] [PubMed]
18. Peng, D.; Zahid, H.F.; Ajlouni, S.; Dunshea, F.R.; Suleria, H.A.R. Lc-esi-qtof/ms profiling of australian mango peel by-product polyphenols and their potential antioxidant activities. *Processes* **2019**, *7*, 764. [CrossRef]
19. Gu, C.; Howell, K.; Dunshea, F.R.; Suleria, H.A.R. Lc-esi-qtof/ms characterisation of phenolic acids and flavonoids in polyphenol-rich fruits and vegetables and their potential antioxidant activities. *Antioxidants* **2019**, *8*, 405. [CrossRef]
20. Tang, J.; Dunshea, F.R.; Suleria, H.A.R. Lc-esi-qtof/ms characterization of phenolic compounds from medicinal plants (hops and juniper berries) and their antioxidant activity. *Foods* **2020**, *9*, 7. [CrossRef] [PubMed]
21. Zhong, B.; Robinson, N.A.; Warner, R.D.; Barrow, C.J.; Dunshea, F.R.; Suleria, H.A.R. Lc-esi-qtof-ms/ms characterization of seaweed phenolics and their antioxidant potential. *Mar. Drugs* **2020**, *18*, 331. [CrossRef] [PubMed]
22. Ma, C.; Dunshea, F.R.; Suleria, H.A.R. Lc-esi-qtof/ms characterization of phenolic compounds in palm fruits (jelly and fishtail palm) and their potential antioxidant activities. *Antioxidants* **2019**, *8*, 483. [CrossRef] [PubMed]
23. Nguyen, N.M.P.; Le, T.T.; Vissenaekens, H.; Gonzales, G.B.; Van Camp, J.; Smagghe, G.; Raes, K. In vitro antioxidant activity and phenolic profiles of tropical fruit by-products. *Int. J. Food Sci. Technol.* **2019**, *54*, 1169–1178. [CrossRef]
24. Escarpa, A.; González, M.C. Approach to the content of total extractable phenolic compounds from different food samples by comparison of chromatographic and spectrophotometric methods. *Anal. Chim. Acta* **2001**, *427*, 119–127. [CrossRef]
25. Shrestha, N.; Shrestha, S.; Bhattarai, A. Determination of Ascorbic Acid in Different Citrus Fruits of Kathmandu Valley. *J. Med. Biol. Sci. Res.* **2016**, *2*, 9–14.

26. Xi, W.; Zhang, G.; Jiang, D.; Zhou, Z. Phenolic compositions and antioxidant activities of grapefruit (citrus paradisi macfadyen) varieties cultivated in china. *Int. J. Food Sci. Nutr.* **2015**, *66*, 858–866. [CrossRef] [PubMed]
27. Irkin, R.; Dogan, S.; Degirmenioglu, N.; Diken, M.E.; Guldas, M. Phenolic content, antioxidant activities and stimulatory roles of citrus fruits on some lactic acid bacteria. *Arch. Biol. Sci.* **2015**, *67*, 1313–1321. [CrossRef]
28. Nurliyana, R.d.; Syed Zahir, I.; Mustapha Suleiman, K.; Aisyah, M.R.; Kamarul Rahim, K. Antioxidant study of pulps and peels of dragon fruits: A comparative study. *Int. Food Res. J.* **2010**, *17*, 367–375.
29. Tenore, G.C.; Novellino, E.; Basile, A. Nutraceutical potential and antioxidant benefits of red pitaya (hylocereus polyrhizus) extracts. *J. Funct. Foods* **2012**, *4*, 129–136. [CrossRef]
30. Sembiring, E.N.; Elya, B.; Sauriasari, R. Phytochemical screening, total flavonoid and total phenolic content and antioxidant activity of different parts of caesalpinia bonduc (l.) roxb. *Polym. J.* **2017**, *10*, 123–127. [CrossRef]
31. Marina, Z.; Noriham, A. Quantification of total phenolic compound and in vitro antioxidant potential of fruit peel extracts. *Int. Food Res. J.* **2014**, *21*.
32. Morais, D.R.; Rotta, E.M.; Sargi, S.C.; Schmidt, E.M.; Bonafe, E.G.; Eberlin, M.N.; Sawaya, A.C.H.F.; Visentainer, J.V. Antioxidant activity, phenolics and uplc–esi(–)–ms of extracts from different tropical fruits parts and processed peels. *Food Res. Int.* **2015**, *77*, 392–399. [CrossRef]
33. Ayala-Zavala, J.F.; Vega-Vega, V.; Rosas-Domínguez, C.; Palafox-Carlos, H.; Villa-Rodriguez, J.A.; Siddiqui, M.W.; Dávila-Aviña, J.E.; González-Aguilar, G.A. Agro-industrial potential of exotic fruit byproducts as a source of food additives. *Food Res. Int.* **2011**, *44*, 1866–1874. [CrossRef]
34. Nogata, Y.; Sakamoto, K.; Shiratsuchi, H.; Ishii, T.; YANO, M.; Ohta, H. Flavonoid composition of fruit tissues of citrus species. *Biosci. Biotechnol. Biochem.* **2006**, *70*, 178–192. [CrossRef] [PubMed]
35. Loh, S.C.; Azlan, A.; Chan, S.H.; Khoo, H.E. Extracts of peel and different parts of md2 pineapple as potent nutraceuticals. *TJPS* **2017**, *41*, 49.
36. Wong, Y.S.; Sia, C.M.; Eng, H.; Ang, Y.K.; Chang, S.K.; Yim, H.S. Influence of extraction conditions on antioxidant properties of passion fruit (passiflora edulis) peel. *Acta Sci. Pol. Technol. Aliment.* **2014**, *13*, 257–265. [CrossRef]
37. Ruiz-Montañez, G.; Ragazzo-Sánchez, J.A.; Calderón-Santoyo, M.; Velázquez-de la Cruz, G.; de León, J.A.; Navarro-Ocaña, A. Evaluation of extraction methods for preparative scale obtention of mangiferin and lupeol from mango peels (mangifera indica l.). *Food Chem.* **2014**, *159*, 267–272.
38. Masibo, M.; He, Q. Major mango polyphenols and their potential significance to human health. *Compr. Rev. Food Sci. Food Saf.* **2008**, *7*, 309–319. [CrossRef]
39. Minatel, I.O.; Borges, C.V.; Ferreira, M.I.; Gomez, H.A.G.; Chen, C.-Y.O.; Lima, G.P.P. *Phenolic Compounds: Functional Properties, Impact of Processing and Bioavailability*; Phenolic Compounds Biological Activity, Ed.; InTech: Rijeka, Croatia, 2017; pp. 1–24.
40. Chang, C.-H.; Lin, H.-Y.; Chang, C.-Y.; Liu, Y.-C. Comparisons on the antioxidant properties of fresh, freeze-dried and hot-air-dried tomatoes. *J. Food Eng.* **2006**, *77*, 478–485. [CrossRef]
41. Ajila, C.M.; Naidu, K.A.; Bhat, S.G.; Rao, U.J.S.P. Bioactive compounds and antioxidant potential of mango peel extract. *Food Chem.* **2007**, *105*, 982–988. [CrossRef]
42. Castro-Vazquez, L.; Alañón, M.E.; Rodríguez-Robledo, V.; Pérez-Coello, M.S.; Hermosín-Gutierrez, I.; Díaz-Maroto, M.C.; Jordán, J.; Galindo, M.F.; Arroyo-Jiménez, M.d.M. Bioactive flavonoids, antioxidant behaviour, and cytoprotective effects of dried grapefruit peels (*citrus paradisi* macf.). *Oxid. Med. Cell. Longev.* **2016**, *2016*, 8915729. [CrossRef]
43. Pal, R.S.; Kumar, V.A.; Arora, S.; Sharma, A.K.; Kumar, V.; Agrawal, S. Physicochemical and antioxidant properties of kiwifruit as a function of cultivar and fruit harvested month. *Braz. Arch. Biol. Technol.* **2015**, *58*, 262–271. [CrossRef]
44. Tremocoldi, M.A.; Rosalen, P.L.; Franchin, M.; Massarioli, A.P.; Denny, C.; Daiuto, É.R.; Paschoal, J.A.R.; Melo, P.S.; Alencar, S.M.d. Exploration of avocado by-products as natural sources of bioactive compounds. *PLoS ONE* **2018**, *13*, e0192577. [CrossRef] [PubMed]
45. Ortega-Arellano, H.F.; Jimenez-Del-Rio, M.; Velez-Pardo, C. Neuroprotective effects of methanolic extract of avocado persea americana (var. Colinred) peel on paraquat-induced locomotor impairment, lipid peroxidation and shortage of life span in transgenic knockdown parkin drosophila melanogaster. *Neurochem. Res.* **2019**, *44*, 1986–1998. [CrossRef]

46. Oboh, G.; Ademosun, A. Characterization of the antioxidant properties of phenolic extracts from some citrus peels. *J. Food Sci. Technol.* **2012**, *49*, 729–736. [CrossRef]
47. Guo, C.; Yang, J.; Wei, J.; Li, Y.; Xu, J.; Jiang, Y. Antioxidant activities of peel, pulp and seed fractions of common fruits as determined by frap assay. *Nutr. Res.* **2003**, *23*, 1719–1726. [CrossRef]
48. Mehmood, T.; Khan, M.R.; Shabbir, M.A.; Zia, M.A. Phytochemical profiling and hplc quantification of citrus peel from different varieties. *Prog. Nutr.* **2018**, *20*, 279–288.
49. Yang, P.; Xu, F.; Li, H.F.; Wang, Y.; Li, F.C.; Shang, M.Y.; Liu, G.X.; Wang, X.; Cai, S.Q. Detection of 191 taxifolin metabolites and their distribution in rats using hplc-esi-it-tof-msn. *Molecules* **2016**, *21*, 1209. [CrossRef] [PubMed]
50. Escobar-Avello, D.; Lozano-Castellon, J.; Mardones, C.; Perez, A.J.; Saez, V.; Riquelme, S.; von Baer, D.; Vallverdu-Queralt, A. Phenolic profile of grape canes: Novel compounds identified by lc-esi-ltq-orbitrap-ms. *Molecules* **2019**, *24*, 3763.
51. Wang, X.Y.; Yan, K.J.; Ma, X.H.; Li, W.; Chu, Y.; Guo, J.H.; Li, S.M.; Zhou, S.P.; Zhu, Y.H.; Liu, C.X. Simultaneous determination and pharmacokinetic study of protocatechuic aldehyde and its major active metabolite protocatechuic acid in rat plasma by liquid chromatography-tandem mass spectrometry. *J. Chromatogr. Sci.* **2016**, *54*, 697–705. [CrossRef]
52. Kim, H.; Choi, H.K.; Moon, J.Y.; Kim, Y.S.; Mosaddik, A.; Cho, S.K. Comparative antioxidant and antiproliferative activities of red and white pitayas and their correlation with flavonoid and polyphenol content. *J Food Sci* **2011**, *76*, C38–C45. [CrossRef]
53. Wang, X.; Liu, J.; Zhang, A.; Sun, H.; Zhang, Y. Chapter 23—systematic characterization of the absorbed components of acanthopanax senticosus stem. In *Serum Pharmacochemistry of Traditional Chinese Medicine*; Wang, X., Ed.; Academic Press: Cambridge, MA, USA, 2017; pp. 313–336.
54. Wang, J.; Jia, Z.; Zhang, Z.; Wang, Y.; Liu, X.; Wang, L.; Lin, R. Analysis of chemical constituents of melastoma dodecandrum lour. By uplc-esi-q-exactive focus-ms/ms. *Molecules* **2017**, *22*, 476. [CrossRef] [PubMed]
55. Sasot, G.; Martinez-Huelamo, M.; Vallverdu-Queralt, A.; Mercader-Marti, M.; Estruch, R.; Lamuela-Raventos, R.M. Identification of phenolic metabolites in human urine after the intake of a functional food made from grape extract by a high resolution ltq-orbitrap-ms approach. *Food Res. Int.* **2017**, *100*, 435–444. [CrossRef] [PubMed]
56. Castro, C.B.; Luz, L.R.; Guedes, J.A.C.; Porto, D.D.; Silva, M.F.S.; Silva, G.S.; Riceli, P.R.V.; Canuto, K.M.; Brito, E.S.; Zampieri, D.S.; et al. Metabolomics-based discovery of biomarkers with cytotoxic potential in extracts of myracrodruon urundeuva. *J. Braz. Chem. Soc.* **2020**, *31*, 775–787. [CrossRef]
57. Wang, Y.F.; Vorsa, N.; Harrington, P.D.; Chen, P. Nontargeted metabolomic study on variation of phenolics in different cranberry cultivars using uplc-im—Hrms. *J. Agric. Food Chem.* **2018**, *66*, 12206–12216. [CrossRef]
58. Zeng, Y.; Lu, Y.; Chen, Z.; Tan, J.; Bai, J.; Li, P.; Wang, Z.; Du, S. Rapid characterization of components in bolbostemma paniculatum by uplc/ltq-orbitrap msn analysis and multivariate statistical analysis for herb discrimination. *Molecules* **2018**, *23*, 1155. [CrossRef]
59. Rychlik, M. Quantification of free coumarin and its liberation from glucosylated precursors by stable isotope dilution assays based on liquid chromatography-tandem mass spectrometric detection. *J. Agric. Food Chem.* **2008**, *56*, 796–801. [CrossRef]
60. Geng, C.A.; Chen, H.; Chen, X.L.; Zhang, X.M.; Lei, L.G.; Chen, J.J. Rapid characterization of chemical constituents in saniculiphyllum guangxiense by ultra fast liquid chromatography with diode array detection and electrospray ionization tandem mass spectrometry. *Int. J. Mass Spectrom.* **2014**, *361*, 9–22. [CrossRef]
61. Yang, S.; Shan, L.; Luo, H.; Sheng, X.; Du, J.; Li, Y. Rapid classification and identification of chemical components of schisandra chinensis by uplc-q-tof/ms combined with data post-processing. *Molecules* **2017**, *22*, 1778. [CrossRef]
62. Stella, L.; De Rosso, M.; Panighel, A.; Vedova, A.D.; Flamini, R.; Traldi, P. Collisionally induced fragmentation of m-h (-) species of resveratrol and piceatannol investigated by deuterium labelling and accurate mass measurements. *Rapid Commun. Mass Spectrom.* **2008**, *22*, 3867–3872. [CrossRef]
63. Reed, K.A. *Identification of Phenolic Compounds from Peanut Skin Using hplc-ms^n*; Virginia Polytechnic Institute and State University: Blacksburg, VA, USA, 2009.
64. Pereira-Netto, A.B. Tropical fruits as natural, exceptionally rich, sources of bioactive compounds. *Int. J. Fruit Sci.* **2018**, *18*, 231–242. [CrossRef]

65. He, S.; Wu, B.; Pan, Y.; Jiang, L. Stilbene oligomers from parthenocissus laetevirens: Isolation, biomimetic synthesis, absolute configuration, and implication of antioxidative defense system in the plant. *J. Org. Chem.* **2008**, *73*, 5233–5241. [CrossRef]
66. Cammell, M.E.; Knight, J.D. Effects of climatic change on the population dynamics of crop pests. In *Advances in Ecological Research*; Begon, M., Fitter, A.H., Macfadyen, A., Eds.; Academic Press: Cambridge, MA, USA, 1992; Volume 22, pp. 117–162.
67. Gang, D.R.; Dinkova-Kostova, A.T.; Davin, L.B.; Lewis, N.G. Phylogenetic links in plant defense systems: Lignans, isoflavonoids, and their reductases. In *Phytochemicals for Pest Control*; Hedin, P.A., Hollingworth, R.M., Eds.; American Chemical Society: Washington, DC, USA, 1997; Volume 658, pp. 58–89.
68. Palafox-Carlos, H.; Yahia, E.M.; González-Aguilar, G.A. Identification and quantification of major phenolic compounds from mango (mangifera indica, cv. Ataulfo) fruit by hplc–dad–ms/ms-esi and their individual contribution to the antioxidant activity during ripening. *Food Chem.* **2012**, *135*, 105–111. [CrossRef]
69. Kim, Y.; Brecht, J.K.; Talcott, S.T. Antioxidant phytochemical and fruit quality changes in mango (mangifera indica l.) following hot water immersion and controlled atmosphere storage. *Food Chem.* **2007**, *1055*, 1327–1334. [CrossRef]
70. Hu, K.; Dars, A.G.; Liu, Q.; Xie, B.; Sun, Z. Phytochemical profiling of the ripening of chinese mango (mangifera indica l.) cultivars by real-time monitoring using uplc-esi-qtof-ms and its potential benefits as prebiotic ingredients. *Food Chem.* **2018**, *256*, 171–180. [CrossRef]
71. Gorinstein, S.; Poovarodom, S.; Leontowicz, H.; Leontowicz, M.; Namiesnik, J.; Vearasilp, S.; Haruenkit, R.; Ruamsuke, P.; Katrich, E.; Tashma, Z. Antioxidant properties and bioactive constituents of some rare exotic thai fruits and comparison with conventional fruits: In vitro and in vivo studies. *Food Res. Int.* **2011**, *44*, 2222–2232. [CrossRef]
72. Rebello, L.P.G.; Ramos, A.M.; Pertuzatti, P.B.; Barcia, M.T.; Castillo-Muñoz, N.; Hermosín-Gutiérrez, I. Flour of banana (musa aaa) peel as a source of antioxidant phenolic compounds. *Food Res. Int.* **2014**, *55*, 397–403. [CrossRef]
73. Li, Y.; Guo, C.; Yang, J.; Wei, J.; Xu, J.; Cheng, S. Evaluation of antioxidant properties of pomegranate peel extract in comparison with pomegranate pulp extract. *Food Chem.* **2006**, *96*, 254–260. [CrossRef]
74. Pal, J.; Prakash, A.; Pandey, G.; Raju, C. Comparatives study of antioxidant activity and total phenolic contents of pomegranate and orange peels extracts. *J. Pharmacogn. Phytochem.* **2017**, *6*, 1359–1362.
75. Sir Elkhatim, K.A.; Elagib, R.A.A.; Hassan, A.B. Content of phenolic compounds and vitamin c and antioxidant activity in wasted parts of sudanese citrus fruits. *Food Sci. Nutr.* **2018**, *6*, 1214–1219. [CrossRef]
76. Li, B.B.; Smith, B.; Hossain, M.M. Extraction of phenolics from citrus peels: I. Solvent extraction method. *Sep. Purif. Technol.* **2006**, *48*, 182–188. [CrossRef]
77. Wolfe, K.; Wu, X.; Liu, R.H. Antioxidant activity of apple peels. *J. Agric. Food Chem.* **2003**, *51*, 609–614. [CrossRef] [PubMed]
78. Russell, W.R.; Labat, A.; Scobbie, L.; Duncan, G.J.; Duthie, G.G. Phenolic acid content of fruits commonly consumed and locally produced in scotland. *Food Chem.* **2009**, *115*, 100–104. [CrossRef]
79. Mihailović, N.; Mihailovic, V.; Kreft, S.; Ciric, A.; Joksović, L.; Djurdjevic, P. Analysis of phenolics in the peel and pulp of wild apples (malus sylvestris (l.) mill.). *J. Food Compos. Anal.* **2017**, *67*.
80. Veberic, R.; Trobec, M.; Herbinger, K.; Hofer, M.; Grill, D.; Stampar, F. Phenolic compounds in some apple (malus domestica borkh) cultivars of organic and integrated production. *J. Sci. Food Agric.* **2005**, *85*, 1687–1694. [CrossRef]
81. López-Cobo, A.; Verardo, V.; Diaz-de-Cerio, E.; Segura-Carretero, A.; Fernández-Gutiérrez, A.; Gómez-Caravaca, A.M. Use of hplc-and gc-qtof to determine hydrophilic and lipophilic phenols in mango fruit (mangifera indica l.) and its by-products. *Food Res. Int.* **2017**, *100*, 423–434.
82. Zain, N.M.; Nazeri, M.A.; Azman, N.A. Assessment on bioactive compounds and the effect of microwave on pitaya peel. *J. Teknol.* **2019**, *81*, 11–19.
83. Silva, L.M.R.d.; Figueiredo, E.A.T.d.; Ricardo, N.M.P.S.; Vieira, I.G.P.; Figueiredo, R.W.d.; Brasil, I.M.; Gomes, C.L. Quantification of bioactive compounds in pulps and by-products of tropical fruits from brazil. *Food Chem.* **2014**, *143*, 398–404. [CrossRef]
84. Singh, S.; Immanuel, G. Extraction of antioxidants from fruit peels and its utilization in paneer. *J. Food Process. Technol.* **2014**, *5*, 1.

85. Schieber, A.; Keller, P.; Carle, R. Determination of phenolic acids and flavonoids of apple and pear by high-performance liquid chromatography. *J. Chromatogr. A* **2001**, *910*, 265–273. [CrossRef]
86. Floegel, A.; Kim, D.-O.; Chung, S.-J.; Koo, S.I.; Chun, O.K. Comparison of abts/dpph assays to measure antioxidant capacity in popular antioxidant-rich us foods. *J. Food Compos. Anal.* **2011**, *24*, 1043–1048. [CrossRef]
87. Pulido, R.; Bravo, L.; Saura-Calixto, F. Antioxidant activity of dietary polyphenols as determined by a modified ferric reducing/antioxidant power assay. *J. Agric. Food Chem.* **2000**, *48*, 3396–3402. [CrossRef] [PubMed]
88. Pękal, A.; Pyrzynska, K. Evaluation of aluminium complexation reaction for flavonoid content assay. *Food Anal. Methods* **2014**, *7*, 1776–1782. [CrossRef]

foods

MDPI

Article

Physicochemical, Sensory, and Cooking Qualities of Pasta Enriched with Oat β-Glucans, Xanthan Gum, and Vital Gluten

Ada Kawęcka, Aldona Sobota * and Emilia Sykut-Domańska

Department of Plant Food Technology and Gastronomy, Division of Engineering and Cereals Technology, University of Life Sciences, Skromna 8 Street, 20-704 Lublin, Poland; ada.krawecka@gmail.com (A.K.); emilia.sykut-domanska@up.lublin.pl (E.S.-D.)

* Correspondence: aldona.sobota@up.lublin.pl

Received: 3 September 2020; Accepted: 1 October 2020; Published: 5 October 2020

check for updates

Abstract: The functional properties of β-glucans derived from oats and barley are confirmed by numerous in vitro and in vivo studies. This study aimed to assess the effect of adding 0, 5, 10, 15, and 20% oat (1,3)(1,4)-β-D-glucans to physicochemical properties, as well as the cooking and sensory qualities of durum wheat pasta. Additionally, to improve the cooking and sensory qualities of pasta, we added 5% of xanthan gum and vital gluten. The present study showed that the addition of β-glucans led to an increase of the water absorption index (WAI), water solubility index (WSI), and viscosity of products. At the same time, an increase in the content of fat, ash, and dietary fiber was observed. The addition of (1,3)(1,4)-β-D-glucans influenced the cooking quality of the pasta, extending the minimum cooking time and increasing the loss of dry matter. At the same time, the color of the product changed. In the case of cooked pasta, the addition of β-glucans decreased the brightness and increased the yellowness and redness. It was found that the products enriched with 10–15% of β-glucans, as well as 5% of xanthan gum and vital gluten would yield functional pasta that may offer health benefits beyond its nutritional value. Further, this could influence high cooking and sensory quality.

Keywords: functional food; fortified pasta; β-glucans; nutritional properties; sensory analysis

1. Introduction

Functional food is part of the fastest growing sector of the global food market and is a response to the growing consumer demand for health products [1]. According to the European Commission, the term "functional food" denotes food that not only has nutritional effects but also exerts a beneficial effect on the physical functions of the body and, in some cases, reduces the risk of specific diseases. These beneficial effects must be confirmed by scientific research. Functional food must have a form easily accessible to the consumer in order to be part of the daily diet [2]. On an industrial scale, these are usually products to which a health-promoting component has been added, its bioavailability has been increased, or an adverse component has been removed [3–5].

A poorly balanced diet has a great impact on the development of chronic non-communicable diseases (NCD). There is still an upward trend in the incidence of type 2 diabetes and cardiovascular diseases [6,7]. An integral part of preventing these diseases is to increase dietary fiber intake [8–10]. Several studies have emphasized that high consumption of cereal-derived fiber is associated with a reduction in the risk of development of type 2 diabetes [11–13]. Dietary fiber has been used for fortification for many years. Its soluble fractions (SDF) are thought to be extremely functional. They increase viscosity in the stomach and delay its emptying. In the intestines, they create a barrier to enzymes and consistently slow down the hydrolysis of nutrients and absorption of glucose and cholesterol from food [14–17]. What is more, soluble fiber is easily fermented by bacteria living in the colon, resulting in the production of short-chain fatty acids (SCFAs), which lower the pH of the environment and stimulate the development of beneficial microflora [13]. The soluble fiber fraction includes (1,3)(1,4)-β-D-glucans, which are polymers of glucose present in the cell walls of cereal grains, especially barley and oats [18]. These ingredients have been documented to exert pro-health effects and may be used as a functional component in food [19]. As reported by Jenkins et al. [20], a 1-g increase in the content of β-glucans in a product reduces a food's glycemic index (GI) by 4 units. A number of scientific studies confirmed that a 4–6-week diet based on products with a low glycemic index significantly reduces the fasting blood glucose level and insulin secretion, in addition to increasing insulin sensitivity. Concurrently, it reduces the level of glycated hemoglobin (HbA1c); hence, it is an effective method to prevent and treat diabetes [21–23]. The consumption of β-D-glucans at the level of 4 g/30 g of digestible carbohydrates present in a meal helps reduce postprandial glucose, while the consumption at a level of 3 g/day helps maintain normal blood cholesterol levels [24].

Pasta produced from semolina durum or common wheat flour is one of the most popular cereal products and can be a suitable food matrix for fortification with functional ingredients. Many studies focus on the possibility of enriching pasta with high-fiber raw materials, including oat flour, β-glucan concentrates [25–31], legume components [32], or pomaces [33,34]. It should be noted that the addition of both insoluble and soluble fractions of dietary fiber can weaken the protein-starch matrix and has a negative effect on the cooking and textural qualities of pasta [17,34,35]. However, some high-fiber materials such as xanthan gum or high-protein material (e.g., vital gluten) can improve dough strength and the cooking and sensory qualities of pasta [36–41]. At the same time, these components may improve the health-promoting value of a product. The effects of adding β-glucans to pasta have been examined by other authors, but there is no research on the possibility of reducing the negative effect of this component addition on pasta sensory and cooking qualities. For this reason, the aim of the study was to determine the possibility of using oat β-glucans and additionally xanthan gum and vital gluten to obtain functional pasta with high quality properties and health benefits.

2. Material and Methods

2.1. Characteristics of Raw Materials

The raw material used in the study was semolina durum (Julia Malom, Kunszállás, Hungary), from which the control sample was produced (i.e., pasta without additives (CON)). Subsequent samples were supplemented with the same level of xanthan gum (5%) (Agnex, Białystok, Poland) and vital wheat gluten (5%) (Polmarkus, Pyskowice, Poland) to the semolina. The oat β-glucans supplement (Brenntag Polska Sp.z o.o., Kędzierzyn-Koźle, Poland) was variable and amounted to 0, 5, 10, 15, and 20% (samples BG0, BG5, BG10, BG15, BG20, respectively). Samples of raw materials weighing 5 kg were moistened to 33% humidity. The detailed model of the experiment is presented in Table 1.

2.2. Pasta Preparation

Fusilli pasta was produced in a semi-technical laboratory scale using a MAC-30S Lab pasta extruder (ItalPast, Parma, Italy). The ingredients of the dough were premixed for 15 min under atmospheric pressure and, subsequently, the dough was mixed and extruded under a vacuum (0.086 MPa). The rotational speed of the screw of the pasta extruder was 48 rpm. The pasta samples were dried at a controlled temperature and humidity in a pasta dryer EAC30-LAB (ItalPast, Parma, Italy) in conditions described previously by Sobota et al. [42].

Table 1. The experimental model.

Samples	Raw Materials (%)				Water Addition (g kg^{-1})	Pressure (MPa)	Barrel Temp. (°C)	Extruder Output (kg h^{-1})
	Semolina Durum	Oat β-Glucans	Vital Wheat Gluten	Xanthan Gum				
CON	100	-	-	-	305.2	8.5	28.9	34.68
BG0	90	0	5	5	305.6	13	29.6	29.88
BG5	85	5	5	5	309	13	29.8	31.56
BG10	80	10	5	5	312.4	13	29.8	32.40
BG15	75	15	5	5	315.8	13.5	29.8	32.40
BG20	70	20	5	5	319.4	13.5	29.8	32.38

CON—control sample; BG—β-glucans.

2.3. Chemical Analysis

The analysis of the chemical composition was performed using American Association of Cereal Chemists Approved Methods (AACC) and Association of Official Analytical Chemists International (AOAC) methods [43,44]. The moisture content was determined with the air-oven method (Method AACC 44-15A). The samples (3 g) were placed in a laboratory dryer and dried at 103 °C ± 1° C to constant weight. After cooling in an exsiccator, the samples were weighed and the moisture contents were calculated. The ash content was determined using AACC method 08–01. The samples were measured into ash dishes at 3 g. Then, the samples were placed in a muffle furnace at 550 °C. They were incinerated until light gray ash or constant weight was obtained (7 h). After cooling, the samples were weighed and the ash contents were calculated. To determine the total protein content, the Kjeldahl method (Method AACC 46-08) and the Kjeltec 2300 (FOSS, Höganäs, Sweden) apparatus were used. The protein content was calculated from the total nitrogen content using converted factor 5.7. The crude fat content was determined via continuous extraction. The SoxtecTM8000 on application AN 310 (FOSS, Höganäs, Sweden) and hexane as a solvent were used. The total dietary fiber (TDF) content, including insoluble dietary fiber (IDF) and soluble dietary fiber (SDF), according to the enzymatic methods (AACC 32-05, AACC 32-21, AOAC 991.43, and AOAC 985.29) was analyzed. Next, 1 g dried samples were subjected to sequential enzymatic digestion using heat-stable α-amylase, protease, and amyloglucosidase. Megazyme enzymes and analytical procedures were used (Megazyme International Ireland Ltd., Wicklow, Ireland). The digestible carbohydrate content was determined by calculating the difference (weight in grams (protein + fat + TDF + ash) in 100 g of dry matter of pasta or raw material).

2.4. Physical Properties

The quality parameters of semolina (i.e., wet gluten content, gluten elasticity, and gluten spreadability) were tested in accordance with the Polish Standard (PN-92/A-74021) [45]. The granulometric composition of durum semolina was determined with the sieve method using a laboratory sieve shaker (Sadkiewicz Instruments—Bydgoszcz) and a set of sieves with the following mesh sizes: 400, 315, 250, 160, 125, and 80 μm. Next, 100 g of samples was sieved for 10 min. The percentage share of each fraction in the semolina and the equivalent diameter (weighted average grain diameter in the sample) were calculated. The details of the method were described by Sobota et al. [35]. In the case of fine-grained raw materials (i.e., vital wheat gluten, xanthan gum, and oat β-glucans), the particle size was examined with microscopic image analysis using an optical microscope (×40 magnification) and DLT CAMViewer 3.7.4043 software. Raw material samples were prepared and the diameters of 70 randomly selected raw material particles were measured, and their average diameter was calculated. The water solubility index (WSI) and the water absorption index (WAI) were investigated according to the centrifuge method (AACC 56-20), with slight modification described by Sobota et al. [35]. Further, 2 g of samples were placed in centrifuge tubes and mixed with 30 mL of distilled water. The suspension was left to rest for 5 min then was centrifuged for 15 min at 2200× g. Next, 10 mL of supernatant was dried to the solid mass and the WSI was calculated using the Formula (1):

$$\text{WSI}\ (\%) = (\text{Weight of dried supernatant} \times \frac{30\ \text{mL}}{10\ \text{mL}} / \text{Dry weight of sample}) \times 100\%. \quad (1)$$

After 10 mL of the supernatant was downloaded to determine the WSI, the remaining supernatant was carefully decanted. The wet samples were weighed and WAI was calculated using the Formula (2):

$$\text{WAI}\ (\%) = ((\text{Weight of wet sample} - \text{Weight of dry sample}) / \text{Dry weight of sample}) \times 100\%. \quad (2)$$

Apparent viscosity was tested in accordance with the method developed by Zarzycki and Sobota [46]. A rotary rheometer RM 180 (Mettler-Toledo AG, Switzerland, software RSI Orchestrator, ver. V6.5.8.) with coaxial cylinders without the bottom cylinder guard (shear rate of 1000 s^{-1}) was used.

A 5% suspension in distilled water at 30 °C was prepared from ground pasta samples. The suspensions were held for 30 min at 30 °C with constant stirring. The suspensions (300 mL) were heated in a laboratory shaker (Elpin type 357, Elpin Plus s.c., Lubawa, Poland). The viscosity was measured at a temperature ranging from 65 to 95 °C; next, they were kept at 95 °C for 20 min, cooled to 50 °C, and kept for 30 min at this temperature. A constant temperature gradient of 1 °C min^{-1} was maintained during the heating and cooling processes. When the viscosity measuring agitation was stopped, the cylinders were immersed in the suspension, and five consecutive readings were taken every 10 s. The measurements were made in 3 replications for each sample.

2.5. Cooking Quality of Pasta Samples

The optimal cooking time (OCT, min) was measured according to Method AACC 16–50 [43]. Next, 50 g of pasta was boiled in 500 mL of distilled water. Every 30 s, the pasta was removed and squeezed between two glass plates until the mealy core disappeared. The time needed for this process was assumed as the optimal cooking time (OCT). The weight increase index (WI) was calculated by dividing the weight of the pasta sample after cooking by the weight of the uncooked pasta sample (50 g) [47]. In order to determine the volume increase index (VII), the volume of the pasta was tested by dipping a 50 g sample of an uncooked product in a measuring cylinder filled with 400 mL of vegetable oil. The volume increase was equal to the volume of the tested pasta sample. A sample of pasta (50 g) was then cooked and the volume of the cooked product was determined in an analogous manner. VII was calculated by dividing the volume of cooked pasta by the volume of an uncooked product. Cooking loss (CL, g/100 g d.m.) was determined by testing the dry matter content in water after cooking a 50 g pasta sample. The dry matter content in water was determined according to the AACC 44-15A method [43].

2.6. Color of Pasta

The color of the cooked and uncooked pasta samples was measured using a colorimeter (X-Rite 8200, Inc., Grand Rapids, MI, USA) with a standard light source (D65), a standard colorimetric observer (10°), and a 12.3 mm diameter hole. White and black calibration references were applied to standardize the instrument before analysis. The following CIE parameters were recorded: L* (lightness, indicates the level of light 100 or dark 0), a* (−a* = indicates greenness, +a* = indicates redness), and b* values (−b* = blue, +b* = yellow). The measurements were performed repeatedly 10 times per each sample.

2.7. Sensory Analysis

The sensory analysis was carried out in accordance with the method described by Sozer et al. [48]. The analysis involved 12 people (8-females and 4-males, 23–48 years old), who had adequate taste sensitivity. The panelists had been previously trained how to evaluate the sensory parameters of pasta: appearance (regularity of shape, lack of deformation, cracks and scratches), color (should be regular and light-yellow), odor and taste (should be characteristic and similar to that of durum semolina pasta), hardness (evaluated as a resistance of cooked pasta to compression by the teeth), adhesiveness (evaluated by placing in the mouth, pressing it against the palate, and determining the force required to remove it with the tongue), springiness (was measured as the degree to which the product returns to its original shape after partial compression). Properly coded samples were cooked for OCT in random order and evaluated within a time no longer than 5 min after cooking.

A 5-point rating scale was used, in which 5 was the maximum score. Assuming that all evaluated parameters had equal weight, the average sensory rating was calculated for each pasta sample.

2.8. Statistical Analysis

The obtained results were subjected to statistical analysis using the statistical program STATISTICA 13.1 (StatSoft ©, Inc. Tulsa, USA). All experimental results were means (± S.D) from at least three assays. One-way analysis of variance (ANOVA) and Tukey's post-hoc test were used to compare the groups. The results were statistically different for p-values < 0.05.

3. Results and Discussion

3.1. Pasta Processing

The addition of β-glucans, vital gluten, and xanthan gum affected the extrusion process of the pasta. The increasing content of β-glucans in the enriched pasta resulted in higher pressure values, compared to the control sample (Table 1). The dough was pushed through a dye under diminished pressure from 8.5 MPa for the control pasta to 13 MPa (BG0, BG5, BG10) and 13.5 MPa (BG15, BG20). The addition of gluten and xanthan gum caused a significant reduction in the efficiency of the pasta extruder. The formation of a strong gluten matrix caused by the addition of vital gluten and xanthan gum limited the flow rate of the dough through the forming holes in the die and thus reduced the extruder's efficiency. Additionally, the high water absorption of xanthan gum and gluten changed the consistency of the dough. It became harder and less plastic. The increase in the β-glucan content increased the plasticity of the dough. The relatively high content of soluble fiber and fat in this raw material made it difficult to build a strong protein-starch matrix in the dough. The flow rate of the dough through the holes in the matrix increased and, consequently, the extruder's performance increased. Pasta samples (uncooked and cooked) obtained in this study are shown in Figure 1.

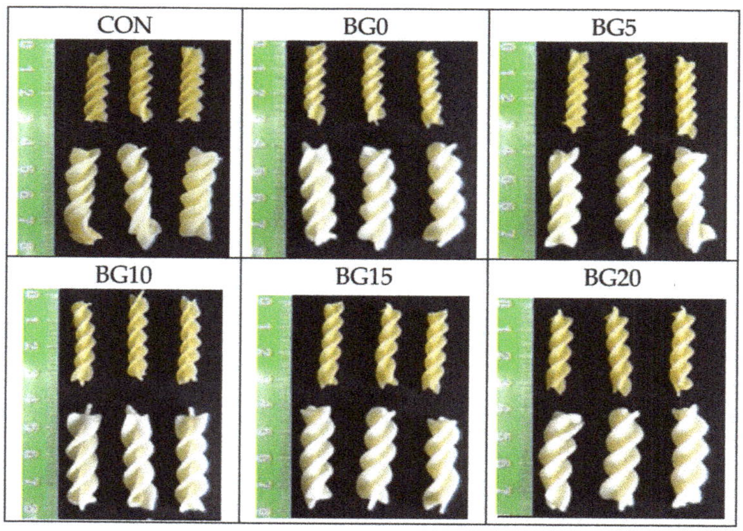

Figure 1. Uncooked and cooked pasta samples. CON—control sample; BG—β-glucans.

3.2. Chemical Analysis

The chemical composition of the raw materials and pasta samples is reported in Table 2. The high content of protein in semolina was responsible for the creation of the strong protein-starch matrix in pasta, which determined the cooking and product quality parameters [22]. In the case of pasta fortified with high-fiber raw materials, weakening of the gluten network and deterioration of cooking quality parameters were most often noted [17,49]. Therefore, vital wheat gluten was used in the β-glucan-fortified pasta samples. The addition of vital wheat gluten, which contained up to 71% protein, caused an increase in the protein content in the sample BG0 and in all samples with β-glucans, compared to the control sample. The addition of β-glucans significantly ($p \leq 0.05$) affected the chemical composition of the pasta. This raw material contained less protein than durum semolina; therefore, increasing the share of β-glucans from 0 to 20% at the expense of semolina led to a decrease in protein content. In addition, xanthan gum was added, which next to β-glucans was a good source of soluble dietary fiber [50]. At the same time, many studies have shown that a small addition of this component had a positive effect on the texture and sensory quality of cooked pasta [36–38,40,41]. As demonstrated by results presented by other authors, the addition of soluble dietary fiber (e.g., xanthan gum, guar gum) meant that, after hydration, non-starch polysaccharides surrounded the protein-starch network, reducing the loss of dry matter [38]. Along with the increase in β-glucans, the ash and fat content increased proportionally. The increase in the ash content resulted from the inclusion of the high-fiber raw materials (e.g., xanthan gum, β-glucans) in the product. The fat content was already high in the β-glucan preparation (Table 2). Both lipids and minerals had a positive effect on the cooking quality of pasta, increasing the stability of the starch-protein matrix [27]. Along with the increase in the share of β-glucans in the pasta, a significant ($p \leq 0.05$) increase in the content of total fiber (TDF) and its soluble fraction (SDF) was noted. The values of TDF and SDF in the BG20 sample were more than four times and six times higher than the control sample, respectively. It should be emphasized that the addition of xanthan gum and gluten also caused a significant ($p \leq 0.05$) increase in dietary fiber content in pasta. Enrichment of the product with β-glucans reinforced these trends. The addition of β-glucans caused a decrease in the share of digestible carbohydrates, which combined with a simultaneous increase in the content of the soluble fiber fraction, which could probably have an impact on reducing the glycemic index of pasta. Soluble fiber (β-glucans) competed with starch granules for water availability and reduced swelling and gelation of starch. As a result, its digestibility and availability decreased [28]. The glycemic index of the product was mainly influenced by the soluble fraction of dietary fiber. It increased the viscosity of the chyme, hindered the access of amylolytic enzymes to starch, reduced the dynamics of starch digestion, and created a sticky film on the intestinal surface impeding glucose absorption into the bloodstream. What is more, fiber stimulated the production of short-chain fatty acids (SCFA) that formed as a result of bacterial fermentation in the large intestine. These acids absorbed by colonocytes through the portal vein entered the liver, where they regulated the metabolism of fatty acids and cholesterol. Propionic acid was of particular importance, as it inhibited the synthesis of fatty acids that reached the bloodstream and regulated adipocytokines (adipokines), i.e., proteins responsible for *inter alia*, glycemic homeostasis, and lipidemia [10].

3.3. Physical Properties

The wet gluten content in the semolina was determined at 28.5%. The spreadability of gluten was 8 mm and the elasticity was estimated at I°. Strong gluten in durum wheat pasta determined its high cooking quality [51]. According to the quality requirements presented in the Polish Standard for Durum Wheat milling products, the minimum content of gluten in semolina durum should be 30% [45]. The raw material did not meet this requirement. Additionally, bearing in mind that the addition of high-fiber raw materials may additionally weaken the gluten network, the addition of vital gluten was used in the mixtures. The particle size of the individual raw materials varied within a wide range, which meant that their mean diameters did not differ significantly, except for the durum semolina (Table 3). The addition of β-glucans affected the water absorption index (WAI), water solubility index (WSI), and viscosity of pasta products (Table 3). The WAI of pasta with added β-glucans was over three times higher, compared to the control sample (CON); however, there were no significant ($p \leq 0.05$) differences between the samples with β-glucans and the BG0 sample.

Table 2. Chemical composition of raw material and pasta samples.

Samples	Moisture	Protein	Fat	Ash	TDF	IDF	SDF	β-Glucans	Digestible Carbohydrate
	(%)						(g/100 g d.m.)		
Raw materials									
Semolina durum	9.50 ^{e} ± 0.04	13.24 ^{c} ± 0.71	1.1 ^{a} ± 0.04	0.76 ^{b} ± 0.01	3.89 ^{abc} ± 0.15	2.07 ^{b} ± 0.09	1.83 ^{ab} ± 0.06	0.18 ^{a} ± 0.05	81.01 ^{e} ± 1.14
Vital wheat gluten	7.31 ^{b} ± 0.05	70.99 ^{g} ± 3.89	1.48 ^{b} ± 0.03	0.57 ^{a} ± 0.01	23.81 ^{e} ± 0.74	22.69 ^{g} ± 0.96	1.12 ^{a} ± 0.22	nd	3.15 ^{b} ± 0.55
Xanthan gum	11.17 ^{h} ± 0.13	5.97 ^{a} ± 0.38	2.03 ^{de} ± 0.16	8.36 ^{g} ± 0.09	83.36 ^{k} ± 4.9	3.17 ^{b} ± 0.28	80.19 ^{f} ± 1.86	nd	0.28 ^{a} ± 0.25
Oat β-glucans	4.19 ^{a} ± 0.07	8.45 ^{b} ± 0.06	3.56 ^{g} ± 0.08	1.99 ^{f} ± 0.02	33.48 ^{f} ± 1.35	0.64 ^{a} ± 0.2	32.85 ^{e} ± 1.15	27.58 ^{f} ± 0.03	52.52 ^{c} ± 1.51
Pasta samples									
CON	8.83 ^{c} ± 0.16	13.57 ^{cd} ± 0.1	1.67 ^{bc} ± 0.05	1.17 ^{c} ± 0.03	4.45 ^{b} ± 0.27	2.34 ^{a} ± 0.19	2.11 ^{b} ± 0.46	0.19 ^{a} ± 0.07	79.14 ^{e} ± 0.37
BG0	9.05 ^{cd} ± 0.10	17.17 ^{f} ± 0.43	1.84 ^{cd} ± 0.07	1.38 ^{d} ± 0.02	15.16 ^{c} ± 0.27	5.97 ^{d} ± 0.19	9.19 ^{c} ± 0.46	0.16 ^{a} ± 0.01	64.45 ^{d} ± 1.47
BG5	10.33 ^{g} ± 0.06	15.7 ^{ef} ± 0.6	2.23 ^{e} ± 0.08	1.39 ^{d} ± 0.02	15.37 ^{c} ± 0.25	5.41 ^{cd} ± 1.59	9.96 ^{cd} ± 0.26	2.03 ^{b} ± 0.23	65.31 ^{d} ± 2.84
BG10	9.93 ^{f} ± 0.18	15.9 ^{ef} ± 0.78	2.55 ^{f} ± 0.09	1.38 ^{d} ± 0.05	14.87 ^{c} ± 0.39	3.9 ^{abc} ± 0.34	10.97 ^{d} ± 0.04	3.31 ^{c} ± 0.44	65.30 ^{d} ± 1.54
BG15	9.32 ^{de} ± 0.15	15.2 ^{de} ± 0.86	2.7 ^{f} ± 0.09	1.49 ^{de} ± 0.05	15.96 ^{cd} ± 0.12	3.5 ^{ab} ± 0.29	12.47 ^{e} ± 0.17	5.44 ^{d} ± 0.32	64.65 ^{d} ± 1.49
BG20	9.17 ^{d} ± 0.03	15.33 ^{de} ± 0.76	2.74 ^{f} ± 0.09	1.55 ^{e} ± 0.05	18.06 ^{d} ± 0.35	4.23 ^{bcd} ± 0.09	13.83 ^{f} ± 0.44	6.22 ^{e} ± 0.25	62.32 ^{d} ± 1.27

Explanation: d.m.—dry matter; IDF—insoluble dietary fiber; SDF—soluble dietary fiber; TDF—total dietary fiber; nd—not detected; CON—control sample; BG—β-glucans. Data are presented as mean (n = 3) ± standard deviation. Data value of each parameter with different superscript letter in the columns are significantly different (Tukey test, $p \leq 0.05$).

Table 3. Physical properties of raw material and pasta samples.

Samples	Equivalent Diameter (μm)	WAI	WSI
		(%)	
Raw materials			
Semolina durum	286.37 ^{b} ± 34	84.7 ^{a} ± 3.7	5.18 ^{a} ± 0.23
Vital wheat gluten	52 ^{a} ± 32	150.3 ^{a} ± 2.9	8.18 ^{bc} ± 0.13
Xanthan gum	25 ^{a} ± 16	2319.3 ^{d} ± 94.7	24.69 ^{f} ± 2.28
Oat β-glucans	83 ^{a} ± 20	900.9 ^{c} ± 17.2	66.26 ^{g} ± 1.9
Pasta samples			
CON	-	100.4 ^{a} ± 2.3	9.5 ^{bcd} ± 0.1
BG0	-	361 ^{b} ± 14.0	7.0 ^{ab} ± 0.1
BG5	-	371.7 ^{b} ± 23.6	9.5 ^{bcd} ± 0.3
BG10	-	376.9 ^{b} ± 4.7	10.9 ^{cde} ± 0.1
BG15	-	377.6 ^{b} ± 4.5	12.2 ^{de} ± 0.1
BG20	-	393.3 ^{b} ± 15.4	12.3 ^{e} ± 0.2

Explanation: WAI—water absorption index; WSI—water solubility index; CON—control sample; BG—β-glucans; Data are presented as mean (n = 3) ± standard deviation. Data value of each parameter with different superscript letter in the columns are significantly different (Tukey test, $p \leq 0.05$).

The high content of dietary fiber (xanthan gum, β-glucans) present in these samples (Table 2) was associated with greater water absorption. As shown in literature reports, distortion of the protein network by the addition of dietary fiber can induce increased water absorption [26]. Foschia et al. [26] observed an increase in water absorption in pasta products when trying to replace semolina with dietary fiber raw materials. High-WAI products quickly satisfy hunger and maintain longer satiety [52]. In the present research, relatively low WSI values were observed for the products enriched with β-glucans. The values obtained may indicate a favorable low level of complex carbohydrate degradation during the production process. This assumption seems to be confirmed by the significantly ($p \leq 0.05$) lower dynamics of digestible carbohydrates in the β-glucan enriched pasta samples (Table 2). Brennan et al. [53] found a positive correlation between WSI values of extruded cereal products enriched with high fiber fungal powder and glycemic response (glycemic index values). Low-WSI cereal products should not generate as high glycemic response as partially degraded starch polymers.

The viscosity of pasta with the addition of β-glucans differed significantly ($p \leq 0.05$) from that in the control sample (CON) (Table 4). The addition of β-glucans into the products meant that, during the suspension heating process, the maximum viscosity was obtained at 75 °C, while the maximum value of this parameter in the case of the control test was only reached at 95 °C. This relationship may be related to the limited swelling of starch granules and the hampered starch pasting in samples containing β-glucans. A significant decrease in viscosity during suspension heating (75–95 °C) may result from partial depolymerization of β-glucans in the process and a decrease in their molecular weight. A significant ($p \leq 0.05$) increase in the viscosity of the BG15 and BG20 products was observed after cooling to 50 °C, which seems to be important in terms of the functional properties of enriched pasta. Increased viscosity in the intestine delays glucose and cholesterol absorption and inhibits bile acid reabsorption [54]. The physicochemical properties of β-glucans can affect the digestibility of starch [17,55]. The increase in the viscosity of oat β-glucans caused a decrease in the digestibility of starch [56]. The functionality of β-glucans as conditioned among others by the molecular weight of β-glucan, which needed to be sufficiently high to be capable to increase the viscosity in intestines [54]. The molecular weight of β-glucan was related to acid degradation [57]. To assess the effect of β-glucan-supplemented pasta on intestinal viscosity, the viscosity of the gastrointestinal content should be examined in simulated in vitro digestion studies.

Table 4. The results of the tests of the apparent viscosity of pasta (Pa s) using the shear rate gradient of 1000 s^{-1}.

Samples	Heating					Cooling	
	Temp. 65 °C	**Temp. 75 °C**	**Temp. 85 °C**	**Temp. 95 °C**	**Temp. 95 °C ***	**Temp. 50 °C**	**Temp. 50 °C ****
Con	$0.006^{aA} \pm 0$	$0.007^{aB} \pm 0$	$0.010^{aC} \pm 0$	$0.012^{aD} \pm 0.001$	$0.014^{aF} \pm 0.001$	$0.015^{aG} \pm 0$	$0.013^{aE} \pm 0$
BG0	$0.012^{bA} \pm 0$	$0.023^{bB} \pm 0$	$0.028^{bC} \pm 0$	$0.031^{cE} \pm 0$	$0.030^{cD} \pm 0$	$0.039^{bG} \pm 0$	$0.035^{bF} \pm 0$
BG5	$0.013^{cA} \pm 0$	$0.034^{dE} \pm 0.001$	$0.031^{cC} \pm 0$	$0.032^{dD} \pm 0$	$0.030^{cB} \pm 0$	$0.040^{cG} \pm 0$	$0.036^{cF} \pm 0$
BG10	$0.014^{dA} \pm 0$	$0.029^{cB} \pm 0$	$0.032^{dC} \pm 0.001$	$0.029^{bB} \pm 0$	$0.029^{bB} \pm 0.001$	$0.039^{bE} \pm 0$	$0.035^{bD} \pm 0$
BG15	$0.017^{eA} \pm 0$	$0.039^{eD} \pm 0$	$0.038^{eC} \pm 0$	$0.038^{fC} \pm 0$	$0.036^{dB} \pm 0$	$0.045^{eF} \pm 0.001$	$0.044^{dE} \pm 0$
BG20	$0.025^{fA} \pm 0.001$	$0.045^{fE} \pm 0.001$	$0.041^{fC} \pm 0.001$	$0.037^{eB} \pm 0$	$0.037^{eB} \pm 0$	$0.048^{dE} \pm 0$	$0.044^{dD} \pm 0$

Explanation: *—measurement after 20 min, **—measurement after 30 min; CON—control sample; BG—β-glucans; Data are presented as mean ± standard deviation. Data value of each parameter with different uppercase superscript letter in the rows are significantly different (Tuckey test, $p \leq 0.05$). Data value of each parameter with different lowercase superscript letter in the columns are significantly different (Tukey test, $p \leq 0.05$).

3.4. Cooking Quality

Studies conducted by other authors [25–31] have indicated that the addition of β-glucans reduced the cooking quality of pasta. Our study focused on the impact of adding β-glucans, xanthan gum, and vital gluten simultaneously. The results demonstrated that the amount of β-glucans addition also significantly ($p \leq 0.05$) affected the cooking quality of pasta samples (Table 5). An increase in the optimal cooking time (OCT) was observed for products enriched with 15% and 20% of β-glucans. The effect of adding soluble fiber (β-glucans/inulin) to extend cooking time has been confirmed by other authors [28,31]. The observed OCT changes may be caused by high water absorption of high-fiber components in products, which compete for water with starch, thereby hindering its swelling and pasting [34,58]. Different results were obtained by Chillo et al. [29], who studied the impact of two commercially available barley β-glucan preparations on the cooking quality of durum wheat spaghetti. They did not notice differences in OCT relative to the control (0% β-glucans) and samples with the variable share of the addition of Glucagel®. In contrast, the OCT of the examined samples with the addition of a Bilans ™ preparation increased with the increase in the share of the preparation. In the 10% β-glucans Bilans ™ assay, the OCT was 1.5 min longer than the control. In the present research, the extension of OCT for samples with the addition of β-glucans may have been the cause of the greater losses of dry matter, compared to the control and BG0, as a result of the passage of part of soluble fiber into the solution during cooking [59].

Table 5. Cooking quality of pasta samples.

Pasta Samples	Optimum Cooking Time (min)	Cooking Loss (% d.m.)	Cooking Weight Increase	Cooking Volume Increase
CON	$9^{a} \pm 0.0$	$4.52^{c} \pm 0.42$	$2.54^{a} \pm 0.04$	$3.16^{a} \pm 0.12$
BG0	$9^{a} \pm 0.0$	$2.95^{a} \pm 0.17$	$2.73^{b} \pm 0.02$	$4.29^{d} \pm 0.18$
BG5	$9^{a} \pm 0.5$	$3.58^{b} \pm 0.13$	$2.77^{b} \pm 0.04$	$3.88^{c} \pm 0.13$
BG10	$9^{a} \pm 0.5$	$3.82^{b} \pm 0.11$	$2.74^{b} \pm 0.02$	$3.46^{b} \pm 0.16$
BG15	$10^{b} \pm 0.0$	$5.03^{d} \pm 0.09$	$2.88^{c} \pm 0.01$	$3.46^{b} \pm 0$
BG20	$11.5^{c} \pm 0.5$	$5.14^{d} \pm 0.09$	$3.09^{d} \pm 0$	$3.38^{b} \pm 0.18$

Explanation: d.m—dry matter; CON—control sample; BG—β-glucans; Data are presented as mean ± standard deviation. Data value of each parameter with different superscript letter in the columns are significantly different (Tukey test, $p \leq 0.05$).

It is worth emphasizing that the almost twice lower dry matter loss in the BG0 sample (compared to BG20) may be related to the highest protein content in this sample (Table 2). Protein has a major impact on maintaining a stable product structure and reducing dry matter loss. The protein-starch matrix is associated with maintenance of better integrity during cooking and improvement of the quality parameters of pasta [51]. Increasing dry matter loss accompanying an increase in the proportion of soluble fiber (guar gum/β-glucans) has been observed [26,27]. The weight values for the BG15 and BG20 samples increased significantly ($p \leq 0.05$) compared to samples with the 0–10% share of β-glucans and the control sample. No statistical differences were observed between the BG0, BG5, and BG10 samples. The increase of the weight of the BG15 and BG20 samples is adequate to the WAI values obtained for these samples (Table 3). Undoubtedly, the ability to absorb water during cooking was related to the extension of the optimal cooking time and the high water capacity of soluble fiber. The reduction of the amount of water available to starch may contribute to its lower pasting capacity and digestibility [26]. In the case of the value increase index, a significant ($p \leq 0.05$) increase was noted in the BG0 pasta (90% semolina, 5% xanthan gum, and 5% gluten) compared to the control sample. Due to its high value of WAI, xanthan gum contributed to a significant ($p \leq 0.05$) increase in the volume of the product during cooking. The increase in the β-glucan content (from 0 to 20%) caused a decrease in this parameter.

3.5. Color of Pasta

The addition of xanthan gum and vital gluten was responsible for the darker color of uncooked pasta (Table 6). An opposite effect was noted with the addition of β-glucans. The product became brighter (higher values of the L* parameter) as the share of this oat component increased. Pasta samples with xanthan gum and vital gluten additionally enriched with 20% of β-glucans (BG20) were characterized by similar brightness as the control sample (CON) (L = 51.10 and 51.52, respectively, for BG20 and CON). Similar results were obtained by Hajji et al. [60] in their analyses of durum wheat pasta enriched with barley β-glucans (at a level of 1–7%). In the case of the uncooked products, no statistical differences ($p \leq 0.05$) were observed between the pasta with β-glucans and the control without β-glucans. The addition of xanthan gum and vital gluten negatively affected the intensity of yellowness (b parameter) and redness (a parameter) of the pasta. The fortification of the pasta with β-glucans reduced the negative changes in the color caused by the addition of vital gluten and xanthan gum, but even samples enriched with 20% of beta-glucan were less yellow and less red than the control sample (CON). The analysis of the color of the cooked pasta demonstrated a different effect of the addition of xanthan gum and vital gluten on product brightness (Table 6). Cooked pasta enriched with xanthan gum, vital gluten, and a small addition of β-glucans (5–10%) had a brighter color compared to the control sample (CON). This was probably due to the intense absorption of water by these products during cooking and the significant ($p \leq 0.05$) increase in the volume index. Consequently, the lower concentration of pigments in the products may have caused the color changes. The increase in the β-glucan addition to the level of 15–20% resulted in a darker color of the cooked pasta; however, it should be emphasized that the values of parameter L for BG15 and BG20 were comparable with those recorded for the control sample (CON). The addition of β-glucans intensify the yellowness of the product; however, pasta samples with up to 20% β-glucans (BG20) were less yellow than the control sample (CON). As reported by Hajji et al. [60], reduction of the intensity of yellowness (b parameter) may be caused by lower protein content in β-glucan compared to that of semolina durum. At the same time, as a function of the increasing proportion of beta-glucans, the volume increase factor declines, which may contribute to a greater concentration of carotenoids in the product - therefore the value of the b parameter and the parameter increases in the case of cooked pasta.

Table 6. Color parameters of cooked and uncooked pasta samples.

Pasta Samples	Uncooked			Cooked		
	L *	a *	b *	L *	a *	b *
CON	51.52 ^{c} ± 1.41	2.11 ^{d} ± 0.15	16.22 ^{c} ± 0.96	76.67 ^{a} ± 0.80	−0.43 ^{b} ± 0.11	17.95 ^{c} ± 1.07
BG0	48.44 ^{a} ± 1.90	1.4 ^{a} ± 0.15	14.49 ^{b} ± 1.36	79.14 ^{b} ± 1.42	−0.66 ^{a} ± 0.09	13.93 ^{a} ± 0.61
BG5	49.78 ^{ab} ± 0.94	1.45 ^{a} ± 0.16	13.84 ^{ab} ± 0.56	79.11 ^{b} ± 1.18	−0.73 ^{a} ± 0.10	13.82 ^{a} ± 0.91
BG10	48.98 ^{a} ± 2.13	1.62 ^{b} ± 0.20	13.58 ^{a} ± 0.88	79.11 ^{b} ± 1.40	−0.43 ^{b} ± 0.12	14.52 ^{ab} ± 0.96
BG15	50.24 ^{b} ± 0.73	1.8 ^{c} ± 0.16	13.81 ^{ab} ± 0.84	76.14 ^{a} ± 1.07	−0.30 ^{b} ± 0.09	14.11 ^{ab} ± 0.59
BG20	51.10 ^{bc} ± 0.70	1.84 ^{c} ± 0.10	13.7 ^{ab} ± 0.34	75.95 ^{a} ± 1.59	−0.31 ^{b} ± 0.08	14.92 ^{b} ± 0.92

Explanation: CON—control sample; BG—β-glucans; * - concern CIE-lab color scale. Data are presented as mean ± standard deviation. Data value of each parameter with different superscript letter in the columns are significantly different (Tukey test, $p \leq 0.05$).

3.6. Sensory Quality

The results of the sensory assessment confirmed that even the 20% addition of β-glucans did not deteriorate the sensory quality of the uncooked products (Table 7). All pasta samples with added β-glucans (5–20%) were approved by the study participants. No significant differences were found between uncooked products enriched with β-glucans, BG0, and the control sample (CON). In the case of cooked pasta, there was a risk that the increase in viscosity in samples with the addition of β-glucans would reduce the tastiness of the product [61]. Although there were no significant differences ($p \leq 0.05$) in the appearance and hardness between the β-glucan-supplemented samples, the BG0 sample, the controls (CON), and the samples supplemented with 20% beta-glucan received significantly lower marks for color, taste, adhesiveness, and springiness, compared to CON and BG0 (Table 7). It should be noted, therefore, that the highest additive of β-glucans that would be acceptable to consumers on a par with the control (CON) is 15% (BG15). Jaworska et al. [62] examined the sensory preferences of consumers of pasta with the addition of oat-β-glucans and reported that pasta with 16% addition of this ingredient turned out to be most often desired among samples with different levels of enrichment (the share of β-glucans was 0, 4, 8, 12, and 16%). The authors revealed that the choice of pasta with the 16% share of β-glucans was determined by the respondents' interest in the inclusion of dietary fiber in the diet and the need to consume functional products.

Table 7. Organoleptic quality of pasta samples.

Pasta Samples	Uncooked			Cooked						
	Apparance	Color	Odor	Apparance	Color	Taste	Odor	Hardness	Adhesiveness	Springness
CON	$4.6^{a} \pm 0.55$	$4.2^{a} \pm 0.45$	$4.8^{a} \pm 0.45$	$3.6^{a} \pm 0.45$	$4.2^{b} \pm 0.45$	$5^{b} \pm 0$	$4.8^{a} \pm 0.45$	$4.6^{a} \pm 0.55$	$5^{c} \pm 0$	$4.8^{b} \pm 0.45$
BG0	$4.8^{a} \pm 0.45$	$4.4^{a} \pm 0.45$	$4.8^{a} \pm 0.45$	$5^{c} \pm 0$	$4.8^{c} \pm 0$	$5^{b} \pm 0$	$5^{a} \pm 0$	$5^{a} \pm 0$	$5^{c} \pm 0$	$5^{b} \pm 0$
BG5	$5^{a} \pm 0$	$4.6^{a} \pm 0$	$5^{a} \pm 0$	$4.6^{bc} \pm 0.35$	$4.6^{bc} \pm 0.55$	$4.6^{ab} \pm 0.55$	$4.8^{a} \pm 0.45$	$4.4^{a} \pm 0.55$	$4.8^{bc} \pm 0.45$	$4.6^{b} \pm 0.55$
BG10	$5^{a} \pm 0$	$4.8^{a} \pm 0.45$	$4.8^{a} \pm 0.45$	$4.6^{bc} \pm 0.35$	$4.6^{bc} \pm 0.55$	$4.6^{ab} \pm 0.55$	$4.8^{a} \pm 0.45$	$4.8^{a} \pm 0.45$	$5^{c} \pm 0$	$4.6^{b} \pm 0.55$
BG15	$5^{a} \pm 0$	$4.83^{a} \pm 0.45$	$4.8^{a} \pm 0.45$	$4.0^{ab} \pm 0.45$	$4.2^{b} \pm 0.45$	$4.2^{ab} \pm 0.45$	$4.4^{a} \pm 0.55$	$4.8^{a} \pm 0.45$	$4.4^{b} \pm 0.15$	$4.2^{ab} \pm 0.45$
BG20	$4.83^{a} \pm 0$	$4.75^{a} \pm 0.45$	$4.83^{a} \pm 0.45$	$3.8^{a} \pm 0.35$	$3.2^{a} \pm 0.45$	$4^{a} \pm 1$	$4.6^{a} \pm 0.55$	$4.4^{a} \pm 0.55$	$3.8^{a} \pm 0.35$	$3.6^{a} \pm 1$

Explanation: CON—control sample; BG—β-glucans; Data are presented as mean ± standard deviation. Data value of each parameter with different superscript letter in the columns are significantly different (Tukey test, $p \leq 0.05$).

4. Conclusions

The obtained results have proved that, by using the appropriate additive β-glucans and simultaneously adding vital wheat gluten and xanthan gum, it is possible to obtain a functional product with high quality properties. The inclusion of xanthan gum and vital wheat gluten improved the cooking quality of the pasta. The addition of β-glucans reduced the negative color changes caused by the addition of xanthan gum and vital wheat gluten in the uncooked products. In contrast, the higher levels of β-glucans (15–20%) in the cooked products contributed to the darkening of the pasta and increased the yellowness, but the pasta was still less yellow than the control. The addition of β-glucans significantly increased the viscosity of the pasta. The highest 20% addition of β-glucans did not cause deterioration of the sensory quality of the uncooked products. In the case of the cooked pasta, the large addition of β-glucans (20%) produced a negative increase in adhesiveness and deteriorated the springiness, color, and taste of the product. In summary, the use of a 10–15% additive of β-glucans and 5% additive of vital wheat gluten and xanthan gum yielded functional pasta containing 3.3–5.5 g β-glucans/100g with high cooking quality and sensory attributes.

Author Contributions: Conceptualization. A.S.; Formal analysis. A.S., A.K. and E.S.-D.; Methodology. A.S.; Project administration. A.K.; Writing – original draft. A.K.; Writing – review & editing. A.K. and A.S. All authors have read and agreed to the published version of the manuscript.

Funding: This research received no external funding.

Conflicts of Interest: The authors declare no conflict of interest.

References

1. Santeramo, F.G.; Carlucci, D.; De Devitiis, B.; Seccia, A.; Stasi, A.; Viscecchia, R.; Nardone, G. Emerging trends in European food, diets and food industry. *Food Res. Int.* **2018**, *104*, 39–47. [CrossRef]
2. Crowe, K.M.; Francis, C. Position of the Academy of Nutrition and Dietetics: Functional Foods. *J. Acad. Nutr. Diet.* **2013**, *113*, 1096–1103. [CrossRef]
3. Alkhatib, A.; Tsang, C.; Tiss, A.; Bahorun, T.; Arefanian, H.; Barake, R.; Khadir, A.; Tuomilehto, J. Functional Foods and Lifestyle Approaches for Diabetes Prevention and Management. *Nutrients* **2017**, *9*, 1310. [CrossRef] [PubMed]
4. Tadesse, S.A.; Emire, S.A. Production and processing of antioxidant bioactive peptides: A driving force for the functional food market. *Heliyon* **2020**, *6*, e04765. [CrossRef] [PubMed]
5. Contreras-Rodriguez, O.; Mata, F.; Verdejo-Román, J.; Ramírez-Bernabé, R.; Moreno, D.; Vilar-Lopez, R.; Soriano-Mas, C.; Verdejo-García, A. Neural-based valuation of functional foods among lean and obese individuals. *Nutr. Res.* **2020**, *78*, 27–35. [CrossRef] [PubMed]
6. Saha, S.; Riemenschneider, H.; Müller, G.; Levin-Zamir, D.; Van den Broucke, S.; Schwarz, P.E.H. Comparative analysis of diabetes self-management education programs in the European Union Member States. *Prim. Care Diabetes* **2017**, *11*, 529–537. [CrossRef]
7. Hammad, M.A.; Syed Sulaiman, S.A.; Abubakar, U.; Mohamed Noor, D.A. The effects of glycemic control on morbidity and survival among diabetic patients. *Diabetes Metab. Syndr. Clin. Res. Rev.* **2019**, *13*, 1035–1040. [CrossRef]
8. McRae, M.P. Dietary Fiber Is Beneficial for the Prevention of Cardiovascular Disease: An Umbrella Review of Meta-analyses. *J. Chiropr. Med.* **2017**, *16*, 289–299. [CrossRef]
9. Xi, P.; Liu, R.H. Whole food approach for type 2 diabetes prevention. *Mol. Nutr. Food Res.* **2016**, *60*, 1819–1836. [CrossRef]
10. Krawęcka, A.; Sobota, A.; Sykut-Domańska, E. Functional Cereal Products in the Diet for Type 2 Diabetes Patients. *Int. J. Food Sci.* **2019**, *2019*, 4012450. [CrossRef]
11. McRae, M.P. Dietary Fiber Intake and Type 2 Diabetes Mellitus: An Umbrella Review of Meta-analyses. *J. Chiropr. Med.* **2018**, *17*, 44–53. [CrossRef] [PubMed]
12. Schulze, M.B.; Schulz, M.; Heidemann, C.; Schienkiewitz, A.; Hoffmann, K.; Boeing, H. Fiber and Magnesium Intake and Incidence of Type 2 Diabetes. *Arch. Intern. Med.* **2007**, *167*, 956. [CrossRef]

13. Davison, K.M.; Temple, N.J. Cereal fiber, fruit fiber, and type 2 diabetes: Explaining the paradox. *J. Diabetes Complications* **2018**, *32*, 240–245. [CrossRef] [PubMed]
14. Othman, R.A.; Moghadasian, M.H.; Jones, P.J. Cholesterol-lowering effects of oat β-glucan. *Nutr. Rev.* **2011**, *69*, 299–309. [CrossRef] [PubMed]
15. Ho, H.V.T.; Sievenpiper, J.L.; Zurbau, A.; Blanco Mejia, S.; Jovanovski, E.; Au-Yeung, F.; Jenkins, A.L.; Vuksan, V. The effect of oat β-glucan on LDL-cholesterol, non-HDL-cholesterol and apoB for CVD risk reduction: A systematic review and meta-analysis of randomised-controlled trials. *Br. J. Nutr.* **2016**, *116*, 1369–1382. [CrossRef] [PubMed]
16. Ciccoritti, R.; Nocente, F.; Sgrulletta, D.; Gazza, L. Cooking quality, biochemical and technological characteristics of bran-enriched pasta obtained by a novel pasta-making process. *LWT Food Sci. Technol.* **2019**, *101*, 10–16. [CrossRef]
17. Cleary, L.; Brennan, C. The influence of a (1→3)(1→4)-β-D-glucan rich fraction from barley on the physico-chemical properties and in vitro reducing sugars release of durum wheat pasta. *Int. J. Food Sci. Technol.* **2006**, *41*, 910–918. [CrossRef]
18. Jayachandran, M.; Chen, J.; Chung, S.S.M.; Xu, B. A critical review on the impacts of β-glucans on gut microbiota and human health. *J. Nutr. Biochem.* **2018**, *61*, 101–110. [CrossRef]
19. Brennan, C.S.; Cleary, L.J. The potential use of cereal (1→3)(1→4)-β-D-glucans as functional food ingredients. *J. Cereal Sci.* **2005**, *42*, 1–13. [CrossRef]
20. Jenkins, A.L.; Jenkins, D.J.A.; Zdravkovic, U.; Würsch, P.; Vuksan, V. Depression of the glycemic index by high levels of β-glucan fiber in two functional foods tested in type 2 diabetes. *Eur. J. Clin. Nutr.* **2002**, *56*, 622–628. [CrossRef]
21. Thomas, A.; Schönauer, M.; Achermann, F.; Schnell, O.; Hanefeld, M.; Ziegelasch, H.J.; Mastrototaro, J.; Heinemann, L. The "glucose pentagon": Assessing glycemic control of patients with diabetes mellitus by a model integrating different parameters from glucose profiles. *Diabetes Technol. Ther.* **2009**, *11*, 399–409. [CrossRef] [PubMed]
22. Rizkalla, S.W.; Taghrid, L.; Laromiguiere, M.; Huet, D.; Boillot, J.; Rigoir, A.; Elgrably, F.; Slama, G. Improved plasma glucose control, whole-body glucose utilization, and lipid profile on a low-glycemic index diet in type 2 diabetic men: A randomized controlled trial. *Diabetes Care* **2004**, *27*, 1866–1872. [CrossRef] [PubMed]
23. Brand-Miller, J.; Hayne, S.; Petocz, P.; Colagiuri, S. Low-glycemic index diets in the management of diabetes: A meta-analysis of randomized controlled trials. *Diabetes Care* **2003**, *26*, 2261–2267. [CrossRef] [PubMed]
24. European Council. *Official Journal of the European Union*; EU Publications: Luxemburg, 2012; Volume 55.
25. Piwińska, M.; Wyrwisz, J.; Kurek, M.; Wierzbicka, A. Effect of oat β-glucan fiber powder and vacuum-drying on cooking quality and physical properties of pasta. *CyTA J. Food* **2016**, *14*, 101–108. [CrossRef]
26. Foschia, M.; Peressini, D.; Sensidoni, A.; Brennan, M.A.; Brennan, C.S. How combinations of dietary fibres can affect physicochemical characteristics of pasta. *Starch/Staerke* **2015**, *66*, 41–46. [CrossRef]
27. Aravind, N.; Sissons, M.; Fellows, C.M. Effect of soluble fibre (guar gum and carboxymethylcellulose) addition on technological, sensory and structural properties of durum wheat spaghetti. *Food Chem.* **2012**, *131*, 893–900. [CrossRef]
28. Aravind, N.; Sissons, M.; Egan, N.; Fellows, C. Effect of insoluble dietary fibre addition on technological, sensory, and structural properties of durum wheat spaghetti. *Food Chem.* **2012**, *130*, 299–309. [CrossRef]
29. Chillo, S.; Ranawana, D.V.; Henry, C.J.K. Effect of two barley β-glucan concentrates on in vitro glycaemic impact and cooking quality of spaghetti. *LWT Food Sci. Technol.* **2011**, *44*, 940–948. [CrossRef]
30. Brennan, C.S.; Kuri, V.; Tudorica, C.M. Inulin-enriched pasta: Effects on textural properties and starch degradation. *Food Chem.* **2004**, *86*, 189–193. [CrossRef]
31. Kudake, D.C.; Pawar, A.V.; Muley, A.B.; Parate, V.R.; Talib, M.I. Enrichment of Wheat Flour Noodles with Oat Flour: Effect on Physical, Nutritional, Antioxidant and Sensory Properties. *Int. J. Curr. Microbiol. Appl. Sci.* **2017**, *6*, 204–213. [CrossRef]
32. Jayasena, V.; Nasar-Abbas, S.M. Development and quality evaluation of high-protein and high-dietary-fiber pasta using lupin flour. *J. Texture Stud.* **2012**, *43*, 153–163. [CrossRef]
33. Tolve, R.; Pasini, G.; Vignale, F.; Favati, F.; Simonato, B. Effect of grape pomace addition on the technological, sensory, and nutritional properties of durum wheat pasta. *Foods* **2020**, *9*, 354. [CrossRef] [PubMed]

34. Sykut-Domańska, E.; Zarzycki, P.; Sobota, A.; Teterycz, D.; Wirkijowska, A.; Blicharz-Kania, A.; Andrejko, D.; Mazurkiewicz, J. The potential use of by-products from coconut industry for production of pasta. *J. Food Process. Preserv.* **2020**, *44*, e14490. [CrossRef]
35. Sobota, A.; Rzedzicki, Z.; Zarzycki, P.; Kuzawińska, E. Application of common wheat bran for the industrial production of high-fibre pasta. *Int. J. Food Sci. Technol.* **2015**, *50*, 111–119. [CrossRef]
36. Milde, L.B.; Chigal, P.S.; Olivera, J.E.; González, K.G. Incorporation of xanthan gum to gluten-free pasta with cassava starch. Physical, textural and sensory attributes. *LWT* **2020**, *131*, 109674. [CrossRef]
37. Raungrusmee, S.; Shrestha, S.; Sadiq, M.B.; Anal, A.K. Influence of resistant starch, xanthan gum, inulin and defatted rice bran on the physicochemical, functional and sensory properties of low glycemic gluten-free noodles. *LWT* **2020**, *126*, 109279. [CrossRef]
38. Brennan, C.S.; Tudorica, C.M. Fresh Pasta Quality as Affected by Enrichment of Nonstarch Polysaccharides. *J. Food Sci.* **2007**, *72*, S659–S665. [CrossRef]
39. Villeneuve, S.; Des Marchais, L.-P.; Gauvreau, V.; Mercier, S.; Do, C.B.; Arcand, Y. Effect of flaxseed processing on engineering properties and fatty acids profiles of pasta. *Food Bioprod. Process.* **2013**, *91*, 183–191. [CrossRef]
40. Mekuria, B.; Sa, E. Effects of Vital Gluten Enrichment on Qualities of Value Added Products. *J. Food Process. Technol.* **2015**, *6*, 1. [CrossRef]
41. Kaur, M.; Sandhu, K.S.; Arora, A.P.; Sharma, A. Gluten free biscuits prepared from buckwheat flour by incorporation of various gums: Physicochemical and sensory properties. *LWT Food Sci. Technol.* **2015**, *62*, 628–632. [CrossRef]
42. Sobota, A.; Wirkijowska, A.; Zarzycki, P. Application of vegetable concentrates and powders in coloured pasta production. *Int. J. Food Sci. Technol.* **2020**, *55*, 2677–2687. [CrossRef]
43. AACC. American Association of Cereal Chemistry Approved Methods, 10th ed.; St. Paul. Available online: http://methods.aaccnet.org/toc.aspx (accessed on 29 July 2020).
44. AOAC. Association of Official Analytical Chemists International. Official Methods, 20th ed.; Rockville. Available online: https://www.aoac.org/official-methods-of-analysis-21st-edition-2019/ (accessed on 29 July 2020).
45. The Polish Committee for Standarization. PKN Polish Standard. Cereal Products. Wheat Flour for Alimentary Pasta. Available online: https://infostore.saiglobal.com/STORE/Details.aspx?ProductID=340448 (accessed on 29 July 2020).
46. Zarzycki, P.; Sobota, A. Effect of storage temperature on falling number and apparent viscosity of gruels from wheat flours. *J. Food Sci. Technol.* **2015**, *52*, 437–443. [CrossRef]
47. Obuchowski, W. *Technologia Przemysłowej Produkcji Makaronu (Technology of Industrial Pasta Production)*; Wydawnictwo Akademii Rolniczej im. Augusta Cieszkowskiego: Poznań, Poland, 1997.
48. Sozer, N.; Dalgiç, A.C.; Kaya, A. Thermal, textural and cooking properties of spaghetti enriched with resistant starch. *J. Food Eng.* **2007**, *81*, 476–484. [CrossRef]
49. Tudoricǎ, C.M.; Kuri, V.; Brennan, C.S. Nutritional and physicochemical characteristics of dietary fiber enriched pasta. *J. Agric. Food Chem.* **2002**, *50*, 347–356. [CrossRef] [PubMed]
50. Chawla, R.; Patil, G.R. Soluble dietary fiber. *Compr. Rev. Food Sci. Food Saf.* **2010**, *9*, 178–196. [CrossRef]
51. Sobota, A.; Dobosz, M. Quality of whole grain pasta available in market (in Polish). *Food. Sci. Technol. Qual.* **2010**, *6*, 83–99.
52. Zarzycki, P.; Sobota, A.; Kuzawińska, E.; Wirkijowska, A.; Sykut-Domańska, E. Estimation of degree of starch gelatinisation in instant pasta using measurements of viscosity and water absorption of ground instant pasta dispersions. *Acta Agrophys.* **2017**, *24*, 625–632.
53. Brennan, M.A.; Derbyshire, E.; Tiwari, B.K.; Brennan, C.S. Enrichment of extruded snack products with coproducts from chestnut mushroom (Agrocybe aegerita) production: Interactions between dietary fiber, physicochemical characteristics, and glycemic load. *J. Agric. Food Chem.* **2012**, *60*, 4396–4401. [CrossRef]
54. Anttila, H.; Sontag-Strohm, T.; Salovaara, H. Viscosity of beta-glucan in oat products. *Agric. Food Sci.* **2004**, *13*, 80–87. [CrossRef]
55. Mäkelä, N.; Brinck, O.; Sontag-Strohm, T. Viscosity of β-glucan from oat products at the intestinal phase of the gastrointestinal model. *Food Hydrocoll.* **2020**, *100*, 105422. [CrossRef]
56. Regand, A.; Chowdhury, Z.; Tosh, S.M.; Wolever, T.M.S.; Wood, P. The molecular weight, solubility and viscosity of oat beta-glucan affect human glycemic response by modifying starch digestibility. *Food Chem.* **2011**, *129*, 297–304. [CrossRef] [PubMed]

57. Sun, T.; Li, J.; Qin, Y.; Xie, J.; Xue, B.; Li, X.; Gan, J.; Bian, X.; Shao, Z. Rheological and functional properties of oat β-glucan with different molecular weight. *J. Mol. Struct.* **2020**, *1209*, 127944. [CrossRef]
58. Zarzycki, P.; Sykut-Domańska, E.; Sobota, A.; Teterycz, D.; Krawęcka, A.; Blicharz-Kania, A.; Andrejko, D.; Zdybel, B. Flaxseed Enriched Pasta—Chemical Composition and Cooking Quality. *Foods* **2020**, *9*, 404. [CrossRef] [PubMed]
59. Sobota, A.; Zarzycki, P. Effect of Pasta Cooking Time on the Content and Fractional Composition of Dietary Fiber. *J. Food Qual.* **2013**, *36*, 127–132. [CrossRef]
60. Hajji, T.; Sfayhi-Terras, D.; El Felah, M.; Rezgui, S.; Ferchichi, A. Incorporation of β-glucans into pasta extracted from two tunisian barley cultivars. *Int. J. Food Eng.* **2016**, *12*, 701–710. [CrossRef]
61. Cassidy, Y.M.; McSorley, E.M.; Allsopp, P.J. Effect of soluble dietary fibre on postprandial blood glucose response and its potential as a functional food ingredient. *J. Funct. Foods* **2018**, *46*, 423–439. [CrossRef]
62. Jaworska, D.; Królak, M.; Przybylski, W.; Jezewska-Zychowicz, M. Acceptance of Fresh Pasta with β-Glucan Addition: Expected Versus Perceived Liking. *Foods* **2020**, *9*, 869. [CrossRef]

foods

MDPI

Article

Nutrient Composition of Different Hazelnut Cultivars Grown in Germany

Anke Katharina Müller [1,2], Ute Helms [2,3], Carsten Rohrer [1,2], Monika Möhler [4], Frank Hellwig [5], Michael Glei [2,6], Tanja Schwerdtle [7,8], Stefan Lorkowski [1,2] and Christine Dawczynski [1,2,3,*]

1 Department of Nutritional Biochemistry and Physiology, Institute of Nutritional Sciences, Friedrich Schiller University Jena, Dornburger Str. 25, 07743 Jena, Germany; anke.katharina.mueller@uni-jena.de (A.K.M.); carsten.rohrer@uni-jena.de (C.R.); stefan.lorkowski@uni-jena.de (S.L.)

2 Competence Cluster for Nutrition and Cardiovascular Health (nutriCARD) Halle-Jena-Leipzig, Dornburger Str. 25, 07743 Jena, Germany; ute.helms@uni-jena.de (U.H.); michael.glei@uni-jena.de (M.G.)

3 Junior Research Group Nutritional Concepts, Institute of Nutritional Sciences, Friedrich Schiller University Jena, Dornburger Str. 29, 07743 Jena, Germany

4 Department of Fruit Growing, Education & Research Institute of Horticulture, Leipziger Straße 75a, 99085 Erfurt, Germay

5 Institute for Systematic Botany, Friedrich Schiller University Jena, Philosophenweg 16, 07743 Jena, Germany; frank.hellwig@uni-jena.de

6 Department of Nutritional Toxicology, Institute of Nutritional Sciences, Friedrich Schiller University Jena, Dornburger Str. 24, 07743 Jena, Germany

7 Department of Food Chemistry, Institute of Nutritional Science, University of Potsdam, Arthur-Scheunert-Allee 114–116, 14558 Nuthetal, Germany; taschwer@uni-potsdam.de

8 Competence Cluster Nutrition Research (NutriAct), Berlin-Potsdam, Arthur-Scheunert-Allee 114–116, 14558 Nuthetal, Germany

* Correspondence: christine.dawczynski@uni-jena.de; Tel.: +49-3641-9-49656

Received: 27 September 2020; Accepted: 31 October 2020; Published: 3 November 2020

check for updates

Abstract: Hazelnuts are rarely cultivated in Germany, although they are a valuable source for macro- and micronutrients and can thus contribute to a healthy diet. Near the present, 15 varieties were cultivated in Thuringia, Germany, as a pilot study for further research. The aim of our study was to evaluate the micro- and macronutrient composition of representative, randomly mixed samples of the 15 different hazelnut cultivars. Protein, fat, and fiber contents were determined using established methods. Fatty acids, tocopherols, minerals, trace elements, and ultra-trace elements were analyzed using gas chromatography, high-performance liquid chromatography, and inductively coupled plasma triple quadrupole mass-spectrometry, respectively. We found that the different hazelnut varieties contained valuable amounts of fat, protein, dietary fiber, minerals, trace elements, and α-tocopherol, however, in different quantities. The variations in nutrient composition were independent of growth conditions, which were identical for all hazelnut varieties. Therefore, each hazelnut cultivar has its specific nutrient profile.

Keywords: *Corylus avellana* L.; nutrient composition; hazelnut cultivars; minerals; tocopherols

1. Introduction

Hazelnuts (*Corylus avellana* L., cobnut, including *Corylus maxima* Mill., Lambert Filbert) are popular tree nuts used in the human diet, which are mainly produced in Turkey followed by Italy and Spain [1]. Commercial hazelnut cultivation in Germany is rare, although hazelnuts are getting more popular and are often an essential ingredient in food production, e.g., in confectionery industry [2]. There are several reasons for the rare production, such as uncertain profitability forecasts for this region, long yield times, and missing organizational structures or practical experience. However, climatic conditions are changing.

Nut consumption is regularly recommended worldwide due to the beneficial health effects of nuts, and nuts are an essential part of the Mediterranean diet [3]. The prevention of cardiovascular diseases by consuming tree nuts, in particular hazelnuts, were investigated widely and indicate positive effects most notably based on improved blood lipid profiles [4]. Hazelnuts contain appreciable amounts of macronutrients, such as fat, protein, and fiber, but also micronutrients, such as minerals and vitamins. Hazelnuts have a high content of monounsaturated fatty acids (MUFA) and contain relatively small amounts of saturated fatty acids (SFA). Hazelnuts are characterized by a particularly high concentration of oleic acid with contents up to 70% of all fatty acids (FA), followed by linolic acid and palmitic acid. Schlörmann et al. measured fiber levels of 8.7% for hazelnuts, which indicate that tree nuts are a good natural source of dietary fiber [5]. This high fiber content is discussed to be partly responsible for the inverse association of nut consumption and gaining weight [6]. Other important components are vitamin E, with α-tocopherol (αTOH) as the most abundant form with up to 40.6 mg/100 g [7], and minerals such as magnesium, calcium, potassium, copper, and iron [5,8]. However, not all nuts have the same contents of these ingredients, so the recommendation is to consume a mixture with a variety of different nuts and intake amounts up to 42.5 g per day [9].

The aim of this pilot study was therefore to evaluate 15 European cultivars grown in Thuringia in Germany regarding their nutritional value and to identify the most useful cultivar for large-scale hazelnut plantings, because we expected variety-dependent nutrient differences. Hence, the protein, fiber, and fat contents of these different hazelnut cultivars were determined. In addition, the FA distribution, ash and mineral content, and data on αTOH amounts are presented.

2. Materials and Methods

2.1. Samples

In this study, varieties from different areas of origin were tested for cultivation. The varieties came from Germany, England, Spain, or Italy. The investigated 15 hazelnut (*Corylus avellana* L.) cultivars (Tonda di Giffoni, Juningia, Ennis, Cosford, Rote Lambert (Red Lambert), Englische Riesen, Webbs Preisnuss (Webb's Prize Cob), Gustav Zeller (Gustav's Zellernuss), Pauetet, Corabel, Hallesche Riesen (Hall's Giant), Wunder aus Bollweiler (Merveille de Bollwiller), Gunslebert (Gunslebener Zellernuss), Emoa-1, Eckige Barceloner (Barcelloner Zellernuss)) were grown in experimental orchards in the region of Thuringia in Germany in 2005 (Figure 1). There is an ongoing debate about the status of *Corylus maxima* Mill. Lambert Filbert, which is either classified as an individual species or is considered as *Corylus avellana*. Since there is no concluding evidence which would favor the separation as an individual species and because of reported hybridization between both taxa, we do not discriminate between both species here. For a recent critical evaluation on *Corylus* taxonomy, we refer to Holstein et al. [10]. For synonymous names of cultivars, the reader is referred to Mehlenbacher [11], NCGR-Corvallis *Corylus* catalogue [12], and information provided by the Food and Agriculture Organization of the United Nations (FAO) [13]. The technical basis for the successful cultivation of hazelnuts in Germany was laid by many years of attempts by the Bavarian Department of Food, Agriculture, and Forestry in Fürth, Germany. To allow for reliable evaluations of the different hazelnut varieties, agronomical conditions were identical for the different cultivars: five trees of each cultivar were grown by the Department of Fruit Growing, Education, and Research Institute of Horticulture (Erfurt, Germany) for 12 years on a single testing ground in Thuringia, Germany. All cultivars were grown on their own roots except for the cultivar Ennis, which was grafted on *Corylus colurna* L. All cultivars were grown under identical treatment with drip irrigation and fertigation. Cultivation took place on comparable soil covered with loess. The mean planting distance between the five trees of a cultivar was about 4.5 m between the rows and about 2.5 m between the trees. Nuts were harvested in the same season of the same year but depending on differences in maturing times of each cultivar as assessed by independent qualified experts. The crop was carried out in 2016 and varied from 0.9 kg (average amount of Ennis) to 7.6 kg (average amount of Barcelloner Zellernuss) per tree (data not shown). Selected data and observations on growth and yield of the hazelnut varieties evaluated here are presented in Supplemental Table S1. After harvesting, nuts were dried at 30 to 35 °C in the dark for at least 7 days and were stored in their shell until use. For all analyses, representative samples of 100 randomly selected and freshly grounded nuts of a cultivar were used with skin.

Figure 1. Representative exemplary pictures of the hazelnut cultivars studied.

2.2. *Quantification of Main Constituents of Hazelnuts*

All chemical analyses of the samples were done in accordance with the official methods of the Association of Official Agricultural Chemists (AOAC) [14]. Hazelnut fat content was examined using a Soxhlet extractor with petroleum ether and crude protein (Nx6.25) content was determined using a Kjeldahl apparatus. The total dietary fiber content of the different fat-free hazelnut varieties was measured according to the AOAC-certified protocol [15], using the Merck total fiber assay kit (Merck, Darmstadt, Germany). To determine the ash content, defined amounts of hazelnuts were dried and then completely incinerated in a muffle furnace at 525 °C.

2.3. *Fatty Acid Composition*

Fatty acid analysis of nuts was performed using gas chromatography (GC-17 V3; Shimadzu Corporation, Kyoto, Japan) equipped with a flame ionization detector and an autosampler (AOC-5000), as described [16]. Fatty acid concentrations were expressed as percentage of the total area of all FA methyl esters (% of total fatty acid methyl esters, FAME) using GC solution software version 2.3 (Shimadzu).

2.4. Tocopherol Determination

All high-performance liquid chromatography grade solvents and TOH standards were purchased from Merck, (Darmstadt, Germany) and LGC (Wesel, Germany), respectively. According to DIN EN 12822, HPLC (LC-20 AT; Shimadzu) was used to measure TOH concentrations of the grounded and homogenized nuts. After saponification with potassium hydroxide and extraction with n-hexane, TOH were separated on an Eurospher 100-5 Diol Vertex Plus Column 250 × 4 mm (Knauer, Berlin, Germany) with a mixture of n-hexane/methyl t-butylether (98/2 v/m) as mobile phase at a flow rate of 1.5 mL/min. Isomers of TOH were determined using a fluorescence detector (λex 295 nm, λem 330 nm; RF-10A XL; Shimadzu) and quantified using external standard calibration curves.

2.5. Quantification of Minerals, Trace and Ultra-Trace Elements

Hazelnut samples were digested with nitric acid in a closed microwave digestion system (Mars 6, CEM, Kamp-Lintfort, Germany) and multi-element quantification was carried out with an ICP-QQQ-MS 8800 mass spectrometer (Agilent, Waldbronn, Germany) [17]. Calcium, magnesium, manganese, iron, copper, zinc, and cadmium were measured on mass, and arsenic and molybdenum were analyzed in the mass-shift mode using oxygen as a reaction gas to eliminate interferences. Rhodium (1 µg/L) was used as internal standard and helium (3 mL/min) as collision gas. For selenium isotope dilution, the analysis was applied as described [18]. The nebulizer gas flow and parameters of lenses, Q1, collision cell, and Q2 were tuned daily for maximum sensitivity (oxide ratio <1.0% (140Ce16O+/140Ce+), double charged ratio <1.5% (140Ce++/140Ce+), background counts <0.1 cps). For quality assurance, the measurement blanks and recalibration check points were determined periodically every 20 samples. For the verification of the applied method, the certified fish muscle reference material ERM-BB422 (Joint Research Centre, European Commission, Geel, Belgium) was successfully analyzed.

2.6. Statistics

Replicates were measured as indicated in Tables 1–4. The results were expressed as means with standard deviation (SD) or indicated otherwise in Tables 1–4.

3. Results and Discussion

The aim of the present study was to comprehensively evaluate the nutrient profile of 15 hazelnut varieties cultivated in Germany by analyzing their protein, ash, fiber, and fat content. In addition, the FA distribution was measured and data on TOH, minerals, and trace elements were collected. The present data are in accordance with previously published data of nutrient profiles in hazelnuts. Though, most of these investigations were carried out in hazelnuts from Turkey. Local variances can be explained by differences in soil composition and weather conditions. For the nuts analyzed in the present study, agricultural conditions did not differ as they were grown on the same ground. Thus, the observed alterations in the nutrient profiles of the hazelnuts studied here likely depend on the cultivar. The data presented here indicate also a high nutritional value of the hazelnuts grown in Germany. Thus, harvesting hazelnuts in Germany, even on commercial scale, could be an interesting option for improving nutrient supply.

3.1. Fat, Crude Protein, Dietary Fiber, Moisture and Ash

Results for the nutritional properties of 15 hazelnut cultivars grown in Germany are shown in Table 1. Fat is the predominant component and the total fat content varied between the 15 cultivars. While the cultivar Red Lambert contained 64.8 g/100 g fat in relation to the fresh weight, the Corabel variety had a fat content of 47.9 g/100 g. This range is comparable with data published previously.

Savage and McNeil compared six varieties grown in New Zealand and described a fat content of 54.6 to 63.2% [19]. Later reports on varietal differences in the fat content of hazelnuts in Turkey and other regions revealed similar results [2,20–23]. Taş and Gökmen reported comparable fat contents in a range of 58.1 to 68.9% for hazelnuts harvested in Turkey in 2014 [24], while another group found somewhat lower total fat contents of 53.4 to 63.5% in six hazelnut cultivars grown in Iran in 2010 [25].

Table 1. Composition [g/100 g] of 15 hazelnut cultivars grown under identical conditions in Thuringia, Germany.

	Fat	Protein	Dietary Fiber	Ash	Moisture
Tonda di Giffoni	62.7 ± 1.0	16.5 ± 0.2	13.8 ± 0.4	2.1 ± 0.0	3.9 ± 0.0
Juningia	62.3 ± 0.7	11.7 ± 0.1	15.3 ± 0.1	1.9 ± 0.0	3.8 ± 0.0
Ennis	59.8 ± 0.6	12.4 ± 0.1	18.9 ± 1.1	2.2 ± 0.1	3.5 ± 0.0
Cosford	52.6 ± 0.8	10.2 ± 0.2	14.9 ± 0.4	2.8 ± 0.1	4.0 ± 0.4
Red Lambert	64.8 ± 0.8	10.6 ± 0.2	16.4 ± 0.8	1.9 ± 0.0	3.3 ± 0.0
Englische Riesen	51.9 ± 0.0	19.7 ± 0.2	14.9 ± 1.2	2.8 ± 0.1	4.5 ± 0.1
Webb's Prize Cob	50.9 ± 0.1	15.9 ± 0.0	22.2 ± 0.8	2.7 ± 0.0	4.5 ± 0.1
Gustav's Zellernuss	60.6 ± 1.3	14.3 ± 0.1	18.1 ± 1.8	2.6 ± 0.1	4.0 ± 0.1
Pauetet	57.7 ± 0.2	16.0 ± 0.0	14.4 ± 0.1	2.2 ± 0.0	3.9 ± 0.1
Corabel	47.9 ± 0.8	22.1 ± 0.1	14.7 ± 0.2	3.1 ± 0.0	4.4 ± 0.0
Hall's Giant	54.1 ± 1.3	18.4 ± 0.1	19.7 ± 0.7	2.5 ± 0.0	4.0 ± 0.0
Merveille de Bollweiler	54.1 ± 1.4	14.2 ± 0.2	19.5 ± 0.2	2.7 ± 0.0	4.3 ± 0.0
Gunslebener Zellernuss	50.3 ± 0.3	17.4 ± 0.2	16.7 ± 0.8	3.2 ± 0.1	4.4 ± 0.0
Emoa-1	56.9 ± 0.3	15.1 ± 0.1	13.9 ± 0.6	2.6 ± 0.0	3.9 ± 0.0
Barcelloner Zellernuss	60.2 ± 0.2	16.1 ± 0.1	13.4 ± 0.5	2.2 ± 0.0	3.9 ± 0.1

Data refer to fresh weight; values are expressed as means ± SD (n = 2).

Variations were also found for other macronutrients. The crude protein content was the highest in the Corabel variety (22.1 g/100 g), whereas the Cosford variety contained the lowest (10.2 g/100 g). These results are in line with that obtained in other studies, where protein contents of 12 to 22% have been found [2,19,21,23,26]. Amaral et al. investigated 19 cultivars grown in Portugal and reported lower protein contents ranging from 9.3 to 12.7% [20].

Compared to published values, we found higher contents of dietary fiber. The Webb's Prize Cob variety had an outstanding content of 22.2 g/100 g, whereas the Barcelloner Zellernuss variety contained only 13.4 g/100 g. Other studies reported values in the range of 9.5 to 13.2% [19,27,28]. With an average of 16.6% for the hazelnuts studied here, and especially for the varieties Webb's Prize Cob, Hall's Giant (19.7 g/100 g), and Merveille de Bollweiler (19.5 g/100 g), we found remarkably higher dietary fiber contents.

The content of ash, which allows an estimation about the mineral content, was in the range of 1.9 g/100 g (Red Lambert) to 3.2 g/100 g (Gunslebener Zellernuss). Previous reports have shown similar results [2,19]. Locatelli et al. reported slightly lower ash contents in a range of 1.30 to 2.75% [23].

3.2. Fatty Acid Composition

The tested varieties showed both differences as well as similarities regarding their FA distribution (Table 2). Palmitic acid (C16:0) accounted for only around 5% in all studied hazelnut varieties. Furthermore, proportions of palmitoleic acid (C16:1c9; data not shown), stearic acid (C18:0), and α-linolenic acid (C18:3c9,c12,c15) did not exceed 2.1%. However, there were noticeable differences in the content of the major FA. Oleic acid (C18:1c9) content varied in the range of 65.1 to 81.7%. The variety Tonda di Giffoni had the highest content (81.7%), while the variety Corabel showed the lowest value (65.1%). Next, linoleic acid (C18:3c9,c12) content differed from 10.3% in the Tonda di Giffoni variety to 26.8% in Corabel. The differences in these FA account for a cultivar-specific pattern of MUFA (66.9 to 83.0%) and polyunsaturated FA (PUFA; 10.4 to 27.0%). The total SFA content did not exceed 7.2%. The total n-3 (omega-3) PUFA content was very low for all hazelnuts (<0.2%). The total n-6 (omega-6) PUFA value varied from 10.3 (Tonda di Giffoni) to 26.8% (Corabel), depending on the cultivar.

The FA distribution of the analyzed hazelnuts were in good agreement with data previously reported [23,25]. Specific characteristics are the slightly lower levels of stearic acid in our varieties, with the highest value of 2.1% for Pauetet, while Locatelli et al. reported contents up to 4.9% [23]. In addition, five varieties (Cosford, Englische Riesen, Webb's Prize Cob, Corabel, and Gunslebener Zellernuss) have oleic acid contents of less than 70%, which are low compared to literature data. This results in an inversely related high content of linoleic acid in these varieties of >20%, which exceeds the values reported in previous publications [20–24,29]. In previous studies, mostly Turkish varieties or nuts cultivated in southern regions were examined. However, an earlier report on the FA distribution of hazelnut cultivars grown in Iran revealed values comparable to ours [25]. In this study, oleic acid content varied from 64.2 to 81.3% and linoleic acid from 10.0 to 21.1%. Bacchetta et al. reported significant differences between two crop years regarding the FA content of 75 hazelnut cultivars from different countries [30]. This indicates that multiple determinants can influence the nutrient composition and especially the proportions of FA.

Table 2. Fatty acid composition (% of total FAME [1]) of 15 hazelnut cultivars grown under identical conditions in Thuringia, Germany.

	C16:0	C18:0	C18:1 n-9	C18:2 n-6 (LA) [2]	C-18:3 n-3 (ALA) [3]	Σ SFA [4]	Σ MUFA [5]	Σ PUFA [6]	Σ n-3	Σ n-6
Tonda di Giffoni	4.5	1.8	81.7	10.3	0.1	6.5	83.0	10.7	0.1	10.3
Juningia	4.7	1.8	81.0	10.5	0.1	6.8	82.6	10.7	0.1	10.5
Ennis	5.2	1.8	77.1	13.9	0.1	7.2	78.8	14.0	0.1	13.9
Cosford	4.7	1.2	68.1	24.1	0.2	6.0	69.7	24.3	0.2	24.1
Red Lambert	4.9	1.9	80.3	11.0	0.1	7.0	81.8	11.2	0.1	11.0
Englische Riesen	4.1	1.0	69.0	24.0	0.2	5.2	70.5	24.3	0.2	24.0
Webb's Prize Cob	4.9	0.8	65.8	26.2	0.2	5.9	67.6	26.5	0.2	26.2
Gustav's Zellernuss	4.4	1.7	76.4	15.6	0.1	6.3	77.9	15.8	0.1	15.6
Pauetet	4.7	2.1	80.8	10.6	0.1	7.0	82.2	10.8	0.1	10.6
Corabel	4.9	1.0	65.1	26.8	0.2	6.1	66.9	27.0	0.2	26.8
Hall's Giant	4.3	1.7	75.2	16.9	0.1	6.3	76.6	17.1	0.1	16.9
Merveille de Bollweiler	4.4	1.4	77.0	15.3	0.1	6.0	78.6	15.5	0.1	15.3
Gunslebener Zellernuss	4.6	1.1	66.2	25.9	0.2	5.9	67.9	26.1	0.2	25.9
Emoa-1	4.6	1.8	72.1	19.5	0.1	6.6	73.6	19.8	0.1	19.5
Barcelloner Zellernuss	4.3	1.8	79.4	12.8	0.1	6.3	80.7	13.0	0.1	12.8

[1] FAME, fatty acid methyl esters. [2] LA, linoleic acid. [3] ALA, α-linolenic acid. [4] SFA, saturated fatty acids. [5] MUFA, monounsaturated fatty acids. [6] PUFA, polyunsaturated fatty acids.

3.3. Tocopherols

Table 3 shows that αTOH is the major form of vitamin E in hazelnuts. Only traces of the vitamers β-, γ-, and δTOH were detected with contents of less than 2 mg/100 g. The variety Juningia contained the highest concentration of αTOH (28.9 mg/100 g), while only 9.9 mg/100 g were found in the variety Hall's Giant. Tocopherol levels found here are in accordance with data from the literature. With a mean of 17.5 mg/100 g, the results herein are comparable to data on hazelnuts from Turkey and Portugal [27,31]. Taş and Gökmen reported the differences between two crop years and described notable decreases in the concentration of TOH for the second year for 14 varieties grown in Turkey [24]. However, another group measured TOH contents in hazelnuts grown in Poland and reported 73.90 mg/100 g αTOH for unroasted hazelnuts [32]. A study comparing nutritional values of hazelnuts mainly collected in Turkey determined total TOH contents in the range from 25.8 to 69.8 mg/100 g in hazelnut kernel oils [33]. Environmental, local, and analytical conditions are possible explanations for these remarkable differences.

Table 3. Vitamin E content [mg/100 g] of 15 hazelnut cultivars grown under identical conditions in Thuringia, Germany.

	α-Tocopherol	β-Tocopherol	γ-Tocopherol	δ-Tocopherol
Tonda di Giffoni	13.5	<0.6	<1.0	<0.6
Juningia	28.9	0.80	<1.0	<0.6
Ennis	21.2	0.60	<1.0	<0.6
Cosford	20.7	<0.6	<1.0	<0.6
Red Lambert	24.8	<0.6	2.00	<0.6
Englische Riesen	16.6	0.60	<1.0	<0.6
Webb's Prize Cob	16.3	<0.6	<1.0	<0.6
Gustav's Zellernuss	13.3	<0.6	<1.0	<0.6
Pauetet	19.2	<0.6	<1.0	<0.6
Corabel	10.9	<0.6	<1.0	<0.6
Hall's Giant	9.9	<0.6	<1.0	<0.6
Merveille de Bollweiler	11.8	<0.6	<1.0	<0.6
Gunslebener Zellernuss	18.6	<0.6	<1.0	<0.6
Emoa-1	15.6	<0.6	<1.0	<0.6
Barcelloner Zellernuss	16.1	<0.6	<1.0	<0.6

Data refer to fresh weight.

3.4. Micronutrients

Nuts are known as a good source for minerals, which can contribute to a healthy diet [8]. We measured the amount of magnesium, calcium, manganese, iron, copper, zinc, cadmium, selenium, and arsenic in the 15 hazelnut cultivars. There are considerable differences between the examined hazelnuts regarding mineral, trace, and ultra-trace element compositions (Table 4). Magnesium contents ranged from 148 ± 3 mg/100 g (Tonda di Giffoni) to 213 ± 5 mg/100 g (Merveille de Bollweiler). Köksal et al. determined 15 cultivars of hazelnuts grown in Turkey and measured similar magnesium levels of 144 to 224 mg/100 g [21]. Data from other studies confirm these results [27,34]. In addition, the nuts contained high amounts of calcium. Calcium contents ranged from 140 ± 2 (Ennis) to 247 ± 2 mg/100 g (Cosford) and fit well into the picture previously published for hazelnuts grown in Turkey [27,34]. The lowest manganese content was found in Red Lambert (0.682 ± 0.001 mg/100 g), and the highest was determined for Gunslebener Zellernuss (3.92 ± 0.02 mg/100 g). Özdemir et al. reported comparable results with a range from 1.4 to 2.6 mg/100 g for commercial Turkish hazelnuts [34]. Juningia had the lowest level of iron with a content of 2.88 ± 0.07 mg/100 g, while the highest amount was found in Gunslebener Zellernuss with 4.67 ± 0.03 mg/100 g. These values are comparable with data provided in the literature [2,21,34–36]. Contents of copper ranged from 0.764 ± 0.011 to 2.17 ± 0.03 mg/100 g with highest amounts in the variety Corabel and lowest in the variety Ennis and are similar to those reported by others [33]. Corabel contained also the highest levels of zinc (3.93 ± 0.06 mg/100 g), while the variety Juningia contained only 2.12 ± 0.003 mg/100 g; both values are comparable with those reported in the literature [21,36]. Until now, less is known about the content of the trace element molybdenum in hazelnuts. The recommended value for an adequate intake is 65 μg/day [37]. Özkutlu et al. measured between 0.09 to 0.31 mg/kg molybdenum in hazelnuts grown in Turkey [38]. Our hazelnut varieties can contribute to a sufficient intake with contents ranging from 0.109 ± 0.003 mg/100 g (Tonda die Giffoni) to 0.515 ± 0.008 mg/100 g (Merveille de Bollweiler). Levels of selenium are low in the examined samples with highest contents in Tonda di Giffoni and Pauetet with 5.10 ± 0.20 μg/100 g and 6.25 ± 0.51 μg/100 g, respectively. Hazelnuts grown in Turkey showed higher amounts of selenium [27,36], but it is important to note that the soil content for selenium is very low in Germany [39]. Cadmium was only detected at very low amounts in Gunslebener Zellernuss (0.91 ± 0.06 μg/100 g), all other samples were under the limit of quantitation (data not shown). Arsenic levels remained lower than 4 μg/100 g in all varieties, which complies with data for foods of terrestrial origin [40].

Table 4. Composition of minerals, trace, and ultra-trace elements of 15 hazelnut cultivars grown under identical conditions in Thuringia, Germany.

	Mg (mg/100 g)	Ca (mg/100 g)	Mn (mg/100 g)	Fe (mg/100 g)	Cu (mg/100 g)	Zn (mg/100 g)	Mo (mg/100 g)	Se (µg/100 g)	As (µg/100 g)
Tonda di Giffoni	148 ± 3	177 ± 4	2.78 ± 0.09	3.02 ± 0.04	1.14 ± 0.02	2.48 ± 0.03	0.109 ± 0.003	5.10 ± 0.20	1.57 ± 0.14
Juningia	155 ± 3	155 ± 1	1.22 ± 0.02	2.88 ± 0.07	0.948 ± 0.013	2.12 ± 0.003	0.310 ± 0.004	4.33 ± 0.28	0.95 ± 0.04
Ennis	162 ± 3	140 ± 2	1.17 ± 0.02	3.21 ± 0.02	0.764 ± 0.011	2.36 ± 0.05	0.414 ± 0.004	3.11 ± 0.11	2.01 ± 0.06
Cosford	178 ± 3	247 ± 2	2.94 ± 0.01	3.42 ± 0.03	1.26 ± 0.01	2.91 ± 0.03	0.256 ± 0.003	4.55 ± 0.45	2.53 ± 0.09
Red Lambert	151 ± 1	176 ± 1	0.682 ± 0.001	3.34 ± 0.03	1.21 ± 0.004	2.48 ± 0.01	0.231 ± 0.002	2.73 ± 0.21	2.57 ± 0.07
Englische Riesen	211 ± 1	241 ± 3	2.71 ± 0.02	3.09 ± 0.04	1.36 ± 0.01	2.87 ± 0.01	0.309 ± 0.002	4.49 ± 0.47	3.58 ± 0.04
Webb's Prize Cob	173 ± 3	235 ± 3	1.40 ± 0.01	3.71 ± 0.05	1.06 ± 0.02	2.92 ± 0.01	0.331 ± 0.006	3.69 ± 0.23	3.47 ± 0.12
Gustav's Zellernuss	206 ± 2	224 ± 1	2.10 ± 0.01	3.90 ± 0.004	1.84 ± 0.06	3.01 ± 0.01	0.479 ± 0.010	3.29 ± 0.27	3.81 ± 0.07
Pauetet	162 ± 2	175 ± 0.4	1.87 ± 0.02	3.73 ± 0.06	1.52 ± 0.02	3.02 ± 0.04	0.117 ± 0.001	6.25 ± 0.51	2.35 ± 0.06
Corabel	188 ± 5	212 ± 5	2.91 ± 0.05	4.26 ± 0.11	2.17 ± 0.03	3.93 ± 0.06	0.280 ± 0.005	4.79 ± 0.45	2.38 ± 0.07
Hall's Giant	182 ± 1	211 ± 14	1.67 ± 0.10	3.67 ± 0.04	1.71 ± 0.05	3.40 ± 0.03	0.297 ± 0.005	2.94 ± 0.23	2.91 ± 0.07
Merveille de Bollweiler	213 ± 5	201 ± 6	1.77 ± 0.04	3.74 ± 0.12	1.85 ± 0.05	2.88 ± 0.05	0.515 ± 0.008	3.68 ± 0.05	2.13 ± 0.05
Gunslebener Zellernuss	209 ± 3	207 ± 1	3.92 ± 0.02	4.67 ± 0.03	1.79 ± 0.04	3.43 ± 0.03	0.351 ± 0.001	4.11 ± 0.56	2.69 ± 0.03
Emoa-1	180 ± 3	225 ± 1	1.94 ± 0.04	3.49 ± 0.05	0.972 ± 0.010	2.81 ± 0.02	0.279 ± 0.004	4.23 ± 0.11	1.99 ± 0.07
Barcelloner Zellernuss	171 ± 6	232 ± 11	2.33 ± 0.10	3.99 ± 0.29	0.779 ± 0.027	2.68 ± 0.11	0.204 ± 0.003	3.83 ± 0.26	2.61 ± 0.16

Data refer to fresh weight; values are expressed as means ± SD ($n = 3$).

4. Conclusions

The data from this pilot study show differences in nutrient profiles depending on the varieties. The observed variations in nutrient composition were independent of growth conditions and climate, which were identical for all hazelnut varieties, as well as year-to-year changes. Just as growth conditions, technical processability, and resistance to pests, the nutrient content is only one of several parameters for assessing the suitability of a variety for commercial cultivation. Based on these parameters and nutrient profiles, Emoa-1, Corabel, Webb's Prize Cob, Barcelloner Zellernuss, and Merveille de Bollweiler were selected for long-term studies which are ongoing. While Corabel was the variety with the highest protein content, Webb's Prize Cob variety showed a high content of dietary fiber and Merveille de Bollweiler had the highest content of magnesium. On the other hand, Emoa-1 conceded reliably good harvests and aromatic nuts and Barcelloner Zellernuss distinguished due to a very high yield. Assessing nutrient profiles, growing behavior, and resistance to pests of these cultivars over time will allow for recommending suitable varieties, rootstocks, and cultivation systems for regional cultivation as well as changes in nutrient profiles from year to year.

Supplementary Materials: The following are available online at http://www.mdpi.com/2304-8158/9/11/1596/s1, Table S1: Agronomic data of and observations for the 15 varieties cultivated in Germany.

Author Contributions: Conceptualization: S.L., C.D.; funding acquisition: S.L., C.D.; investigation: M.M., U.H., C.R., F.H., T.S., A.K.M.; project administration: S.L., C.D.; resources: M.M., S.L.; supervision: M.G., S.L., C.D.; validation: A.K.M., S.L., C.D.; visualization: A.K.M.; writing—original draft: A.K.M.; writing—review and editing: U.H., C.R., M.M., F.H., T.S., M.G., S.L., C.D. All authors have read and agreed to the published version of the manuscript.

Funding: The work of C.D. and S.L. is supported by grants from the Federal Ministry of Education and Research (01EA1411A; 01EA1708), and the German Ministry of Economics and Technology (AiF 16642 BR) via AiF (German Federation of Industrial Research Associations), FEI (Research Association of the German Food Industry), and by the Free State of Thuringia and the European Social Fund (2016 FGR 0045). The work of T.S. is supported by grants of the Federal Ministry of Education and Research (01EA1408B), the German Research Foundation (DFG) (SCHW903/10-1) and TraceAge (DFG research unit on the interactions of essential trace elements in healthy and diseased elderly, Potsdam-Berlin-Jena, FOR 2558/1).

Acknowledgments: We thank the staff of SYNLAB Analytics and Services Germany, Jena, especially Anne Arnold and Lars Müller, for their support to analyze tocopherols. The technical assistance of Sören Meyer from the Department of Food Chemistry for measuring minerals, trace, and ultra-trace elements is highly appreciated.

Conflicts of Interest: The authors declare no conflict of interest.

References

1. FAOSTAT. *Value of Agricultural Production*; The Food and Agriculture Organization (FAO): Rome, Italy, 2016.
2. Ozdemir, F.; Akinci, I. Physical and nutritional properties of four major commercial Turkish hazelnut varieties. *J. Food Eng.* **2004**, *63*, 341–347. [CrossRef]
3. Ros, E. Nuts and CVD. *Br. J. Nutr.* **2015**, *113*, S111–S120. [CrossRef] [PubMed]
4. Perna, S.; Giacosa, A.; Bonitta, G.; Bologna, C.; Isu, A.; Guido, D.; Rondanelli, M. Effects of Hazelnut Consumption on Blood Lipids and Body Weight: A Systematic Review and Bayesian Meta-Analysis. *Nutrients* **2016**, *8*, 747. [CrossRef] [PubMed]
5. Schlormann, W.; Birringer, M.; Böhm, V.; Löber, K.; Jahreis, G.; Lorkowski, S.; Müller, A.; Schöne, F.; Glei, M. Influence of roasting conditions on health-related compounds in different nuts. *Food Chem.* **2015**, *180*, 77–85. [CrossRef] [PubMed]
6. Jackson, C.L.; Hu, F.B. Long-term associations of nut consumption with body weight and obesity. *Am. J. Clin. Nutr.* **2014**, *100*, 408S–411S. [CrossRef] [PubMed]
7. Stuetz, W.; Schlörmann, W.; Glei, M. B-vitamins, carotenoids and α-/γ-tocopherol in raw and roasted nuts. *Food Chem.* **2017**, *221*, 222–227. [CrossRef] [PubMed]
8. Ros, E. Health Benefits of Nut Consumption. *Nutrients* **2010**, *2*, 652–682. [CrossRef]
9. FDA. *Qualified Health Claims: Letter of Enforcement Discretion Nuts and Coronary Heart Disease (Docket No 02p-0505)*; Administration, F.D., Ed.; Food and Drug Administration: Silver Spring, MD, USA, 2003.

10. Holstein, N.; El Tamer, S.; Weigend, M. The nutty world of hazel names—a critical taxonomic checklist of the genus Corylus (Betulaceae). *Eur. J. Taxon.* **2018**. [CrossRef]
11. Mehlenbacher, S.A. *Betulaceae Corylus*; CAB International: Wallingford, UK, 2008.
12. NCGR-Corvallis. Ncgr-Corvallis Corylus Catalog. Available online: https://www.ars.usda.gov/ARSUserFiles/20721500/catalogs/corcore.html (accessed on 13 December 2018).
13. Köksal, A.I. *Inventory of Hazelnut Research, Germplasm and References*; FAO Regional Office for Europe, Interregional (REU/RNE), Cooperative Research Network on Nuts (ESCORENA): Rome, Italy, 2000.
14. AOAC. *Official Methods of Analysis of Aoac International*; Association of Official Analytical Chemists: Gaithersburg, MD, USA, 2012.
15. Lee, S.C.; Prosky, L.; De Vries, J.W. Determination of Total, Soluble, and Insoluble Dietary Fiber in Foods—Enzymatic-Gravimetric Method, MES-TRIS Buffer: Collaborative Study. *J. AOAC Int.* **1992**, *75*, 395–416. [CrossRef]
16. Dawczynski, C.; Schubert, R.; Jahreis, G. Amino acids, fatty acids, and dietary fibre in edible seaweed products. *Food Chem.* **2007**, *103*, 891–899. [CrossRef]
17. Meyer, S.; Markova, M.; Pohl, G.; Marschall, T.A.; Pivovarova, O.; Pfeiffer, A.F.; Schwerdtle, T. Development, validation and application of an ICP-MS/MS method to quantify minerals and (ultra-)trace elements in human serum. *J. Trace Elements Med. Biol.* **2018**, *49*, 157–163. [CrossRef] [PubMed]
18. Marschall, T.A.; Kroepfl, N.; Jensen, K.B.; Bornhorst, J.; Meermann, B.; Kuehnelt, D.; Schwerdtle, T. Tracing cytotoxic effects of small organic Se species in human liver cells back to total cellular Se and Se metabolites. *Metallomics* **2017**, *9*, 268–277. [CrossRef] [PubMed]
19. Savage, G.; McNeil, D.L. Chemical composition of hazelnuts (Corylus avellana L.) grown in New Zealand. *Int. J. Food Sci. Nutr.* **1998**, *49*, 199–203. [CrossRef]
20. Amaral, J.S.; Casal, S.; Citová, I.; Santos, A.; Seabra, R.M.; Oliveira, B.P.P. Characterization of several hazelnut (Corylus avellana L.) cultivars based in chemical, fatty acid and sterol composition. *Eur. Food Res. Technol.* **2005**, *222*, 274–280. [CrossRef]
21. Köksal, A.I.; Artik, N.; Şimşek, A.; Güneş, N. Nutrient composition of hazelnut (Corylus avellana L.) varieties cultivated in Turkey. *Food Chem.* **2006**, *99*, 509–515. [CrossRef]
22. Alasalvar, C.; Pelvan, E.; Topal, B. Effects of roasting on oil and fatty acid composition of Turkish hazelnut varieties (Corylus avellanaL.). *Int. J. Food Sci. Nutr.* **2010**, *61*, 630–642. [CrossRef]
23. Locatelli, M.; Coïsson, J.D.; Travaglia, F.; Bordiga, M.; Arlorio, M. Impact of Roasting on Identification of Hazelnut (Corylus avellana L.) Origin: A Chemometric Approach. *J. Agric. Food Chem.* **2015**, *63*, 7294–7303. [CrossRef] [PubMed]
24. Taş, N.G.; Gökmen, V. Profiling triacylglycerols, fatty acids and tocopherols in hazelnut varieties grown in Turkey. *J. Food Compos. Anal.* **2015**, *44*, 115–121. [CrossRef]
25. Rezaei, F.; Bakhshi, D.; Fotouhi Ghazvini, R.; Javadi Majd, D.; Pourghayoumi, M. Evaluation of fatty acid content and nutritional properties of selected native and imported hazelnut (corylus avellana l.) varieties grown in iran. *J. Appl. Bot. Food Qual.* **2014**, *87*, 104–107. [CrossRef]
26. Bonvehí, J.S. A chemical study of the protein fractions of Tarragona hazelnuts (Corylus avellana). *Eur. Food Res. Technol.* **1995**, *201*, 371–374. [CrossRef] [PubMed]
27. Alasalvar, C.; Shahidi, F.; Liyanapathirana, C.M.; Ohshima, T. Turkish Tombul Hazelnut (Corylus avellanaL.). 1. Compositional Characteristics. *J. Agric. Food Chem.* **2003**, *51*, 3790–3796. [CrossRef]
28. Dodevska, M.; Sobajic, S.; Djordjevic, B. Fibre and polyphenols of selected fruits, nuts and green leafy vegetables used in Serbian diet. *J. Serbian Chem. Soc.* **2015**, *80*, 21–33. [CrossRef]
29. Ciemniewska-Żytkiewicz, H.; Verardo, V.; Pasini, F.; Bryś, J.; Koczoń, P.; Caboni, M.F. Determination of lipid and phenolic fraction in two hazelnut (Corylus avellana L.) cultivars grown in Poland. *Food Chem.* **2015**, *168*, 615–622. [CrossRef]
30. Bacchetta, L.; Aramini, M.; Zini, A.; Di Giammatteo, V.; Spera, D.; Drogoudi, P.; Rovira, M.; Silva, A.P.; Solar, A.; Botta, R. Fatty acids and alpha-tocopherol composition in hazelnut (Corylus avellana L.): A chemometric approach to emphasize the quality of European germplasm. *Euphytica* **2013**, *191*, 57–73. [CrossRef]
31. Amaral, J.S.; Casal, S.; Alves, M.R.; Seabra, R.M.; Oliveira, M.B.P.P. Tocopherol and Tocotrienol Content of Hazelnut Cultivars Grown in Portugal. *J. Agric. Food Chem.* **2006**, *54*, 1329–1336. [CrossRef]

32. Marzocchi, S.; Pasini, F.; Verardo, V.; Ciemniewska-Żytkiewicz, H.; Caboni, M.F.; Romani, S. Effects of different roasting conditions on physical-chemical properties of Polish hazelnuts (Corylus avellana L. var. Katalońoski). *LWT* **2017**, *77*, 440–448. [CrossRef]
33. Matthäus, B.; Özcan, M.M. The comparison of properties of the oil and kernels of various hazelnuts from Germany and Turkey. *Eur. J. Lipid Sci. Technol.* **2012**, *114*, 801–806. [CrossRef]
34. Özdemir, M.; Açkurt, F.; Kaplan, M.; Yıldız, M.; Loker, M.; Gürcan, T.; Biringen, G.; Okay, A.; Seyhan, F.G. Evaluation of new Turkish hybrid hazelnut (Corylus avellana L.) varieties: Fatty acid composition, α-tocopherol content, mineral composition and stability. *Food Chem.* **2001**, *73*, 411–415. [CrossRef]
35. Açkurt, F.; Ozdemir, M.; Biringen, G.; Loker, M. Effects of geographical origin and variety on vitamin and mineral composition of hazelnut (Corylus avellana L.) varieties cultivated in Turkey. *Food Chem.* **1999**, *65*, 309–313. [CrossRef]
36. Simsek, A.; Aykut, O. Evaluation of the microelement profile of Turkish hazelnut (Corylus avellanaL.) varieties for human nutrition and health. *Int. J. Food Sci. Nutr.* **2007**, *58*, 677–688. [CrossRef]
37. EFSA Panel on Dietetic Products, Nutrition, and Allergies (NDA). Scientific Opinion on Dietary Reference Values for molybdenum. *EFSA J.* **2013**, *11*, 3333. [CrossRef]
38. Özkutlu, F.; Ziya Doğru, Y.; Özenç, N.; Yazici, G.; Turan, M.; Akçay, F. The importance of turkish hazelnut trace and heavy metal contents for human nutrition. *J. Soil Sci. Environ. Manag.* **2011**, *2*, 25–33.
39. Utermann, J.; Fuchs, M.; Düwel, O. *Flächenrepräsentative Hintergrundwerte für arsen, Antimon, Beryllium, Molybdän, Kobalt, Selen, Thallium, Uran und Vanadium in Böden Deutschlands aus Länderübergreifender Sicht*; Bundesanstalt für Geowissenschaften und Rohstoffe: Hannover, Germary, 2008.
40. Molin, M.; Ulven, S.M.; Meltzer, H.M.; Alexander, J. Arsenic in the human food chain, biotransformation and toxicology—Review focusing on seafood arsenic. *J. Trace Elements Med. Biol.* **2015**, *31*, 249–259. [CrossRef]

Publisher's Note: MDPI stays neutral with regard to jurisdictional claims in published maps and institutional affiliations.

foods

MDPI

Article

Effects of Genotype, Storage Temperature and Time on Quality and Compositional Traits of Cherry Tomato

Miriam Distefano †, Elena Arena †, Rosario Paolo Mauro *, Selina Brighina, Cherubino Leonardi, Biagio Fallico and Francesco Giuffrida

Dipartimento di Agricoltura, Alimentazione e Ambiente (Di3A), University of Catania, 95123 Catania, Italy; miriam.distefano@unict.it (M.D.); earena@unict.it (E.A.); selina.brighina@unict.it (S.B.); cherubino.leonardi@unict.it (C.L.); biagio.fallico@unict.it (B.F.); francesco.giuffrida@unict.it (F.G.)

* Correspondence: rosario.mauro@unict.it

† Both authors contributed equally to this work.

Received: 30 October 2020; Accepted: 20 November 2020; Published: 25 November 2020

check for updates

Abstract: The experiment addressed the effects of two storage temperatures, namely 10 (T_{10}) and 20 °C (T_{20}), on main quality and functional traits of three cherry tomato cultivars ('Eletta', 'Sugarland' and 'Ottymo'), after 0 (S_0), 7 (S_7) and 14 (S_{14}) days of storage. At T_{10} both fruit weight and firmness were better retained during storage. At S_{14}, T_{10} promoted fruit Chroma and overall fruit color deviation (ΔE^*_{ab}). Total polyphenols content (TPC) of fruits peaked at S_7 (4660 mg GAE kg^{-1} DW) then declined at S_{14} (by 16%), with the highest values recorded at T_{10}. Lycopene showed a similar trend, but with a higher average concentration recorded at T_{20} (488 mg kg^{-1} DW). β-carotene content peaked at S_{14}, irrespective of the storage temperature. At S_{14}, the concentrations of phytoene and phytofluene were higher at T_{20} (48.3 and 40.9 mg kg^{-1} DW, respectively), but the opposite was found at S_7. 'Sugarland' and 'Ottymo' showed the highest ΔE^*_{ab} along storage, with the former cultivar proving the highest TPC and lycopene content, whereas 'Eletta' did so for phytoene and phytofluene. Our results suggest that unravelling the possible functional interactions among these three carotenoids would allow for a better orientation of breeding programs, targeting the phytochemical evolution of tomatoes during refrigerated storage.

Keywords: *Solanum lycopersicum* L.; refrigerated storage; fruit quality; carotenoids; total polyphenols content

1. Introduction

Tomato (*Solanum lycopersicum* L.) is one of most important vegetable crops throughout the world, with an estimated production of 182 Mt from more than 4.8 Mha cropland [1]. In the Mediterranean basin it is the primary field and greenhouse vegetable crop [2], since tomato strongly characterizes the Mediterranean diet, hence its consumption is widely spread around this macroarea [3]. Fresh tomatoes commercialization is often characterized by significant temporal gaps among production and consumption. This implies the optimization of quality maintenance of the product along the distribution chain, in order to match the consumers' sensorial and nutritional demands [4]. Indeed, fresh vegetables are perishable commodities, whose postharvest decay represents a primary matter of social concern in terms of economic (loss of capital, fuel and manpower) and environmental costs (due to landfilling), associated to losses of valuable phytonutrients [5]. In this context, temperature is a key factor to extend quality of fresh horticultural products along the distribution chain [6]. Because of its sensitivity to chilling injuries [7], the optimization of tomatoes cold storage implies a compromise between temperatures low enough to slow down the ripening process but high enough to generate either no or tolerable side effects on the main organoleptic and nutritional traits [8]. Similarly to other plant foods [9], tomato is a source of many valuable phytonutrients having potential health benefits, including minerals, vitamins C and E, organic acids, polyphenols and carotenoids [10]. Carotenoids represent by far the most studied phytochemical fraction of tomatoes [11], which are considered the main dietary source of lycopene [12], i.e., the prevailing constituent conferring the typical pigmentation to red-ripe fruits. From a nutritional viewpoint, lycopene is a powerful antioxidant, whose intake has been linked to reduced frequency and severity of several types of cancer and heart diseases [13]. Moreover, it has been indicated as the most effective singlet oxygen quencher among all known carotenoids [14]. β-carotene is the second main carotenoid constituent of tomato fruits [15]. It is a red-orange pigment having strong chemoprotective functions and the highest provitamin A activity in the human metabolism, and its deficiency can result in xerophthalmia, blindness, and even premature death [16]. Although both carotenoids can be specifically ingested through dietary supplements, scientific evidences seem to point out stronger health benefits associated to their direct assumption from tomato matrices, likely as a consequence of synergistic effects involving other naturally occurring compounds [17]. Among these, the colorless carotenoids phytoene and phytofluene have been supposed to have biological activity, as in the case of skin protection from UV-induced erythema or in the protection of human lipoproteins from oxidation [18].

Over the last decades, cherry tomato has been intensively targeted in breeding programs of many seed companies, in order to match the evolving standards in tomato production, commercialization and consumption [19]. Consequently, the currently available cultivars are characterized by better functional profile than the past [20], wide compositional variability [21] and rapid temporal turnovers. Such diversification and dynamism represent a challenging task to optimize the product management along the distribution chain, since postharvest quality modifications are strongly affected by both storage conditions and genotype [22]. Hence, to address the growing demands for tomatoes with high quality and functional profiles, it is appropriate to in-depth the knowledge about whole patterns of change in these properties, as a function of the storage conditions applied to the emerging germplasm.

Due to this, the aim of the present work was to investigate the postharvest modifications on main quality variables of three recently widespread cherry tomato cultivars in a Mediterranean environment induced by different thermal regimes (10 and 20 °C) and storage time (up to 14 days).

2. Materials and Methods

2.1. Experimental Site and Plant Material

A greenhouse experiment was carried out from February to June 2019, at the experimental farm of the University of Catania (Sicily, South Italy: 37°24′27″ N, 15°03′36″ E, 6 m a.s.l.). The climate of the area is semi-arid Mediterranean, with mild winters and hot, dry summers. An 800 m^2, multi-aisle cold greenhouse was used, having a steel tubular structure with adjustable windows on the roof and along the sides, and covered with polycarbonate slabs. Three cherry tomato cultivars, namely 'Eletta', 'Ottymo' and 'Sugarland', recently diffused in the reference area, were grown in the experiment, chosen on the basis of their different main carpometric traits (Table 1). To this end, data were previously acquired from different local farms operating in comparable growth conditions.

Table 1. Main information and fruit characteristics related to the cultivars selected for the study.

	'Eletta'	'Ottymo'	'Sugarland'
Seed company	TSI Italia srl, Foggia (FG), Italy	Vilmorin Italia srl, Funo (BO), Italy	Rijk Zwaan Italia srl, Bologna (BO), Italy
Fruit color	Deep red	Red	Deep red
Average fruit diameter (mm)	15 ± 2	18 ± 2	12 ± 1
Average fruit weight (g)	15.0 ± 1.5	20.5 ± 2.5	12.0 ± 1.0

2.2. Growth Conditions, Fruit Sampling and Storage

Plants were transplanted on 11th February 2019 within the greenhouse at the stage of two true leaves, in an open soilless cultivation system using 5 L plastic pots (20 cm height, 19 cm width), with perlite as growing medium (particle size 2–6 mm). Before transplanting, plantlets were selected for uniform size and health appearance, whereas pots were arranged in simple rows, adopting a 0.40 × 1.00 m rectangular format (center to center) and 1 plant per pot (2.5 plants m^{-2}). Plants were grown at single stem up to the 8th cluster, whereas all clusters were pruned leaving 12 fruits, whose setting was allowed by using bumblebee hives. Each net experimental unit contained 12 plants. During the cycle, the crop was uniformly fertigated with a standard nutrient solution [23], adopting a leaching fraction of at least 75%, to avoid root zone salinization [24].

From 14 to 16 May, tomatoes belonging to the 4th cluster were harvested by hand at the red stage (stage F) according to Gautier et al. [25]. This was done to allow tomatoes to reach stage G (deep red) during postharvest, as it is usual among local growers. Soon after harvest, fruits were transported to the laboratory and processed for further analysis. Overall, 72 clusters were collected (8 clusters × 3 cultivars × 3 replicates) and divided in 3 batches for the characterization of fruits after 0 (harvest date), 7 and 14 days of storage (hereafter S_0, S_7 and S_{14}, respectively), either at 10 ± 0.5 (T_{10}) or 20 ± 0.5 °C (T_{20}) and 85% relative humidity (RH). The lowest thermal regime was chosen since it represents a mild stressing conditions frequently adopted during transportation and storage of cherry tomatoes, whereas T_{20} was comparatively chosen as it simulates storage at room temperature [26]. Before storage, fruits were detached from rachis, selected for absence of defeats and uniform appearance within each genotype, washed with deionized water and dried with paper for further analysis. Fifteen to twenty-two fruits per replicate were placed in common commercial trays, i.e., transparent PET trays Mod. C500/41p (190 × 115 × 41 mm) covered with a perforated PET LC32 lid (Carton Pack s.p.a., Rutigliano, Italy) for a final net weight of 250 ± 8 g, then stored at the abovementioned conditions.

2.3. Carpometric Determinations

At each time point, fruit fresh weight was determined on 10 fruits per tray, then their firmness was determined through a Digital Texture Analyser mod. TA-XT2 (Stable Micro Systems, Godalming, UK) and defined as the force (N) needed to impress a 2 mm fruit deformation along the polar axis, between two steel plates.

2.4. Cherry Tomato Quality Variables

For each sample, ~50 g of cherry tomato were homogenized up to a puree in a home blender (La Moulinette, Groupe SEB, Écully, France) and immediately analyzed for: soluble solids content, dry matter, pH, total acidity (TA), reducing sugars, total polyphenols and carotenoids profile and content.

The soluble solids content (SSC) was estimated with an Abbe refractometer 16531 (Carl Zeiss, Oberkochen, Germany) at 20 °C and the results were expressed as °Brix. The dry matter was determined by gravimetric analysis. An aliquot of cherry tomato puree were placed in an oven at 70 °C (Thermo Fisher Scientific, Waltham, MA, USA) until the constant weight [27]. The pH was measured using a pHmeter (Mettler Toledo, MP 220), and tritatable acidity (TA) was determined by titrating an aliquot of the puree sample with 0.05 N NaOH to pH 8.1. TA was expressed as g kg^{-1} of cherry tomato fresh weight (FW), as citric acid.

Reducing sugars (fructose and glucose), were estimated using Fehling's method according to the official Italian method of analysis (D.M. 3.2.1989, GU n.168/1989). An aliquot of the puree sample (20 g) were transferred into a volumetric flask (50 mL) and neutralized with 1 N NaOH. Subsequently sample was cleared by the addition of 10 mL saturated sodium sulphate decahydrate and 5 mL saturated basic lead acetate. The samples were diluted up to 50 g with distilled water, mixed and centrifuged for 10 min at 10,000 rpm. The supernatant was filtered through a filter paper (Whatman No 1, Whatman

International, Maidstone, UK) and used to completely reduce in hot condition a mixed of the Fehling's solution using methylene blue solution as indicator. The Fehling solution was prepared as follow: 5 mL of each stock Fehling solution A and B were mixed with 40 mL of distilled water immediately before the determination. Results were expressed as g of reducing sugars kg^{-1} of dry weight (DW) and all analysis were conducted in triplicate.

2.5. Fruit Chromatic Coordinates

The fruit chromatic coordinates were measured as described by McGuire [28] on the equatorial axis of whole fruits (two measurements per fruits), through a tristimulus Minolta Chroma meter (model CR-200, Minolta Corp.) calibrated with a standard white tile (UE certificated) with illuminant D65/10°, measuring color in terms of lightness (L^*), green-red axis (a^*) and blue-yellow axis (b^*). Fruit color was described as $(a^*/b^*)^2$, Chroma [as $(a^{*2} + b^{*2})^{1/2}$], tomato color index [$TCI = 2000\ a^*/L^*(a^{*2} + b^{*2})^{1/2}$] and total color difference [$\Delta E^*_{ab} = (\Delta L^{*2} + \Delta a^{*2} + \Delta b^{*2})^{1/2}$], this last describing the color deviation recoded at S_7 and S_{14}.

2.6. Total Polyphenols Content

The extraction of polyphenol compounds was performed according to Atanasova et al. [29] with some modifications. An aliquot of cherry tomato puree sample (1 g) was mixed and shacked with 40 mL of acetone (80% solution in distilled water) and left in the dark, overnight at room temperature. After that, each sample was filtered (0.45 μm Albet) and the supernatant was collected for determination of total polyphenols content (TPC). This was determined according to Gahler et al. [30] using the Folin-Ciocâlteu reagent and measuring spectrophotometrically the absorbance at 725 nm using a Perkin Elmer lambda 25 Uv-Vis spectrometer. Gallic acid was used as standard (standard curve, 0.29–8.18 mg kg^{-1}; $R^2 = 1.00$) and TPC was expressed as mg gallic acid equivalents (GAE) kg^{-1} on a dry weight (DW) basis. All analyses were carried out in triplicate.

2.7. Carotenoids Extraction and HPLC Analysis

Tomato carotenoids were extracted using the method of Siracusa et al. [31]. An aliquot of the cherry tomato puree sample (0.5 g) was transferred into a vial and 5 mL of a n-hexane/acetone/ethanol (2:1:1) solution were added. The vial was left shaking for 40 min. in the dark at room temperature. Subsequently 1 mL of H_2O (HPLC grade) was added and a further 2 min. agitation was applied. The resulting heterogeneous mixtures were left decanting until phases separation. The apolar coloured layers were transferred into an amber vial and analysed.

Quantitative analyses were carried out on an HPLC (Shimadzu USA Manufacturing Company Inc., Class VPLC-10 Dvp, Canby, OR, USA) equipped with a DAD (Shimadzu SPD-M10Avp). The column was a Gemini NX C18 (150 × 4.6 mm; 3 μm particle size; Phenomenex, Italy), fitted with a guard cartridge packed with the same stationary phase. The flow rate was 0.7 mL/min. and the injector volume was 20 μL. Carotenoids were eluted with the following gradient of A (Methanol: H_2O 75:25) and B (Ethyl acetate): T0 30% B; T15 82% B; T25 30% B. All reagents used were HPLC purity grade: water, methanol and Ethyl acetate were obtained from Merck. The wavelength range was 220–660 nm, and the chromatograms were monitored at 473 nm for lycopene; at 453 nm for β-carotene; at 348 nm for phytofluene and at 288 nm for phytoene. Carotenoids were identified by splitting the peak of the carotenoids from the tomato-solution sample with a standard of β-carotene and lycopene; ($p \geq 95\%$ and $p \geq 98\%$, Sigma-Aldrich, St. Louis, MO, USA) and by comparing retention times and UV spectra with those of standards. Quantification of β-carotene and lycopene was performed using external calibration curves; for phytofluene and phytoene the calibration curve of β-carotene was used. Linearity was checked for β-carotene between 3.36 and 21 mg kg^{-1} ($R^2 = 1.00$) and lycopene between 2.56 and 40.0 mg kg^{-1} ($R^2 = 1.00$). All analyses were performer in triplicate, including the extraction procedure, and the results were expressed as mg kg^{-1} DW.

2.8. Statistical Procedures

All data were subjected to Shapiro-Wilk and Levene's test, in order to check for normality and homoscedasticity, respectively, then to a factorial 'storage temperature × genotype × storage time' (T × G × S) analysis of variance (ANOVA), according to the experimental layout adopted in the experiment. Percentage data were Bliss transformed before the ANOVA (untransformed data are reported and discussed), whereas multiple means comparisons were performed through Tukey's honestly significant difference (HSD) test ($p \leq 0.05$). All calculations were performed using Excel version 2016 (Microsoft Corporation, Redmond, WA, USA) and Minitab version 16.1.1 (Minitab Inc., State College, PA, USA).

3. Results

In the present study, the significance resulting from the ANOVA related to storage temperature (T), genotype (G) and storage time (S) and their first order interactions is reported in Table 2 (Fisher-Snedecor F-test), whereas their effects on variable means are reported in Tables 3–6 and Figures 1–3.

Table 2. *F*-test values of the main factors and their first order interactions related to observed variables, with the significance resulting from the analysis of variance.

Variable	Source of Variation					
	Storage Temperature	Genotype	Storage Time	(T) × (G)	(T) × (S)	(G) × (S)
Average fruit weight	13.6 ***	140.8 ***	65.6 ***	3.8 *	NS	8.1 ***
Fruit dry matter	10.6 **	88.4 ***	29.7 ***	3.5 *	NS	8.7 ***
Fruit firmness	8.7 **	29.4 ***	11.5 ***	3.9 *	NS	NS
Reducing sugars content	NS	50.2 ***	10.2 ***	NS	NS	3.8 *
SSC/TA	NS	265.2 ***	4.7 *	NS	NS	12.4 ***
Fruit pH	NS	19.6 ***	NS	NS	NS	NS
$(a^*/b^*)^2$	NS	38.0 ***	6.3 **	NS	NS	NS
Chroma	39.4 ***	544.7 ***	30.9 ***	NS	13.2 ***	NS
Tomato color index	NS	39.9 ***	4.9 *	NS	NS	NS
ΔE^*_{ab}	8.1 **	15.1 ***	5.7 *	6.0 **	NS	5.0 *
Total polyphenols content	20.9 ***	9.5 ***	56.4 ***	17.2 ***	4.7*	32.5 ***
Phytoene content	NS	82.8 ***	7.5 **	NS	3.6 *	15.1 ***
Phytofluene content	NS	44.3 ***	6.5 **	6.3 **	3.9 *	6.8 ***
Lycopene content	33.8 ***	1462.3 ***	138.5 ***	3.6 *	9.8 ***	121.4 ***
β-carotene content	NS	17.4 ***	23.1 ***	NS	NS	11.7 ***

SSC: soluble solids content; TA: tritatable acidity. (T): storage temperature; (G): genotype; (S): storage time. NS: not significant; *, ** and ***: significant at $p \leq 0.05$, 0.01 and 0.001, respectively.

3.1. Carpometric Traits

Average fruit weight showed a significant 'T × G' interaction since, passing from T_{10} to T_{20}, 'Ottymo' and 'Sugarland' showed the highest reduction (−9%, on average) (Table 3). Moreover, both cultivars proved the highest decline of fruit weight at the end of the storage period, as this variable was reduced by 28%, on the average of both cultivars (Figure 1A).

Fruit dry matter, proved a higher value at T_{20} than at T_{10}, reaching the highest rise among the thermal regimes in 'Ottymo' (+15%) and 'Sugarland' (+12%) (Table 3). Both genotypes highlighted the highest rise during the storage period, as their fruit dry matter increased by 44% on average, passing from S_0 to S_{14} (Figure 1B). Differently, at T_{20} fruit firmness was significantly reduced, with 'Ottymo' showing the strongest decline passing from T_{10} to T_{20} (−19%) (Table 3). For this variable, a decreasing trend was recorded along the storage period, since, by comparison with the initial value, fruit firmness was reduced by 22% at S_{14} (Table 3).

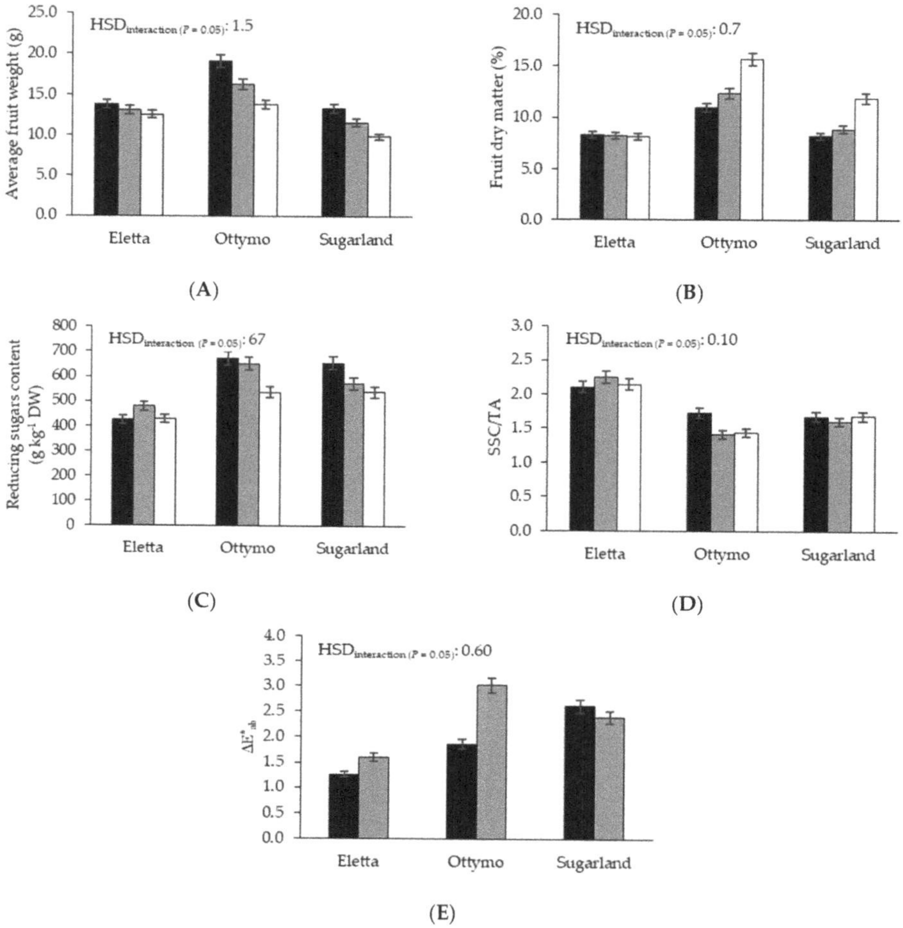

Figure 1. Average fruit weight (**A**), fruit dry matter (**B**), reducing sugars content (**C**), SSC/TA (**D**) and ΔE^*_{ab} (**E**) as affected by 'genotype × storage time' interaction. Black bars: S_0; grey bars: S_7; white bars S_{14}.

Table 3. Carpometric variables of cherry tomato as affected by the main factors.

Variable		Genotype			Storage Time			Storage Temperature Mean
		'Eletta'	'Ottymo'	'Sugarland'	S_0	S_7	S_{14}	
Average fruit weight (g)	T_{10}	13.3	17.2	12.0	15.3	14.1	12.9	14.1 [a]
	T_{20}	13.0	15.5	11.1	15.5	13.6	12.0	13.2 [b]
	Mean	13.2 [b]	16.4 [a]	11.6 [c]	15.4 [a]	13.9 [b]	12.5 [c]	
$HSD_{interaction}$			0.8			0.8		
Fruit dry matter content (%)	T_{10}	8.2	12.1	9.1	9.1	9.4	10.9	9.8 [b]
	T_{20}	8.2	13.9	10.2	9.2	10.3	12.9	10.8 [a]
	Mean	8.2 [c]	13.0 [a]	9.7 [b]	9.2 [b]	9.8 [b]	11.9 [a]	
$HSD_{interaction}$			1.1			NS		
Fruit firmness (N)	T_{10}	14.1	19.7	12.8	16.5	16.1	14.1	15.6 [a]
	T_{20}	13.6	16.0	11.6	16.6	12.9	11.7	13.7 [b]
	Mean	13.8 [b]	17.9 [a]	12.2 [c]	16.6 [a]	14.5 [b]	12.9 [c]	
$HSD_{interaction}$			2.2			NS		

Different letters among factor means indicate significance at Tukey's HSD test ($p \leq 0.05$). Interaction values ($p = 0.05$) related to 'storage temperature × genotype' and 'storage temperature × storage time' are reported. NS: not significant.

3.2. Cherry Tomato Quality Variables

On the average of the other factors, 'Ottymo' and 'Sugarland' proved the highest reducing sugars content (581 g kg^{-1} DW, on average), whereas the former cultivar proved the lowest pH; differently, 'Sugarland' minimized the SSC/TA ratio (Table 4). Both reducing sugars content and SSC/TA declined passing from S_0 to S_{14} (by 14 and 4%, respectively). For the former variable, the strongest reduction along the storage period was noticed in 'Ottymo' and 'Sugarland' (−19%, on average) (Figure 1C), whereas for SSC/TA the only significant reduction within the S_0–S_{14} period was found in 'Ottymo' (−17%) (Figure 1D).

Table 4. Compositional variables related to fruit taste of cherry tomato as affected by the main factors.

Variable		Genotype			Storage Time			Storage Temperature Mean
		'Eletta'	'Ottymo'	'Sugarland'	S_0	S_7	S_{14}	
Reducing sugars content (g kg^{-1} DW)	T_{10}	434	595	552	557	555	463	527 ^{a}
	T_{20}	420	597	579	565	536	499	529 ^{a}
	Mean	427 ^{b}	596 ^{a}	565 ^{a}	561 ^{a}	545 ^{a}	481 ^{b}	
$HSD_{interaction}$			NS			NS		
SSC/TA (adimensional)	T_{10}	2.15	1.65	1.52	1.80	1.77	1.71	1.77 ^{a}
	T_{20}	2.18	1.66	1.54	1.86	1.74	1.80	1.79 ^{a}
	Mean	2.16 ^{a}	1.65 ^{b}	1.53 ^{c}	1.83 ^{a}	1.76 ^{b}	1.76 ^{b}	
$HSD_{interaction}$			NS			NS		
Fruit pH	T_{10}	4.59	4.21	4.33	4.37	4.35	4.42	4.38 ^{a}
	T_{20}	4.61	4.22	4.38	4.38	4.42	4.42	4.40 ^{a}
	Mean	4.60 ^{a}	4.21 ^{b}	4.36 ^{ab}	4.37 ^{a}	4.38 ^{a}	4.42 ^{a}	
$HSD_{interaction}$			NS			NS		

Different letters among factor means indicate significance at Tukey's HSD test ($p \leq 0.05$). Interaction values ($p = 0.05$) related to 'storage temperature × genotype' and 'storage temperature × storage time' are reported. NS: not significant.

3.3. Chromatic Variables

Among the chromatic variables, Chroma and ΔE^*_{ab} showed a similar response to storage temperature, as they were both increased at T_{10} (by 4% and 27%, respectively) (Table 5). For Chroma, the increase under cold storage was particularly evident passing from S_7 (24.1) to S_{14} (26.0, +8%) (Table 5). Among the studied genotypes, 'Eletta' showed the highest $(a^*/b^*)^2$ and Chroma (0.81 and 26.4, respectively) and the lowest ΔE^*_{ab} (1.43), whereas the lowest TCI was found in 'Sugarland' (31.3) (Table 5). All the chromatic variables significantly increased between S_7 and S_{14}, but for ΔE^*_{ab} such temporal rise was more prominent in 'Ottymo' (by 61%) (Figure 1E).

Table 5. Chromatic variables of the epicarp of cherry tomato as affected by the main factors.

Variable		Genotype			Storage Time			Storage Temperature Mean
		'Eletta'	'Ottymo'	'Sugarland'	S_0	S_7	S_{14}	
$(a^*/b^*)^2$	T_{10}	0.79	0.74	0.59	0.72	0.67	0.73	0.71 ^{a}
	T_{20}	0.84	0.71	0.63	0.72	0.67	0.78	0.73 ^{a}
	Mean	0.81 ^{a}	0.73 ^{b}	0.61 ^{c}	0.72 ^{a}	0.67 ^{b}	0.75 ^{a}	
$HSD_{interaction}$			NS			NS		
Chroma	T_{10}	26.8	26.5	21.1	24.5	24.1	26.0	24.8 ^{a}
	T_{20}	26.1	25.2	20.3	24.3	23.1	24.1	23.8 ^{b}
	Mean	26.4 ^{a}	25.9 ^{b}	20.7 ^{c}	24.4 ^{b}	23.6 ^{c}	25.1 ^{a}	
$HSD_{interaction}$			NS			0.7		
TCI	T_{10}	33.9	34.9	31.1	33.1	32.9	33.8	33.3 ^{a}
	T_{20}	34.1	34.2	31.6	33.1	32.6	34.0	33.3 ^{a}
	Mean	34.0 ^{a}	34.5 ^{a}	31.3 ^{b}	33.1 ^{ab}	32.7 ^{b}	33.9 ^{a}	
$HSD_{interaction}$			NS			NS		
ΔE^*_{ab}	T_{10}	1.37	2.65	3.11	-	2.18	2.57	2.38 ^{a}
	T_{20}	1.49	2.36	1.77	-	1.64	2.10	1.87 ^{b}
	Mean	1.43 ^{b}	2.50 ^{a}	2.44 ^{a}	-	1.91 ^{b}	2.34 ^{a}	
$HSD_{interaction}$			0.6			NS		

Different letters among factor means indicate significance at Tukey's HSD test ($p \leq 0.05$). Interaction values ($p = 0.05$) related to 'storage temperature × genotype' and 'storage temperature × storage time' are reported. NS: not significant; -: no data.

3.4. Total Polyphenols Content

Total polyphenols content (TPC) was significantly higher at T_{10} (4327 mg GAE kg^{-1} DW) that at T_{20} (4034 mg GAE kg^{-1} DW) (Table 6), but with strong interactive effects with genotype and storage time. Indeed, while 'Eletta' showed no differences among the 2 thermal regimes, TPC was strongly promoted by the lowest thermal regime in 'Sugarland' (+20%), followed by 'Ottymo' (+9%) (Table 6). As regards its temporal trend, TPC significantly increased passing from S_0 to S_7 (+17%) then sharply declined at S_{14} (−16%), with a steeper rise in the S_0–S_7 period recorded at T_{10} (+22%) than at T_{20} (+12%) (Table 6). Moreover, the studied genotypes displayed different time-courses of TPC along the storage period, since 'Sugarland' proved the highest TPC rise passing from S_0 to S_7 (+37%) followed by the strongest decline at S_{14} (−33%) (Figure 3A).

3.5. Carotenoids Content

Figure 2 shows the HPLC carotenoids profile extracted from cherry tomato 'Sugarland'. At harvest date, the level of lycopene ranged from 68.1 to 582.5 mg kg^{-1} DW in 'Ottymo' and 'Sugarland', respectively, followed by β-carotene, which varied from 72.8 to 82.17 mg kg^{-1} DW, in 'Ottymo and 'Eletta', respectively. Among the genotype tested, 'Eletta' proved the highest levels of both phytoene and phytofluene (54.2 and 50.7 mg kg^{-1} DW, respectively). The levels determined in 'Sugarland' and 'Ottymo' varied from 31.0 to 38.2 mg kg^{-1} DW, for phytoene and from 36.1 to 39.9 mg kg^{-1} DW, for phytofluene, respectively.

The phytoene content of the studied genotypes proved different time courses among the 2 thermal regimes, as it significantly increased passing from S_7 to S_{14} when the T_{20} storage was considered (from 41.2 to 48.3 mg kg^{-1} DW, +14%) (Table 6). Among the genotypes, 'Sugarland' proved the highest phytoene rise passing from S_0 to S_7 (from 36.2 to 46.1 mg kg^{-1} DW, +28%), whereas in 'Ottymo' a significant increase was recorded between S_7 (35.1 mg kg^{-1} DW) and S_{14} (45.3 mg kg^{-1} DW, +29%) (Figure 3B).

Table 6. Nutraceutical variables of cherry tomato as affected by the main factors.

Variable		Genotype			Storage Time			Storage Temperature Mean
		'Eletta'	'Ottymo'	'Sugarland'	S_0	S_7	S_{14}	
TPC (mg GAE kg^{-1} FW)	T_{10}	4087	4531	4364	3997	4869	4116	4327 ^{a}
	T_{20}	4287	4167	3647	3982	4452	3668	4034 ^{b}
	Mean	4187 ^{b}	4349 ^{a}	4006 ^{c}	3989 b	4660 a	3892 b	
$HSD_{interaction}$			303			303		
Phytoene content (mg kg^{-1} FW)	T_{10}	50.4	41.5	41.2	43.3	45.5	45.1	44.3 ^{a}
	T_{20}	53.5	38.5	41.3	43.1	41.2	48.3	44.5 ^{a}
	Mean	51.9 ^{a}	40.0 ^{b}	41.2 ^{b}	43.2 ^{b}	43.3 ^{b}	46.7 ^{a}	
$HSD_{interaction}$			NS			3.2		
Phytofluene content (mg kg^{-1} FW)	T_{10}	42.1	38.7	32.3	39.6	37.7	35.6	37.7 ^{a}
	T_{20}	46.5	34.6	33.9	39.8	33.4	41.9	38.3 ^{a}
	Mean	44.3 ^{a}	36.6 ^{b}	33.1 ^{c}	39.7 ^{a}	35.5 ^{c}	38.8 ^{b}	
$HSD_{interaction}$			3.5			3.5		
Lycopene content (mg kg^{-1} FW)	T_{10}	488	185	662	392	526	404	445 ^{b}
	T_{20}	556	207	701	416	577	484	488 ^{a}
	Mean	552 ^{b}	196 ^{c}	682 ^{a}	404 ^{c}	552 ^{a}	444 ^{b}	
$HSD_{interaction}$			35			35		
β-carotene content (mg kg^{-1} DW)	T_{10}	92.0	85.2	83.4	77.0	89.7	94.0	86.9 ^{a}
	T_{20}	96.0	81.7	77.1	76.7	85.9	92.3	84.9 ^{a}
	Mean	94.0 ^{a}	83.5 ^{b}	80.3 ^{b}	76.8 ^{c}	87.8 ^{b}	93.1 ^{a}	
$HSD_{interaction}$			NS			NS		

Different letters among factor means indicate significance at Tukey's HSD test ($p \leq 0.05$). Interaction values ($p = 0.05$) related to 'storage temperature × genotype' and 'storage temperature × storage time' are reported. NS: not significant.

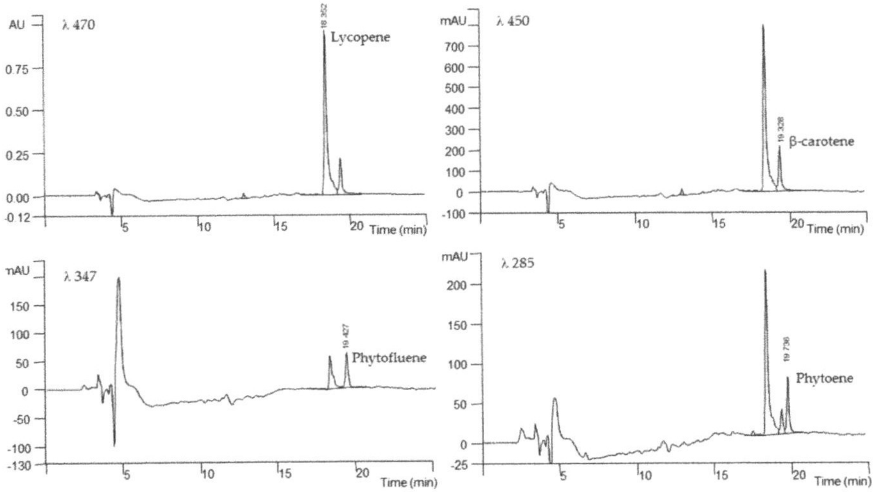

Figure 2. HPLC profile of carotenoids extracted from cherry tomato 'Sugarland' at harvest date (S_0).

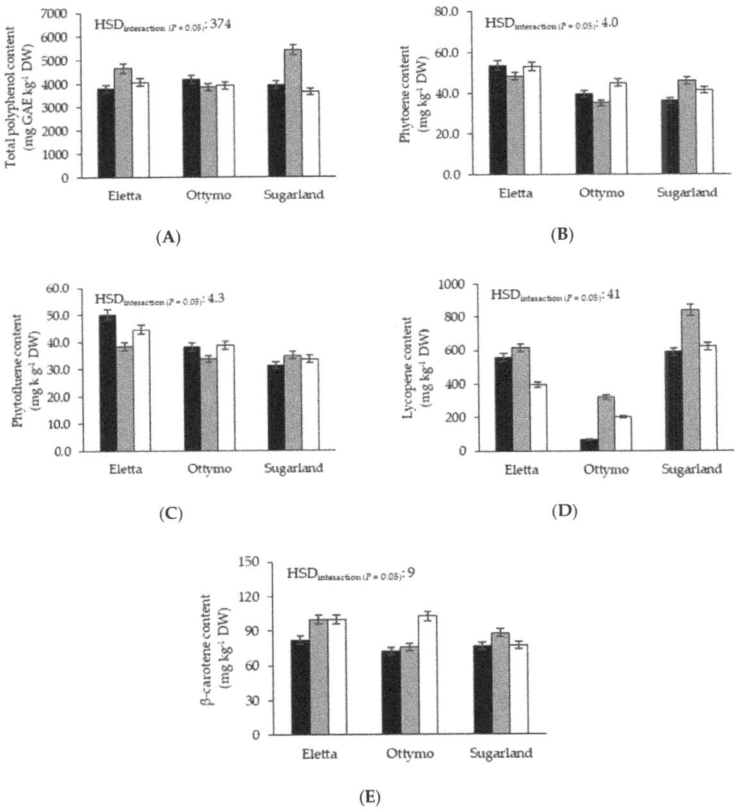

Figure 3. Total polyphenols (**A**), phytoene (**B**), phytofluene (**C**), lycopene (**D**) and β-carotene (**E**) content as affected by 'genotype × storage time' interaction. Black bars: S_0; grey bars: S_7; white bars S_{14}.

Regarding phytofluene, the lowest storage temperature showed a depressive effect in 'Eletta' (in which it was reduced by 9%) and the opposite in 'Ottymo' (in which it increased by 10%) (Table 6). Phytofluene content proved also wider temporal oscillations at T_{20}, as the initial value was reduced by 6.4 mg kg^{-1} DW at S_7 (−9%), then increased by 6.4 mg kg^{-1} DW at S_{14} (+19%) (Table 6). Such temporal oscillations proved to be genotype-dependent too, since 'Eletta' showed the highest reduction passing from S_0 (50 mg kg^{-1} DW) to S_7 (38.4 mg kg^{-1} DW, −23%), then the sharpest rise at S_{14} (44.5 mg kg^{-1} DW, +16%) (Figure 3C).

Lycopene was significantly affected by the storage temperature, as it was lower at T_{10} than at T_{20} (445 vs. 488 mg kg^{-1} DW), and this reduction was more marked for 'Eletta' (−12%) and 'Sugarland' (−6%) (Table 6). Moreover, T_{20} promoted a sharper lycopene rise than T_{10} passing from S_0 to S_7 (from 416 to 577 mg kg^{-1} DW, +39%) followed by a milder decrease at S_{14} (484 mg kg^{-1} DW, −16%) (Table 6). All the studied cultivars showed a significant decrease in lycopene content between S_7 and S_{14} (ranging from 119 to 221 mg kg^{-1} DW in 'Ottymo and 'Eletta', respectively), with 'Ottymo' and 'Sugarland' proving also a higher lycopene increase between S_0 and S_7 (by 252 mg kg^{-1} DW, on average) (Figure 3D).

β-carotene concentration proved to be not sensitive to the storage temperature, and was higher in 'Eletta' (94.0 mg kg^{-1} DW) than in the other genotypes (81.9 mg kg^{-1} DW, on average) and, over the storage period, increased up to 93.1 mg kg^{-1} DW at S_{14} (Table 6). However, such temporal increase was more marked in 'Eletta' within the S_0–S_7 period (from 82.1 to 100.0 mg kg^{-1} DW) and in 'Ottymo' in the S_7–S_{14} one (from 75.9 to 102.3 mg kg^{-1} DW) (Figure 3E).

4. Discussion

The fruits stored at 10 °C showed a higher fruit weight and a lower dry matter content as compared to those stored at 20 °C, indicating that fruit transpiration and water loss were the main processes affected by storage temperature. As a consequence, at 20 °C tomatoes proved a higher loss of fruit firmness over time. The transpiration-driven softening of tomatoes during postharvest is a major problem, as it increases their susceptibility to damages along the distribution chain [32]. Moreover, fruit firmness is considered a key indicator of tomato freshness, able to influence the purchasing behavior of consumers [33]. However, despite cold storage is commonly practiced for reducing postharvest softening of tomatoes, the opposite effect can be found when too low storage temperatures are used, because of the tropical origin of the plant [8]. For this reason, storage temperature over 11–12 °C are advised for storing tomatoes, depending on fruit typology and ripening stage [32–35]. Nonetheless, the differences in terms of fruit weight and firmness we found among the 2 thermal regimes showed that storage at 10 °C was a suitable way to extend these main characteristics of tomato fruits. Among the studied cultivars, both 'Sugarland' (small-fruited) and 'Ottymo' (large-fruited) showed the highest fruit weight reduction during storage, consistent with their steeper rise in dry matter content. Differently, 'Eletta' (medium-fruited) proved the highest temporal stability in relation to both variables. Hence our results suggest that the genotypic attitude of cherry tomato to retain fruit weight and firmness during postharvest, is dependent from factors other than simply the fruit size (i.e., the ratio among berry volume and its external transpiring surface) [36], and likely due to the functional traits of the epicarp. Indeed, it has been reported that the dynamics of fruit water loss and consequent tissue collapse are influenced by genotypic differences in structural characteristics of the cuticle, whose alteration over time is an intrinsic feature of the genetically-programmed ripening process [37].

In tomato, the ethylene-driven ripening and senescence lead to the alteration of the carbon substrates content [38], as they are energy-requiring processes whose kinetic is influenced by the ambient temperature [24]. In our experiment, reducing sugars content, the ratio SSC/TA and fruit pH were not affected by the storage temperature, proving instead to be genotype-dependent. Despite their higher increase in dry matter, 'Sugarland' and 'Ottymo' highlighted the steepest drop in reducing sugars content at the end of storage period (by 19%, on average), denoting within the 10–20 °C range a temperature-insensitive acceleration of their autocatalytic metabolism. This demonstrates that no chilling disturbance in reducing sugars metabolism occurred in the experiment [34]. To this end, while the cultivars did not show appreciable pH variations during storage, 'Ottymo' proved the highest SSC/TA reduction over time, denoting its lowest suitability to keep unchanged the taste peculiarities. Indeed, the SSC/TA ratio is a pivotal organoleptic descriptor, as it is related to the overall balance in the perceived sweetness (SSC) and sourness (TA) of tomatoes [39].

Color is one of most important and widely used parameters to define the quality of tomato and tomato products [40]. When fresh tomato fruits are concerned, it is linked to fruit ripeness and firmness and is generally associated by consumers to tomatoes eating quality. In the present experiment, we used an array of chromatic variables summarizing the main color modifications occurring in tomato epicarp. Chroma, $(a^*/b^*)^2$ and tomato color index have been related to quality traits of tomato [41,42], whereas ΔE^*_{ab} has been successfully used to monitor the quality maintenance of potato sticks during refrigerated storage [43]. All these variables showed a certain variability among the studied cultivars, with two of them, namely Chroma and ΔE^*_{ab}, increasing at T_{10}, overall indicating a higher deviation toward more vivid fruit colors. In particular, after 14 days of storage, a higher reduction of Chroma was recorded at 20 °C, a condition which matched the strongest decrease in fruit weight and firmness experienced by the studied cultivars. 'Sugarland' and 'Ottymo' proved higher ΔE^*_{ab} variations during storage. According to Dattner and Bohn [44], independently from the deviation formula, two colors can be optically distinguished if $\Delta E > 1$. The ΔE^*_{ab} differences attained by 'Sugarland' and 'Ottymo' (2.47 units, on average) and 'Eletta' (1.43) indicate for the former cultivars a higher perceivable color deviation along the storage period, consistent with their higher qualitative decline in terms of fruit weight and turgor.

When phytochemical composition was concerned, total polyphenols, lycopene and β-carotene contents found in our experiment were substantially in line with those reported by Fernandes et al. [45] for cherry tomato 'Moscatel RZ' grown in hydroponic or semi-hydroponic systems. On the other hand, phytoene and phytofluene contents were very similar to those found in cherry tomato by Mapelli-Brahm et al. [46]. Plant polyphenols are a large group of phytochemicals involved in the regulation of plant growth, reproduction and response to the environmental stressors [47]. From a nutraceutical viewpoint, they have strong antioxidant properties probably implicated in the decreased incidence of cardiovascular diseases and certain forms of cancer [48]. Both thermal regimes promoted a bell-shaped postharvest trend of TPC, consisting in their sharp rise at S_7, followed by a decrease at S_{14}, this last indicating the onset of metabolic senescence processes [49]. However, such increase was more marked at 10 °C, suggesting the occurrence of a cold-adaptive response in up-regulating the polyphenols expression during postharvest storage. Indeed, several phenolic compounds typically accumulate in plant cells subjected to cold stress, as they contribute to the homeostasis of cold-induced reactive oxygen species (ROS) and to enhance the thickness of the cell wall, so preventing lipid peroxidation and cell collapse [47]. This would explain the best retention of fruit firmness recorded at 10 °C, indicating at the same time, the improvement of tomato phenolic profile as a benefit induced by a mild cold stress. Thus, although polyphenols have not been considered a priority target in tomato breeding programs, our results suggest that they could represent a sensitive target for improving the functional profile of the tomato, mostly during postharvest cold storage.

Regarding the carotenoid fraction, we recorded variable effects, resulting from different time-course response to storage temperature and duration. Lycopene displayed a bell-shaped temporal trend too since, under both storage temperatures, this carotenoid sharply increased at S_7 then declined at S_{14}. This trend substantially differed from that of β-carotene which continuously increased until S_{14}, so confirming the higher stability of its postharvest accumulation in tomato [50]. According to Rodriguez-Amaya [51], carotenoids accumulation can continue during postharvest transport or storage, provided that the integrity of the fruit is maintained, so preserving the enzymatic activity responsible for carotenogenesis. Lycopene plays a paramount function in protecting the photosynthetic apparatus and plant lipid membranes, as its acyclic polyene structure (11 conjugated double bonds) increases its affinity for singlet oxygen and radical scavenging activity beyond the other carotenoids [52]. For this reason, it has been reported that oxidation is the main cause for lycopene degradation [14]. This could partly explain the depressive effect on lycopene content we recorded upon storing tomatoes in a stressing, ROS-inducing environment (10 °C). In this view, it is interesting to note the contrasting effect of cold storage on tomato compositional traits, resulting in a higher polyphenols accumulation in case of a lower lycopene content. This suggests the existence of a fine tuning among different classes of compounds in response to cold stress. However, by comparing the temporal trend of lycopene with that of its colorless precursors phytoene and phytofluene, clear time-dependent temperature effects on carotenogenesis where noticeable. Indeed, at S_7, the highest lycopene content recorded at 20 °C matched the strongest reduction of both phytoene and phytofluene. In other words, the lowest the lycopene concentration the highest the accumulation of its precursors and vice versa. This implies that reduced transformation kinetics of both phytoene and phytofluene represented the earliest metabolic constraints recorded in response to the imposed cold stress. According to Dumas et al. [53] the over-expression of phytoene desaturase (leading to lycopene synthesis by desaturating both phytoene and phytofluene) is the most important upstream metabolic step in increasing the lycopene content of tomato fruits at harvest. Our results bear this out in postharvest conditions too, as they indicate that, under mild cold stress storage conditions, desaturation of phytoene and phytofluene represents the earliest metabolic bottleneck in lycopene synthesis of cherry tomatoes, hence a possible priority target to modulate the postharvest evolution of their nutraceutical profile. On the other hand, to which extent this implies a mid-term modification of the overall nutraceutical profile of tomato represents an interesting topic, taking into account that, despite they are not effective antiradicals as lycopene, phytoene and phytofluene are among the prevailing carotenoids found in human plasma and tissues, and their bioaccessibility following gastro-intestinal digestion of tomato juice has been found ~3–4 fold higher than that of lycopene [54,55].

Among the studied cultivars 'Sugarland' proved the highest lycopene and total polyphenols content, whereas 'Eletta' overcame the other cultivars for phytoene and phytofluene. Excepting β-carotene, which over time increased more sharply in 'Eletta' and 'Ottymo', these differences were still noticeable at the end of the storage period, regardless of the storage temperature. This highlights, beyond the environmental influence, the existence of a strong genetic component determining the stoichiometric relationships among lycopene and its precursors. Unravelling the possible interactive effects among these three carotenoids in generating the antioxidative health benefits [16,49] will allow for a better orientation of breeding programs toward the most convenient phytochemical evolution of tomatoes during refrigerated storage.

5. Conclusions

The results of the present experiment highlighted complex postharvest modifications of cherry tomatoes in response to the studied factors. By storing them under mild stressing conditions (10 °C) it was possible to improve the stability over time of carpometric traits (mainly fruit weight, firmness and Chroma) having commercial relevance, without alterations of compositional traits related to taste perception (reducing sugars content, SSC/TA and pH). Moreover, when compared to 20 °C, storing at 10 °C boosted the accumulation of total polyphenol and, at least in the short term (within 7 days of storage), the concentration of both phytoene and phytofluene, probably inhibiting their enzymatic desaturation leading to lycopene. This suggests their possible usefulness in modulating the nutraceutical evolution of cold stored cherry tomatoes during postharvest. This idea is reinforced by the stable varietal differences we found in terms of stoichiometric relationships among lycopene, phytoene and phytofluene. Regarding the varietal attitude to postharvest storage, the stability over time of fruit weight, dry matter content, SSC/TA and ΔE^*_{ab} proved to be highly discriminant among cultivars, indicating the lowest ability of 'Ottymo' and 'Eletta' to maintain their fruit peculiarities over time. Thus, our results suggest the use of these variables to screen for cherry tomato germplasm suited to periods of postharvest storage.

Author Contributions: Conceptualization, R.P.M., M.D. and E.A.; methodology, R.P.M., M.D. and S.B.; validation, F.G., R.P.M. and E.A.; formal analysis, M.D. and S.B.; investigation, M.D., R.P.M. and S.B.; data curation, M.D. and S.B.; writing—original draft preparation, R.P.M., E.A. and M.D.; writing—review and editing, F.G., C.L. and B.F.; supervision, F.G., C.L. and B.F. All authors have read and agreed to the published version of the manuscript.

Funding: This research received no external funding.

Conflicts of Interest: The authors declare no conflict of interest.

References

1. Food and Agriculture Organization of the United Nations. Available online: http://www.fao.org/faostat/en/#home (accessed on 26 September 2020).
2. Mauro, R.P.; Lo Monaco, A.; Lombardo, S.; Restuccia, A.; Mauromicale, G. Eradication of Orobanche/Phelipanche spp. seedbank by soil solarization and organic supplementation. *Sci. Hortic.* **2015**, *193*, 62–68. [CrossRef]
3. Capurso, C.; Vendemiale, G. The Mediterranean diet reduces the risk and mortality of the prostate cancer: A narrative review. *Front. Nutr.* **2017**, *4*, 38. [CrossRef] [PubMed]
4. Cainelli, N.; Ruperti, B. Biochemistry and molecular biology in fruits during cold storage. *Annu. Plant Rev. Online* **2019**, *2*, 659–688.
5. Mauro, R.P.; Agnello, M.; Rizzo, V.; Graziani, G.; Fogliano, V.; Leonardi, C.; Giuffrida, F. Recovery of eggplant field waste as a source of phytochemicals. *Sci. Hortic.* **2020**, *261*, 109023. [CrossRef]
6. Petric, T.; Kiferle, C.; Perata, P.; Gonzali, S. Optimizing shelf life conditions for anthocyanin-rich tomatoes. *PLoS ONE* **2018**, *13*, e0205650. [CrossRef]
7. Affandi, F.Y.; Verdonk, J.C.; Ouzounis, T.; Ji, Y.; Woltering, E.J.; Schouten, R.E. Far-red light during cultivation induces postharvest cold tolerance in tomato fruit. *Postharvest Biol. Technol.* **2020**, *159*, 111019. [CrossRef]
8. Tadesse, T.N.; Ibrahim, A.M.; Abtew, W.G. Degradation and formation of fruit color in tomato (*Solanum lycopersicum* L.) in response to storage temperature. *Am. J. Food Technol.* **2015**, *10*, 147–157. [CrossRef]
9. Farzaneh, V.; Gominho, J.; Pereira, H.; Carvalho, I.S. Screening of the antioxidant and enzyme inhibition ootentials of portuguese Pimpinella anisum L. seeds by GC-MS. *Food Anal. Methods* **2018**, *11*, 2645–2656. [CrossRef]
10. Park, M.H.; Sangwanangkul, P.; Baek, D.R. Changes in carotenoid and chlorophyll content of black tomatoes (*Lycopersicon esculentum* L.) during storage at various temperatures. *Saudi J. Biol. Sci.* **2018**, *25*, 57–65. [CrossRef]
11. Martí, R.; Roselló, S.; Cebolla-Cornejo, J. Tomato as a source of carotenoids and polyphenols targeted to cancer prevention. *Cancers* **2016**, *8*, 58. [CrossRef]
12. Story, E.N.; Kopec, R.E.; Schwartz, S.J.; Keith Harris, G. An update on the health effects of tomato lycopene. *Annu. Rev. Food Sci. Technol.* **2010**, *1*, 189–210. [CrossRef] [PubMed]
13. Peters, U.; Leitzmann, M.F.; Chatterjee, N.; Wang, Y.; Albanes, D.; Gelmann, E.P.; Friesen, M.D.; Riboli, E.; Hayes, R.B. Serum lycopene, other carotenoids, and prostate cancer risk: A nested case-control study in the prostate, lung, colorectal, and ovarian cancer screening trial. *Cancer Epidemiol. Biomarkers Prev.* **2007**, *16*, 962–968. [CrossRef] [PubMed]
14. Srivastava, S.; Srivastava, A.K. Lycopene; chemistry, biosynthesis, metabolism and degradation under various abiotic parameters. *J. Food Sci. Technol.* **2015**, *52*, 41–53. [CrossRef]
15. Liu, L.; Shao, Z.; Zhang, M.; Wang, Q. Regulation of carotenoid metabolism in tomato. *Mol. Plant* **2015**, *8*, 28–39. [CrossRef] [PubMed]
16. Gul, K.; Tak, A.; Singh, A.K.; Singh, P.; Yousuf, B.; Wani, A.A. Chemistry, encapsulation, and health benefits of beta-carotene—A review. *Cogent Food Agric.* **2015**, *1*. [CrossRef]
17. Basu, A.; Imrhan, V. Tomatoes versus lycopene in oxidative stress and carcinogenesis: Conclusions from clinical trials. *Eur. J. Clin. Nutr.* **2007**, *61*, 295–303. [CrossRef]
18. Engelmann, N.J.; Clinton, S.K.; Erdman, J.W.J. Nutritional aspects of phytoene and phytofluene, carotenoid precursors of lycopene. *Adv. Nutr.* **2011**, *2*, 51–61. [CrossRef]
19. Passam, H.; Karapanos, I.; Bebeli, P.; Savvas, D. A review of recent research on tomato nutrition, breeding and post-harvest technology with reference to fruit quality. *Eur. J. Plant Sci. Biotechnol.* **2007**, *1*, 1–21.

20. Raiola, A.; Rigano, M.M.; Calafiore, R.; Frusciante, L.; Barone, A. Enhancing the health-promoting effects of tomato fruit for biofortified food. *Mediators Inflamm.* **2014**, *2014*, 1–16. [CrossRef]
21. Kavitha, P.; Shivashankara, K.S.; Rao, V.K.; Sadashiva, A.T.; Ravishankar, K.V.; Sathish, G.J. Genotypic variability for antioxidant and quality parameters among tomato cultivars, hybrids, cherry tomatoes and wild species. *J. Sci. Food Agric.* **2014**, *94*, 993–999. [CrossRef]
22. Alenazi, M.M.; Shafiq, M.; Alsadon, A.A.; Alhelal, I.M.; Alhamdan, A.M.; Solieman, T.H.I.; Ibrahim, A.A.; Shady, M.R.; Al-Selwey, W.A. Improved functional and nutritional properties of tomato fruit during cold storage. *Saudi J. Biol. Sci.* **2020**, *27*, 1467–1474. [CrossRef]
23. Mauro, R.P.; Agnello, M.; Distefano, M.; Sabatino, L.; San Bautista Primo, A.; Leonardi, C.; Giuffrida, F. Chlorophyll fluorescence, photosynthesis and growth of tomato plants as affected by long-term oxygen root zone deprivation and grafting. *Agronomy* **2020**, *10*, 137. [CrossRef]
24. Giuffrida, F.; Agnello, M.; Mauro, R.P.; Ferrante, A.; Leonardi, C. Cultivation under salt stress conditions influences postharvest quality and glucosinolates content of fresh-cut cauliflower. *Sci. Hortic.* **2018**, *236*, 166–174. [CrossRef]
25. Gautier, H.; Diakou-Verdin, V.; Bénard, C.; Reich, M.; Buret, M.; Bourgaud, F.; Poëssel, J.L.; Caris-Veyrat, C.; Génard, M. How does tomato quality (sugar, acid, and nutritional quality) vary with ripening stage, temperature, and irradiance? *J. Agric. Food Chem.* **2008**, *56*, 1241–1250. [CrossRef] [PubMed]
26. Khairi, A.N.; Falah, M.A.F.; Pamungkas, A.P.; Takahashi, N. Optimization of storage temperatures to maintain Lycopene content of tomato from moderate water stress irrigated greenhouse. *IOP Conf. Ser. Mater. Sci. Eng.* **2018**, *403*, 011001. [CrossRef]
27. Carli, P.; Barone, A.; Fogliano, V.; Frusciante, L.; Ercolano, M.R. Dissection of genetic and environmental factors involved in tomato organoleptic quality. *BMC Plant Biol.* **2011**, *11*, 58. [CrossRef]
28. McGuire, R.G. Reporting of objective color measurements. *HortScience* **1992**, *27*, 1254–1255. [CrossRef]
29. Atanasova, A.H.; Denev, P.N.; Tringovska, I.I.; Grozeva, S.Y.; Ganeva, D.G.; Kratchanova, M.G.; Panchev, I.N. Optimization of the key parameters for extraction of polyphenol compounds from tomato fruits (*Solanum lycopersicum* L.). Kinetics of the process. *Bulg. Chem. Commun.* **2014**, *46*, 65–70.
30. Gahler, S.; Otto, K.; Böhm, V. Alterations of vitamin C, total phenolics, and antioxidant capacity as affected by processing tomatoes to different products. *J. Agric. Food Chem.* **2003**, *51*, 7962–7968. [CrossRef]
31. Siracusa, L.; Patanè, C.; Rizzo, V.; Cosentino, S.L.; Ruberto, G. Targeted secondary metabolic and physico-chemical traits analysis to assess genetic variability within a germplasm collection of "long storage" tomatoes. *Food Chem.* **2018**, *244*, 275–283. [CrossRef]
32. Batu, A. Determination of acceptable firmness and colour values of tomatoes. *J. Food Eng.* **2004**, *61*, 471–475. [CrossRef]
33. Bui, H.T.; Makhlouf, J.; Ratti, C. Postharvest ripening characterization of greenhouse tomatoes. *Int. J. Food Prop.* **2010**, *13*, 830–846. [CrossRef]
34. Beckles, D.M. Factors affecting the postharvest soluble solids and sugar content of tomato (*Solanum lycopersicum* L.) fruit. *Postharvest Biol. Technol.* **2012**, *63*, 129–140. [CrossRef]
35. Tadesse, T.N.; Abtew, W.G. Effect of hot water treatment on reduction of chilling injury and keeping quality in tomato (*Solanum lycopersicum* L.) fruits. *J. Stored Prod. Postharvest Res.* **2016**, *7*, 61–68.
36. Leonardi, C.; Baille, A.; Guichard, S. Effects of fruit characteristics and climatic conditions on tomato transpiration in a greenhouse. *J. Hortic. Sci. Biotechnol.* **1999**, *74*, 748–756. [CrossRef]
37. Saladié, M.; Matas, A.J.; Isaacson, T.; Jenks, M.A.; Goodwin, S.M.; Niklas, K.J.; Xiaolin, R.; Labavitch, J.M.; Shackel, K.A.; Fernie, A.R.; et al. A reevaluation of the key factors that influence tomato fruit softening and integrity. *Plant Physiol.* **2007**, *144*, 1012–1028. [CrossRef]
38. Anton, D.; Bender, I.; Kaart, T.; Roasto, M.; Heinonen, M.; Luik, A.; Püssa, T. Changes in polyphenols contents and antioxidant capacities of organically and conventionally cultivated tomato (*Solanum lycopersicum* L.) fruits during ripening. *Int. J. Anal. Chem.* **2017**, *2017*, 6–10. [CrossRef]
39. Mauro, R.P.; Rizzo, V.; Leonardi, C.; Mazzaglia, A.; Muratore, G.; Distefano, M.; Sabatino, L.; Giuffrida, F. Influence of harvest stage and rootstock genotype on compositional and sensory profile of the elongated tomato cv. Sir Elyan. *Agriculture* **2020**, *10*, 82. [CrossRef]
40. Ganje, M.; Jafari, S.M.; Farzaneh, V.; Malekjani, N. Kinetics modelling of color deterioration during thermal processing of tomato paste with the use of response surface methodology. *Heat Mass Transf.* **2018**, *54*, 3663–3671. [CrossRef]

41. Clément, A.; Dorais, M.; Vernon, M. Multivariate approach to the measurement of tomato maturity and gustatory attributes and their rapid assessment by vis-NIR spectroscopy. *J. Agric. Food Chem.* **2008**, *56*, 1538–1544. [CrossRef]
42. Goisser, S.; Wittmann, S.; Fernandes, M.; Mempel, H.; Ulrichs, C. Comparison of colorimeter and different portable food-scanners for non-destructive prediction of lycopene content in tomato fruit. *Postharvest Biol. Technol.* **2020**, *167*, 111232. [CrossRef]
43. Licciardello, F.; Lombardo, S.; Rizzo, V.; Pitino, I.; Pandino, G.; Strano, M.G.; Muratore, G.; Restuccia, C.; Mauromicale, G. Integrated agronomical and technological approach for the quality maintenance of ready-to-fry potato sticks during refrigerated storage. *Postharvest Biol. Technol.* **2018**, *136*, 23–30. [CrossRef]
44. Dattner, M.; Bohn, D. Characterization of print quality in terms of colorimetric aspects. In *Printing on Polymers: Fundamentals and Applications*; Elsevier Inc.: Amsterdam, The Netherlands, 2015; pp. 329–345. ISBN 9780323374682.
45. Fernandes, I.; Leça, J.M.; Aguiar, R.; Fernandes, T.; Marques, J.C.; Cordeiro, N. Influence of crop system fruit quality, carotenoids, fatty acids and phenolic compounds in cherry tomatoes. *Agric. Res.* **2020**. [CrossRef]
46. Mapelli-Brahm, P.; Stinco, C.M.; Meléndez-Martínez, A.J. Comparative study of the bioaccessibility of the colorless carotenoids phytoene and phytofluene in powders and pulps of tomato: Microstructural analysis and effect of addition of sunflower oil. *Food Funct.* **2018**, *9*, 5016–5023. [CrossRef] [PubMed]
47. Sharma, A.; Shahzad, B.; Rehman, A.; Bhardwaj, R.; Landi, M.; Zheng, B. Response of phenylpropanoid pathway and the role of polyphenols in plants under abiotic stress. *Molecules* **2019**, *24*, 2452. [CrossRef]
48. Holst, B.; Williamson, G. Nutrients and phytochemicals: From bioavailability to bioefficacy beyond antioxidants. *Curr. Opin. Biotechnol.* **2008**, *19*, 73–82. [CrossRef]
49. Mirdehghan, S.H.; Valero, D. Bioactive compounds in tomato fruit and its antioxidant activity as affected by incorporation of Aloe, eugenol, and thymol in fruit package during storage. *Int. J. Food Prop.* **2017**, *20*, 1798–1806.
50. Brashlyanova, B.P.; Pevicharova, G.T. Effects of cold storage and ripening on antioxidant components in tomatoes. *Acta Hortic.* **2009**, *830*, 349–353. [CrossRef]
51. Rodriguez-Amaya, D.B. Nature and distribution of carotenoids in foods. In *Shelf Life Studies of Foods and Beverages Chemical, Biological, Physical and Nutritional Aspects*; Charalambous, G., Ed.; Elsevier Science: Amsterdam, The Netherlands, 1993; ISBN 0-444-89459-4.
52. Gruszecki, W.I.; Strzałka, K. Carotenoids as modulators of lipid membrane physical properties. *Biochim. Biophys. Acta Mol. Basis Dis.* **2005**, *1740*, 108–115. [CrossRef]
53. Dumas, Y.; Dadomo, M.; Di Lucca, G.; Grolier, P. Effects of environmental factors and agricultural techniques on antioxidant content of tomatoes. *J. Sci. Food Agric.* **2003**, *83*, 369–382. [CrossRef]
54. Meléndez-Martínez, A.J.; Mapelli-Brahm, P.; Benítez-González, A.; Stinco, C.M. A comprehensive review on the colorless carotenoids phytoene and phytofluene. *Arch. Biochem. Biophys.* **2015**, *572*, 188–200. [CrossRef] [PubMed]
55. Mapelli-Brahm, P.; Corte-Real, J.; Meléndez-Martínez, A.J.; Bohn, T. Bioaccessibility of phytoene and phytofluene is superior to other carotenoids from selected fruit and vegetable juices. *Food Chem.* **2017**, *229*, 304–311. [CrossRef] [PubMed]

foods

Review

Analytical Strategies for Fingerprinting of Antioxidants, Nutritional Substances, and Bioactive Compounds in Foodstuffs Based on High Performance Liquid Chromatography–Mass Spectrometry: An Overview

Dario Donno *, Maria Gabriella Mellano, Giovanni Gamba, Isidoro Riondato and Gabriele Loris Beccaro

Dipartimento di Scienze Agrarie, Forestali e Alimentari, Università degli Studi di Torino, 10095 Grugliasco, Italy; gabriella.mellano@unito.it (M.G.M.); giovanni.gamba@unito.it (G.G.); isidoro.riondato@unito.it (I.R.); gabriele.beccaro@unito.it (G.L.B.)

* Correspondence: dario.donno@unito.it; Tel.: +39-011-670-8751

Received: 29 October 2020; Accepted: 23 November 2020; Published: 25 November 2020

Abstract: New technology development and globalisation have led to extreme changes in the agri-food sector in recent years that need an important food supply chain characterisation from plant materials to commercial productions. Many analytical strategies are commonly utilised in the agri-food industry, often using complementary technologies with different purposes. Chromatography on-line coupled to mass spectrometry (MS) is one of the most selective and sensitive analytical methodologies. The purpose of this overview is to present the most recent MS-based techniques applied to food analysis. An entire section is dedicated to the recent applications of high-resolution MS. Covered topics include liquid (LC)– and gas chromatography (GC)–MS analysis of natural bioactive substances, including carbohydrates, flavonoids and related compounds, lipids, phenolic compounds, vitamins, and other different molecules in foodstuffs from the perspectives of food composition, food authenticity and food adulteration. The results represent an important contribution to the utilisation of GC–MS and LC–MS in the field of natural bioactive compound identification and quantification.

Keywords: food analysis; High Performance Liquid Chromatography–Mass Spectrometry (HPLC–MS) techniques; Orbitrap; High Resolution Mass Spectrometry (HRMS); natural substances; antioxidant molecules

1. Introduction

New technology advancement and globalisation have led to extreme changes in the agri-food sector in recent years that need an important food supply chain characterisation from plant materials to commercial productions. Moreover, consumers call for more food assurances and information on geographical origin, safety, and quality of the used final products [1]. Testing should also be performed to gather additional information that regulatory agencies periodically require not only for a general evaluation of chemical composition but also to accurately identify and quantify all the compounds of interest. Given its importance, the characterisation of complex food matrices is becoming more and more essential [2].

Many analytical strategies are utilised in the agri-food industry, often using complementary technologies with different purposes. Chromatography on-line coupled to mass spectrometry (MS) or diode array detector (DAD) is one of the most selective and sensitive techniques in chemical analysis.

Agri-food products are multicompound matrices; for this reason, sensitivity and selectivity are very important for the identification of components in trace [3]. Among the combinatorial approaches, high performance liquid chromatography–mass spectrometry (HPLC–MS) and gas chromatography–mass spectrometry (GC–MS) are the fastest-growing analytical strategies, already widely used in food analysis (Figure 1). In the last decade, the manuscripts describing GC–MS food analysis doubled, while papers that applied an HPLC–MS approach quadrupled.

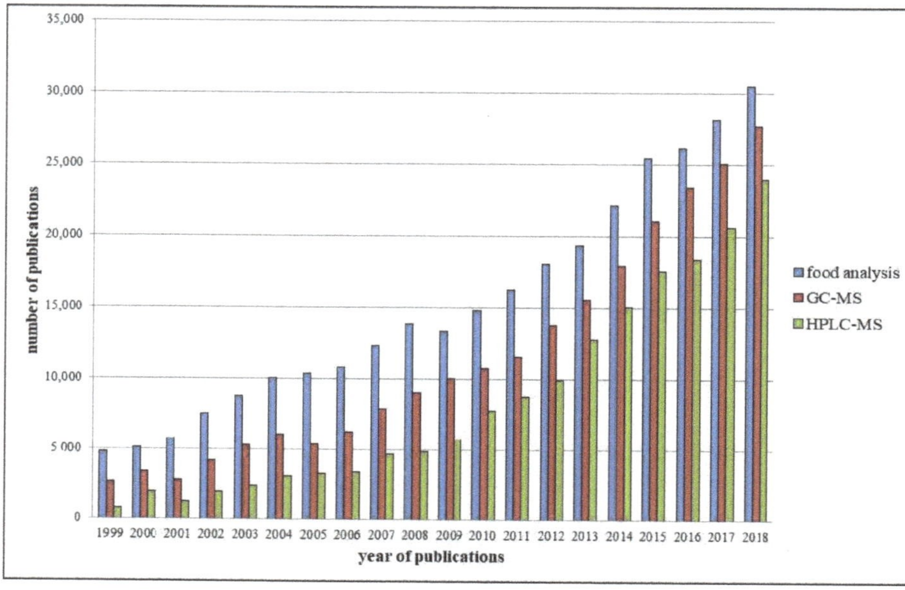

Figure 1. Evaluation of the manuscripts linked to food analysis, and those using GC–MS (20-fold) or HPLC–MS methods (50-fold). The *x*-axis represents the year of publications, while the *y*-axis represents the number of publications. Data derived from the Scopus database using the search term "food analysis" coupled (using "AND" as relationship term) with "HPLC–MS" and "GC–MS", respectively. HPLC–MS = high performance liquid chromatography–mass spectrometry; GC–MS = gas chromatography–mass spectrometry.

In the last period (about 20 years), several important step-changes were developed in MS analysis. In LC–MS, the advances in ionisation interfaces (e.g., atmospheric pressure chemical ionisation—APCI and electrospray ionisation—ESI), coupled to the developments in MS instruments, allowed to apply LC–MS to the wider scientific area [4]. LC–MS techniques were surpassed by LC–tandem MS (MS/MS or MS^2) ones, currently the standard approach, both for more sophisticated configurations and simple lab instruments. An important improvement has been also the improvement of high-resolution LC–MS technology (e.g., Orbitrap can reach resolutions of 10,000–200,000) and time-of-flight (ToF) instruments (as LC–ToF–MS in different configurations) [5–7] as shown in Figure 2.

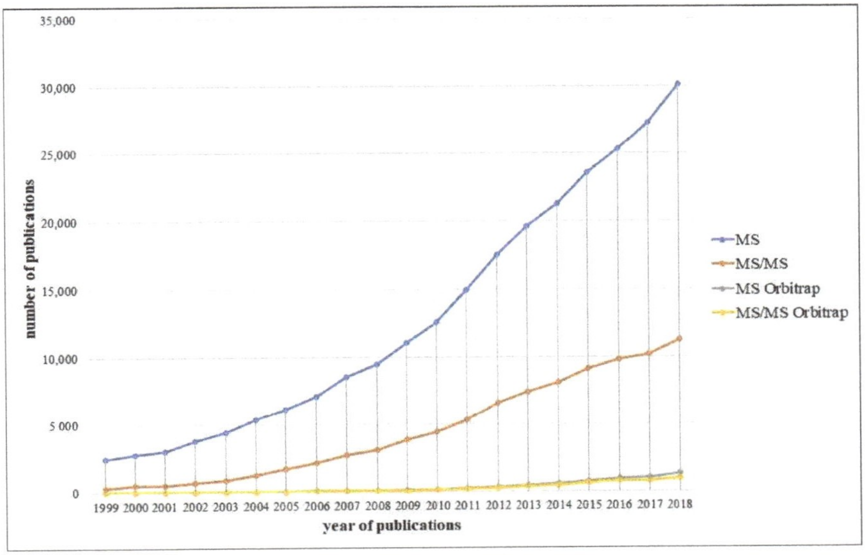

Figure 2. Evaluation of the manuscripts linked to food analysis together to (i) MS, (ii) MS/MS, (iii) MS Orbitrap, and (iv) MS/MS Orbitrap methods. The *x*-axis represents the year of publications, while the *y*-axis represents the number of publications. Data derived from the Scopus database using the search term "food analysis" coupled (using "AND" as relationship term) with "mass spectrometry", "tandem mass spectrometry", "mass spectrometry + Orbitrap", and "tandem mass spectrometry + Orbitrap", respectively. MS = mass spectrometry; MS/MS = tandem mass spectrometry.

The purpose of this overview is to present the most recent MS-based techniques applied in food analysis. An entire section is dedicated to the recent applications of high-resolution mass spectrometry (HRMS). Covered topics include liquid (LC)– and gas chromatography (GC)–MS analysis of natural bioactive substances. The applications of the GC/LC–MS techniques and relative interface systems in food analysis were described in this overview. The results represent an important contribution to the utilisation of GC–MS and LC–MS in the field of natural bioactive compound identification and quantification. The present review includes MS analysis of natural bioactive substances, including carbohydrates, flavonoids and related compounds, lipids, phenolic compounds, vitamins, and other different molecules in foodstuffs.

2. Analysis of Various Molecular Classes

Foods provide nutritional substances and phytochemicals useful for humans and their well-being and health conditions. Moreover, since the past, it was also known that specific foods ensure additional health-promoting benefits to people as prevention and/or treatment of several diseases [8]. For this reason, people achieved a better life quality by eating specific foods (e.g., meat, dairy products, fruits, vegetables, etc.) and relative derived-products (e.g., juices, jams, etc.) or taking dietary supplements or nutraceuticals. Moreover, regional, national, and international regulatory institutions have promoted intense research to identify and characterise new biologically active molecules to be used to formulate, develop, and improve new nutraceuticals and functional foods [9]. In the last years, companies, licensed professionals, marketers, and manufacturers have matured due to the increasing demands for bioactive substances, phytonutrients, nutraceuticals, and their therapeutic services [10]. These substances and compounds are secondary metabolites, namely molecules produced by the plants but not directly involved in the normal development, growth, or reproduction of a natural organism. They may act as "defence" molecules against biotic and abiotic stresses (e.g., parasites, diseases, ultraviolet radiation and oxidants, predators, etc.) for interspecies competition and to facilitate and improve the reproductive processes (e.g., they may also serve as colouring agents and attractive smells) [11]. These same compounds possess important health-promoting effects in animals and man; indeed, scientific evidence from clinical trials, epidemiological studies, and in vitro/in vivo tests has demonstrated that a diet rich in specific foods may reduce the risk of many diseases (e.g., cancer, obesity and diabetes, cardiovascular and inflammatory complications) [12].

Foodstuffs are represented by several substances with different nutritional value, biological properties, and structural traits. Some are main or bulk components (e.g., triglycerides or starch), some trace or minor components (e.g., phenolic compounds), and some are pesticides, mycotoxins, or undesirable impurities. The same compound could be a main or minor substance, in relation to the type of considered food [1].

The quantitative and qualitative studies on the characterisation of biomolecules are firstly focused on the selection of good extraction methods [13]. Extraction is the first stage in the study of chemical composition, and it is very important for the outcomes and final results. The main aims of extraction in foods are (i) to extract targeted bioactive or nutritional substances from a complex matrix, (ii) to increase the selectivity of the analytical method, (iii) to increase the sensitivity of the selected test or assay by increasing the quantity of targeted molecules, (iv) to convert the nutritive o active substances into more suitable compounds to be detected and separated, and (v) to allow a more reproducible and strong method that should be not related to variability in the original sample [14]. For this reason, extraction should be carefully performed and optimised. Polarities of different molecules and substances, that may significantly vary because of their relationship with the matrix and their conjugation status, influence the extraction solvents to be selected and used [15]. The selection of solvents for compound extraction should also involve other factors as use of cosolvent, environmental safety, financial feasibility, human toxicity, mass transfer, and molecular affinity between solute and solvent [14]. The optimisation of parameters for compound extraction may improve the extraction efficiency of the molecule of interest and it may also reduce the solvent consumption and waste production in order to allow a more environmentally friendly extraction process [16,17]. Sometimes, specific pretreatment processes are necessary to not modify the raw materials and better extract the substances (e.g., pasteurisation, use of depolymerising enzymes, chemical and microbial acidification storage under specifically modified atmospheres); moreover, additional secondary procedures (e.g., sonication, stirring, rotary shaking, etc.) can be applied to improve the extraction [18]. The main traditional extraction methods are reported in Table 1.

Table 1. Analytical methods for the extraction of the main molecules in foodstuffs [15,19–23].

Method	Used Solvents	Time	The Volume of Requested Solvent (mL)
Accelerated solvent extraction, static (ASE)	Methanol	20–40 min	20–40
Microwave assisted extraction (MAE)	Ethanol, methanol, or a mixture of water and alcohol	10–40 min	20–50
Pressurised hot water extraction (PHWE)	Water with/without 10–30% ethanol	40–50 min	40–45
Pressurised liquid extraction, dynamic (PLE)	Methanol	20–40 min	20–30
Sonication	Ethanol, methanol, or a mixture of water and alcohol	1 h	50–100
Soxhlet extraction	Ethanol, methanol, or a mixture of water and alcohol	3–18 h	150–200
Supercritical fluid extraction (SFE)	Carbon dioxide with/without modifiers (e.g., methanol)	30–100 min	Not applicable
Surfactant assisted PHWE	Water with surfactants (e.g., SDS or Triton X100)	40–50 min	40–45

For characterisation (identification and quantification) of nutritional and bioactive molecules, the most important step is the selection of the right analytical strategies to be applied. Spectroscopic protocols usually allow a total identification and quantification of the substances in the food, but in most examples, the specific phytochemical composition may remain not fully observed. In any case, spectrophotometry is usually used for the evaluation of antioxidant capacity [24]. For this reason, more detailed information on the chemical composition requires additional analytical techniques, as chromatography and mass spectrometry, to fully describe the structures of the compounds in the food matrices [14]. Technological improvement and innovation in the analytical instrumentation have allowed more and more sophisticated evaluations (quantitative and qualitative) of the nutritional and health-promoting chemical composition in food [25]. Chromatography is mainly utilised to analyse natural food and relative derived-products. Chromatographic techniques offer very high separation ability because chemical ingredients in a complex extract can be divided into many relatively simple parts (subfractions). Moreover, the last analytical approaches, as the application of hyphenated spectrometry and chromatography (e.g., capillary electrophoresis-diode array detection (CE-DAD), high performance liquid chromatography–diode array detection (HPLC–DAD), gas chromatography/high performance liquid chromatography–mass spectroscopy (GC/HPLC–MS), and high performance liquid chromatography–nuclear magnetic resonance (HPLC–NMR)), may allow to obtain additional spectral information. These techniques are very important for on-line structural elucidation and qualitative analysis [26,27]. In particular, high performance liquid chromatography is now the most widely utilised analytical technique, hyphenated with different detectors (mass spectrometry, fluorescence, and diode arrays) [28]. It is referred as the most important standard strategy in the characterisation and authentication of foodstuffs thanks to its reproducibility, sensitivity, and precision despite large organic solvents consumption, relatively long analytical time, and high instrumental cost [29]. Quality control and product authentication of the food can be carried out by the application of a specific approach based on marker compounds; a marker substance is a chemical ingredient of raw material, preparation, beverage, or food product that is used for characterisation and/or quality control aims, in particular if the nutritive or active constituents are not known or previously identified [30].

Due to this high number of variables, a specific analytical technique is not able to cover all the food analysis aspects. If an analytical method is developed (e.g., for flavonoid analysis), it may be applied for several foods (e.g., soft drinks, fruits, wines, or vegetables). Examples of MS coupled to different chromatographic techniques are shown in the following sections in order to characterise different classes of substances. Some molecules present a closely similar structure (e.g., carbohydrates), while other compounds are defined by their similar biological properties (e.g., vitamins), despite potentially large differences in structure and polarity; for this reason, analytical conditions and sample preparations could be very different depending on the considered molecules.

2.1. Proteins

Proteins are a very important dietary ingredient for the survival of humans and animals. They are essential in nutrition to supply enough amounts of specific amino acids, that may be divided into different categories according to their rates of protein in vivo synthesis: (i) indispensable (e.g., valine, tryptophan, threonine, phenylalanine, methionine, lysine, leucine, isoleucine, and histidine); (ii) conditionally indispensable (e.g., tyrosine, cysteine, arginine); (iii) dispensable (e.g., serine, proline, glycine, glutamic acid, glutamine, aspartic acid, asparagine, alanine) [31]. The main studies are focused on physico-chemical traits and structural characteristics of food proteins to elucidate their molecular structure responsible for their actions and functionalities and consequently their relative structure–property relationships. The total nitrogen analysis by Kjeldahl system is the main reference method for the evaluation of proteins in food products and now it is used for validation and/or calibration of alternative methods for protein analysis [32]. Molecules that easily ionise in aqueous solutions (e.g., nitrogen compounds mostly as proteins and single peptides) are potentially suitable for LC–MS techniques. Consequently, LC–MS is fundamental in proteomics and peptidomics, which are branches of biology that assess the full set of peptides or proteins in a single sample. MS-based proteomics and peptidomics are increasingly used for food analysis [33,34].

2.2. Lipids

Food derived from plants and animals contain lipids, an essential substance involved in maintaining many life activities. A high number of fat molecules with different functional traits and chemical proprieties are provided by nature. Lipidic molecules significantly contribute to the sensory and nutritional traits of food. Some lipids are structurally simple compounds (e.g., fatty acids), while others show a variable and complex structure (e.g., steroids, sphingomyelins, phospholipids, glycosphingolipids, acylglycerols) [35]. Extract preparation is a crucial step, for analysing lipids and lipid trace substances in lipid-rich systems (e.g., fats derived from animal matrices). HPLC–MS is the main tool used for the characterisation of lipid substances in foods. Two-dimensional HPLC is commonly on-line coupled to MS to improve selectivity for the analysis of single compound structure because of the complexity of these molecules (triacylglycerols, fatty acids, and phospholipids) [36]. In the past, GC–MS techniques were used to analyse triacylglycerols, but this approach required complex derivatisation and extraction procedures. HPLC–MS analysis has been implemented in recent years to simplify sample preparation. The use of MS/MS has enabled MS structural elucidation of individual triacylglycerols [37,38].

2.3. Carbohydrates

Carbohydrates are among the most common nutrients in foods and they show many nutritional and physiological roles. For example, celluloses are structural elements in the plant kingdom, and they are defined as an excellent dietary fibre source. Starch and simple sugars (e.g., fructose and glucose) are the main sources of energy in humans and animals. Other carbohydrates are conjugated to several macromolecules, as flavonoids or proteins, and they present several health-promoting and functional roles [39]. It is often very difficult to analyse structural carbohydrates by MS, but these compounds may be targeted by HPLC–MS analysis. Glycoproteins are specialised components of many foodstuffs (e.g., cereals, milk, and colostrum) and possess high functional and nutritional value [40–42]. GC–MS, after derivatisation, is traditionally used to analyse small oligosaccharides and monosaccharides. A less common technique for oligosaccharide analysis is MALDI-TOF/TOF; in particular, it is very useful for oligosaccharide structure elucidation, as previously reported by Hotchkiss et al. (2015) [43]. Molecular conjugates and oligosaccharides can be identified and quantified by HPLC–MS, that also allows the analysis of small oligosaccharides and unknown carbohydrates without time-consuming derivatisation, purification, and fractionation. ESI, rather than APCI, is usually preferred as ionisation mode due to the carbohydrate low volatility and high polarity. However, carbohydrates might have low sensitivity in ESI, even if a cationisation may avoid this problem [44].

2.4. Vitamins

Health-promoting capacities and nutritional properties of foods were not only associated with the presence of bioactive compounds, as phenolic acids, terpenes, or flavonoids, but they were also attributed to other molecules, such as vitamins. They are very important trace components with several structural features. Today, research in food composition is also involved in the characterisation of antioxidant vitamins (quantity and quality evaluation) and the percentage of their daily requirements. Indeed, thanks to the high vitamin content and the ease of consumption compared to other foods with similar biological properties, consumption of some specific functional foods may be targeted to specific people sectors (e.g., the elderly, sportsmen, pregnant women, and children) [45].

Most of the vitamins are easy to oxidise and thermally-labile [46–48] as vitamin C, vitamin B complex, vitamin E, and carotenoids; for this reason, HPLC, in particular if hyphenated to MS techniques, is often used for their identification and quantification. Typically, ESI or, sometimes, APCI ionisation is used when HPLC–MS is employed. Many examples were reported in previous literature for vitamin identification and quantitation in food products by HPLC–MS [49,50] or HPLC-DAD [51,52]. In the last years, HPLC–MS has largely been replaced by ultra-(U)HPLC–MS, providing higher throughput and better chromatographic resolution. Due to their several structural isomeric forms, carotenoids are preferentially analysed by (U)HPLC–MS. APCI (or also atmospheric pressure photoionisation (APPI)) may be more useful than ESI (now commonly used) because most of the carotenoids are apolar compounds [53].

2.5. Phenolic Compounds

Phenolics are secondary plant metabolites with many potential health-promoting properties that present a large structural variety. There are many active studies aimed at identifying new chemical molecules and characterising the polyphenolic content of several vegetables, plants, and fruits. Quality and quantity of phenolics in natural foods may significantly vary in accordance with several extrinsic and intrinsic factors (e.g., growing conditions and soil composition, maturity stage, genetics, and post-harvest conditions) [54]. Dietary intake of polyphenolic compounds is greatly influenced by the preferences of single people and eating habits [55]. Simple phenolics (e.g., flavonoids and phenolic acid conjugates) are important ingredients of vegetables, fruits, and relative derived-products as juices, beverages, and jams. These molecules present a wide range of health-promoting properties as the antioxidant capacity and anti-inflammatory activity [56]; for example, they may exert protective effects against cancer, cardiovascular problems, and other major diseases [57].

Polyphenolic compounds may be divided into many several classes [58]; cinnamic acid, benzoic acids, catechins, and flavonols are the most abundant in foods [59,60]. Low quantities of polyphenols are usually observed in complex plant systems and it is very difficult to isolate them in high amounts [61,62]. HPLC–MS is the main techniques for flavonoid analysis with good sensitivity in ESI systems [27]. In particular, MS/MS (or MS^2) is often utilised in HPLC–MS flavonoids analysis to allow high structural information and mass resolution and increase selectivity after an HPLC-DAD screening [63–65].

Regarding structure analysis, the common phenolic fragmentation pathways do not depend on the ionisation mode (APCI, ESI, or matrix-assisted laser desorption ionisation) and the applied analyser. However, the used different instrumentation may significantly influence and change relative fragment abundances [66]. For this reason, it is preferred to detect the presence/absence of specific MS fragments rather than to use relative intensity changes [67].

Information on the glycan part stereochemistry in flavonoid glycosides are not usually provided by HPLC–MS analysis. The sugar part may still be easily identified because glycosidic fragments, commonly present in flavonoid glycosides, show different m/z ions for hexosides (e.g., glucose, galactose), deoxyhexosides (e.g., rhamnose), and pentosides (e.g., xylose, arabinose). Specific MS/MS analysis (MS^n spectra) may be used to identify the aglycone in a flavonoid O-glycoside; this result may be also obtained by the comparison with literature data or the corresponding spectra of reference compounds [68]. Enough information for C-glycoside identification may be provided by a specific MS^3 spectra analysis. Alternatively, the MS/MS aglycone fragments, produced by higher fragmentation energy, may be utilised. Structural characterisation and identification of many flavonoids in different foodstuffs can be achieved by specific mass spectrometric scanning techniques [69].

2.6. Allergens

Allergies derived from food are one of the most emergent issues in food science and technology studies. The mainly considered allergens are peanuts, wheat, egg, milk, tree nuts, soybeans, shellfish and fish. In any case, several suspected food allergens have not yet been characterised despite the recent advances in biomedical sciences and immunopathology [1]. The determination of allergens in foods can be carried out by many methods based on the biological immune response, often studied by screening-based strategies, such as enzyme-linked immunosorbent assays (ELISA). However, proteomic techniques and HPLC–MS offer different and unique strengths. The MS high sensitivity and selectivity in protein analysis could overcome some problems associated with appropriate allergen identification [70]. For example, the main peanut allergens in food matrices were successfully identified and quantified at a low ppm level by HPLC–MS/MS [71,72].

2.7. Food Additives

Additives in foods are very limited because they should be explicitly authorised before using in foodstuffs. Foodstuffs are monitored for additives and analytical controls are, therefore, an area of increasing importance and concern. Indeed, it is very important to check foodstuffs for additives and HPLC–MS is often the preferred method. Some additives are small compounds (as benzoic acid used for conservation), while other additives are macromolecules; moreover, some additives are synthetic products (composed by a 'single' substance), while others are natural extracts (with many and variable components) [1]. Food additives are molecules with different chemical structure grouped according to their application; for this reason, it is difficult to select the best method to be used for their identification and quantification. Most of these analytical strategies are molecule-specific (without an estimation of the total amount of the considered chemical class), while other protocols can be utilised for the characterisation of groups of compounds. Many techniques for the real-time extraction of different agri-food additives and residues in several matrices are used [73].

3. Quality and Authentication of Food

Food quality, perceived by consumers, is a critical factor for their final economic value [74]. The consumer often links quality with (i) a specific production system; (ii) the utilisation of particular ingredients; (iii) the food origin (also a particular region); (iv) the product authenticity. The European Union has defined different brands as Protected Geographical Indication (PGI), Traditional Speciality Guaranteed (TSG), and Protected Designation of Origin (PDO) to ensure quality and valorise authentic labelling of food and agricultural products.

Quality is a multifunctional data pattern including physical, chemical, microbiological, technological, and sensorial food traits [75]. Food quality may be defined as "the totality of features and traits of a service or product that bear on its ability to satisfy stated or implied needs" (ISO 9000) [76]. Therefore, quality should not be only addressed by HPLC–MS-based techniques. However, HPLC–MS methods can provide important information on nutritional value, origin, safety, sensory attributes, molecular composition, and unique traits. In the last years, these analytical techniques have become the preferred methods when specific and sensitive molecular characterisation of complex systems is required [77].

The identification of selected molecular markers can be used to assess food quality parameters [78,79]. The utilisation of these biomarker compounds should be defined in terms of quantity or absence/presence, together with their organoleptic, biochemical, and chemical role [80,81]. Moreover, their main metabolic pathway should also be known. Increasing consumer awareness is linked to the development of increasingly sophisticated techniques. Numerous cases of food adulteration have been detected using MS-based techniques [82], as shown in Table 2.

Table 2. Cases of food adulterations solved by the application of mass spectrometry coupled to chromatographic techniques.

Food	Adulteration	Analytical Technique	Reference
Meat, milk	Melamine and its metabolites (cyanuric acid, ammelide, and ammeline) in animal feed, meat, milk and infant formulations, and other processed productions	HPLC–MS/MS	[83]
Food colourants	Harmful colourants as Sudan I–IV dyes (lipophilic azo dyes, used in scientific and industrial applications, even if banned as food colourants because of their carcinogenicity)	HPLC–MS	[84–86]
Wine	(i) Addition of sugar even if forbidden; (ii) illegal mixing of different cultivars; (iii) origin falsification; (iv) flavouring or colouring wines by fruit extracts (e.g., elderberry);	HPLC–MS (anthocyanin profiles)	[87–89]

HPLC–MS, high performance liquid chromatography–mass spectrometry; HPLC–MS/MS, high performance liquid chromatography–tandem mass spectrometry.

When the aim is to characterise the quality of a specific agri-food production (i.e., cheese, ham, or fruit), several issues should be considered, and many analytical strategies are required. Safety problems to be addressed may include the identification of mycotoxins or agrochemicals, while nutritional traits may require the determination of bulk components (e.g., carbohydrates, proteins, and lipids) plus other minor compounds in trace amounts (e.g., vitamins) [1]. The identification and quantification of compounds with functional or health-promoting properties, as anti-hypertensive peptides or conjugated linoleic acid, is also increasingly required as well as the characterisation of molecular markers for food quality evaluation. The utilisation of chromatographic techniques and relative high separation ability hyphenated to MS molecular identification capability could solve the main analytical problems in food characterisation [67]. Moreover, portable and micro gas-chromatography (Micro GC) is a fast-developing and, in the last years, more mature technologies have been well developed and successfully commercialised with great potential in food and environmental applications [90–92]. Multiple Mini MS instruments were also recently developed to be used in food analysis [93].

4. Considerations on Statistical Data Analysis

In food analysis, a consistent method should be developed, then optimised, and finally validated by a multivariate statistical experimental design, the Design of Experiments (DoE) [26,94]. In the multivariate statistics, several trials are selected and carried out following a well-designed experimental approach to simultaneously evaluate several factors. The elements that present a specific influence on a chromatographic analytical strategy are defined by a screening design; in food HPLC analysis, these factors may include sample concentration, buffer concentration and relative pH, column temperature and type, mobile phase composition, injection volume, and detection wavelength [95,96].

Moreover, to evaluate the large number of data derived from the hyphenation between chromatographic techniques, as HPLC and GC, and mass spectrometry (e.g., MS, MS^2, HRMS, etc.), a large range of analytical strategies showed to be versatile and useful tools for the visualisation, extraction, and interpretation of the chemical information. In particular, pattern recognition methods, as cluster analysis (CA) and principal component analysis (PCA), proved to allow a better representation of the information included in the HPLC/GC–MS fingerprints. The original variables are converted into new variables summarising the systemic patterns of the variability among the food samples; indeed, an exploratory data analysis is easier to study if it is shown as a multivariate data table rather than a low-dimensional plane [97].

5. HRMS in Food Analysis

For a long time, HRMS techniques have traditionally been limited to the most important applications due to their high instrumental complexity. This situation is now changed thanks to modern HRMS instruments, such as Orbitrap, Fourier-transform ion cyclotron resonance (FT–ICR), and ToF. FT–ICR and Orbitrap MS are high-resolution instruments applied in quality and safety food analysis thanks to their high mass precision (1–2 ppm, high discrimination between ions of interest and isobaric interferences) and high resolution (typically 100,000–1,000,000 FWHM) [98]. In particular, Orbitrap instruments have a resolution higher than 200,000. A recent hybrid mass spectrometer combines a state-of-the-art segmented quadrupole for high-performance precursor ion selection with a high-resolution (HR) and accurate-mass (AM) ultra-high-field Orbitrap mass analyser with a final resolution value of 240,000 [99]. The resolution for these instruments is usually reported at 200 Da and it decreases with the increase of the square root of the *m/z* value and it is proportional to the acquisition time. Instead, TOF instruments have a resolving power that changes very little with the mass [100]. Low-resolution mass analysers show a resolving power of <10,000 (mass accuracy > 5 ppm), while high-resolution ones present a resolving power of 10,000–100,000 (mass accuracy < 5 ppm). In target analysis, specificity may be improved by changing from low to high resolution, but it cannot be clearly defined by comparing LC–HRMS and LC–MS/MS performance [5] and the choice may be more influenced by additional tests, such as retrospective data analysis by LC–HRMS [4].

For many years HRMS coupled to GC has been the main technology used for food analysis. Currently, LC–MS/MS and LC–MS are extensively accepted, and the switch from GC–MS to LC–MS was encouraged by many analysts [101]. Moreover, many commercially available HRMS detectors can be easily hyphenated to LC systems rather than GC systems. LC–MS is preferred to GC–MS thanks to its additional selectivity (due to lower detection limits and less use of an extensive sample clean-up) and its familiarity with LC–MS/MS [102,103].

In the last few years, LC–HRMS has become the most used technique in food analysis, as shown in Table 3, due to a very high number of bioactive compounds and relative derived-degradation products to be analytically monitored by a single protocol [104].

Table 3. Use of high-resolution MS in several agri-food analyses.

Analyte	Other Used Methods	Analytical Problems	Mass Spectrometry Methods [1]	Reference
Marine biotoxins	Mice assay	A high number of false-negative and false-positive findings	LC–MS LC–MS/MS	[105,106]
Mycotoxins	LC [1]-based detection (e.g., electrochemical Kobra® cell or fluorescence) or immunoassay tests	Need for a good clean-up and high-sensitivity detection and use of expensive and time-consuming but very specific immunoassay	LC–ESI/MS, LC–MS/MS	[107–109]
Veterinary drug residues	Immunoassay tests or GC [1]-fluorescence detection	Need for derivatisation before injection into the GC due to high molecular weight and polarity of these compounds	LC–MS/MS UHPLC–MS/MS	[110,111]
Organic contaminants	GC [1]-electron capture detection	Low limits of detection and complexity of the matrix	GC–MS/MS LC–MS/MS UHPLC–QToF–MS	[112,113]
Bioactive compounds (e.g., proteins with allergenic potential)	Bioassays (e.g., enzyme-linked immunosorbent assay)	Expensive and time-consuming tests	LC–MS/MS	[114]

[1] LC–MS, liquid chromatography–mass spectrometry; LC–MS/MS, liquid chromatography–tandem mass spectrometry; LC, liquid chromatography; LC–ESI/MS, liquid chromatography–electrospray ionisation/mass spectrometry; GC, gas chromatography; UHPLC–MS/MS, ultra-high performance liquid chromatography–tandem mass spectrometry; GC–MS/MS, gas chromatography–tandem mass spectrometry; UHPLC–QToF–MS, ultra-high performance liquid chromatography–quadrupole-time of flight–mass spectrometry.

HRMS is a tool for nontargeted or semitargeted screening, that allows to detect suspected peaks and identify and confirm the relative chemical structure. Screening is useful for a no–yes answer in relation to the detection of a set of several molecules or a single compound in many samples. Speed and cost-effectiveness are the main considered parameters. Although quantification may be required, screening analysis is focused on the absence of false-negative data and the number of false-positive findings at specific concentrations [102]. MS techniques (if compared to GC–flame ionisation detection or LC–UV detection) provide a high selectivity that allows verifying lower analyte concentrations in complex matrices without too many false-positive findings. MS/MS-based targeted multiresidue screening techniques allow an even higher selectivity [111,115]. Modern MS/MS instruments are capable of selected reaction monitoring (SRM) at trace concentrations with shorter dwell times (5–10 ms). Therefore, many molecules can be monitored (up to 1000) by the application of retention time-window-based SRM traces and high-end instruments [116]. HRMS is now used as the main screening tool to distinguish between potential target molecules, but without reference standards (unlike LC–MS/MS) [4].

6. HRMS Related to Adulteration and Authenticity

Adulteration is the voluntary addition of one or more nonauthorised substances to food products for economic gain. Adulteration is often related to food quality, with significant differences in the price between lower and higher grade food products [117].

Adulteration is closely linked to authenticity [118,119], although authenticity is commonly related to the voluntary mislabelling as economic deception [120]. Rice, honey, olive oil, wine, and fruit-derived products (as jams or juices) are the main products that relate their authenticity to botanical species, floral type, and geographical origin [121–124]. Table 4 reports data on 137 identified single incidents of economically motivated adulteration in several food categories, based on media reports journal and articles from 1980 to 2013 [120]:

Table 4. Incidents of economically motivated adulteration in 11 food categories from 1980 to 2013.

Food Category	Number of Incidents
Seafood and fish	24
Dairy products	15
Fruit juices	12
Fats and oils	12
Grain products	11
Natural sweeteners and honey	10
Herbal extracts and spices	8
Alcoholic beverages and wine	7
Infant products	5
Proteins by plant material	5
Other food products	28

Targeted analyses (i.e., adulterant is known) and untargeted ones (i.e., adulteration involves not known substances) are related to the detection of adulteration of beverages and foods [125]. For example, in dietary supplement analysis, the matrix is already under suspicion because the new adulterant is often structurally linked to an already known molecule or his chemical analogue [126]. Authenticity confirmation is more demanding than adulteration one; for this reason, LC–HRMS can be used as an effective fingerprinting tool, that, coupled to chemometrics, allows to obtain a probability (sometimes a percentage) of authenticity for a foodstuff [4].

In the last years, the use of MS-Orbitrap interfaced with an ion trap mass analyser has continuously increased for fingerprint applications. The developments in chromatography and relative interface techniques and the advances in MS technologies (e.g., hybrid MS instruments composed by two or more mass analysers) will further increase the potential of analytical technologies in food analysis [127].

7. Conclusions

In the last years, the agri-food industry started to apply innovative analytical strategies for a full characterisation of food productions because a molecular characterisation was required by regulatory agencies and consumers. These requests shifted the food sector to the pharmaceutical one. Therefore, validated analytical strategies using high-performance systems were improved to ensure food authenticity, safety, and quality. Chromatographic techniques coupled to suitable detection strategies produce an effective tool to separate the single molecules and develop a specific profile of the food sample, called "fingerprint"; the combination of a chromatographic separation system with a spectroscopic detector (mass spectrometry) has become the most used approach for the characterisation and/or confirmation of the identity of selected and/or unknown phytochemical substances. If chromatography, hyphenated with mass spectrometry or other detection systems, is further combined with chemometric techniques, clearer patterns might be developed for analytical fingerprints.

The literature related to MS in the analysis of food quality and safety has recently increased. In this review, a brief overview of the main MS and HRMS application to the food analysis was provided.

Indeed, accurate, reliable, and rapid MS techniques for analyte characterisation in food products are indispensable for food quality and safety control. New (U)HPLC–MS(MS) systems allowed to characterise many food products at the molecular level routinely identifying and quantifying desirable and undesirable molecules in different foods. HPLC–MS is often the analytical technique with the highest performance level because it provides a simultaneous assessment of several substances in complex systems (e.g., food products). For this reason, it is now an effective tool for the certification of food authenticity, quality, and safety.

The potential of HPLC–MS is due to the hyphenation of the HPLC separation power and the molecular structure identification provided by MS. The sample-preparation procedures have been minimised and the selectivity for quali-quantitative analyses of complex matrices (e.g., food samples) increased using coupled MS techniques (LC–MS and GC–MS) and MS/MS. Moreover, the developments of HRMS for FT–ICR–MS, Orbitrap, and quadrupole-ToF allowed higher specificity and sensitivity for detection of potential unknown compounds in food (e.g., biomolecules, toxins, and pollutants) and routine food analysis. The analysis of substances with different polarity ranges was allowed by the availability of different ionisation techniques (ESI, APCI, APPI).

Another challenge for food quality and safety control is to promptly provide (often within 1 day) accurate results avoiding damage to food samples. The recently developed ionisation techniques have shown potential, thanks to the ease of automation, fast acquisition, and no-sample treatment. For this reason, the development of reliable, robust, and simple portable MS instruments for accurate, rapid, and in situ characterisation of molecules in food samples remains one of the main future targets. The growing interest in LC-Orbitrap applications for food analysis is associated with important improvements in the data handling software and to reductions in the instrumentation costs. These advances will stimulate high interest in Foodomics for (i) verification of authenticity, (ii) detection of adulteration, (iii) fingerprinting of foods, and (iv) detection of illegal veterinary drugs.

Finally, three main trends could become increasingly important for food analysis: (i) HPLC–MS methods will be further improved and used to solve new issues in the agri-food industry; (ii) analytical protocols will be developed to reduce detection limits for harmful molecules or to improve the multi-compound analysis; (iii) HPLC–MS will be applied in regulatory agencies and food industry, mainly in routine quality control at the production and commercialisation steps.

Author Contributions: Conceptualisation, G.L.B. and D.D.; methodology, D.D. and M.G.M.; data curation, D.D., I.R. and G.G.; writing—original draft preparation, D.D.; writing—review and editing, D.D. and G.L.B.; supervision, G.L.B. All authors have read and agreed to the published version of the manuscript.

Funding: This research received no external funding.

Conflicts of Interest: The authors declare no conflict of interest.

References

1. Di Stefano, V.; Avellone, G.; Bongiorno, D.; Cunsolo, V.; Muccilli, V.; Sforza, S.; Dossena, A.; Drahos, L.; Vékey, K. Applications of liquid chromatography spectrometry for food analysis. *J. Chromatogr. A* **2012**, *1259*, 74–85. [CrossRef] [PubMed]
2. Fibigr, J.; Šatínský, D.; Solich, P. Current trends in the analysis and quality control of food supplements based on plant extracts. *Anal. Chim. Acta* **2018**, *1036*, 1–15. [CrossRef] [PubMed]
3. Careri, M.; Bianchi, F.; Corradini, C. Recent advances in the application of mass spectrometry in food-related analysis. *J. Chromatogr. A* **2002**, *970*, 3–64. [CrossRef]
4. Senyuva, H.Z.; Gökmen, V.; Sarikaya, E.A. Future perspectives in Orbitrap-high-resolution mass spectrometry in food analysis: A review. *Food Addit. Contam. Part A* **2015**, *32*, 1568–1606. [CrossRef] [PubMed]
5. Holčapek, M.; Jirásko, R.; Lísa, M. Recent developments in liquid chromatography–mass spectrometry and related techniques. *J. Chromatogr. A* **2012**, *1259*, 3–15. [CrossRef] [PubMed]
6. Zubarev, R.A.; Makarov, A. Orbitrap mass spectrometry. *Anal. Chem.* **2013**, *85*, 5288–5296. [CrossRef]

7. Bourgou, S.; Bettaieb, I.; Mkadmini, K.; Isoda, H.; Ksouri, R.; Ksouri, W.M. LC-ESI-TOF-MS and GC-MS profiling of Artemisia herba-alba and evaluation of its bioactive properties. *Food Res. Int.* **2017**, *99*, 702–712. [CrossRef]
8. Schieber, A. Functional foods and nutraceuticals. *Food Res. Int.* **2012**, *46*, 437. [CrossRef]
9. Pisanello, D. *Chemistry of Foods: EU Legal and Regulatory Approaches*; Springer Science and Business Media LLC: Berlin, Germany, 2014.
10. Bagchi, D.; Preuss, H.G.; Kehrer, J.P. Nutraceutical and functional food industries: Aspects on safety and regulatory requirements. *Toxicol. Lett.* **2004**, *150*, 1–2. [CrossRef]
11. Kaur, S.; Das, M. Functional foods: An overview. *Food Sci. Biotechnol.* **2011**, *20*, 861–875. [CrossRef]
12. Cheung, P.C.K.; Mehta, B.M. *Handbook of Food Chemistry*; Springer: Berlin, Germany, 2015.
13. Smith, R.M. Before the injection-modern methods of sample preparation for separation techniques. *J. Chromatogr. A* **2003**, *1000*, 3–27. [CrossRef]
14. Nedović, V.; Raspor, P.; Lević, J.; Tumbas Šaponjac, V.; Barbosa-Cánovas, G.V. (Eds.) *Emerging and Traditional Technologies for Safe, Healthy and Quality Food*; Springer Science and Business Media LLC: Berlin, Germany, 2016.
15. Sasidharan, S.; Chen, Y.; Saravanan, D.; Sundram, K.; Latha, L.Y. Extraction, isolation and characterization of bioactive compounds from plants' extracts. *Afr. J. Tradit. Complement. Altern. Med.* **2011**, *8*, 1–10. [CrossRef] [PubMed]
16. Chemat, F.; Rombaut, N.; Sicaire, A.-G.; Meullemiestre, A.; Fabiano-Tixier, A.-S.; Abert-Vian, M. Ultrasound assisted extraction of food and natural products. Mechanisms, techniques, combinations, protocols and applications. A review. *Ultrason. Sonochem.* **2017**, *34*, 540–560. [CrossRef] [PubMed]
17. Chemat, F.; Abert-Vian, M.; Cravotto, G. Green extraction of natural products: Concept and principles. *Int. J. Mol. Sci.* **2012**, *13*, 8615–8627. [CrossRef] [PubMed]
18. Transformation of vegetable waste into value added products:(A) the upgrading concept; (B) practical implementations. *Fuel Energy Abstr.* **2003**, *44*, 272. [CrossRef]
19. Azmir, J.; Zaidul, I.; Rahman, M.; Sharif, K.; Mohamed, A.; Sahena, F.; Jahurul, M.; Ghafoor, K.; Norulaini, N.A.N.; Omar, A. Techniques for extraction of bioactive compounds from plant materials: A review. *J. Food Eng.* **2013**, *117*, 426–436. [CrossRef]
20. Domínguez-Rodríguez, G.; Marina, M.L.; Plaza, M. Strategies for the extraction and analysis of non-extractable polyphenols from plants. *J. Chromatogr. A* **2017**, *1514*, 1–15. [CrossRef]
21. Gil-Chávez, G.J.; Villa, J.A.; Ayala-Zavala, J.F.; Heredia, J.B.; Sepulveda, D.; Yahia, E.M.; Velderrain-Rodríguez, G.R. Technologies for extraction and production of bioactive compounds to be used as nutraceuticals and food ingredients: An overview. *Compr. Rev. Food Sci. Food Saf.* **2013**, *12*, 5–23. [CrossRef]
22. Osorio-Tobón, J.F. Recent advances and comparisons of conventional and alternative extraction techniques of phenolic compounds. *J. Food Sci. Technol.* **2020**, *57*, 4299–4315. [CrossRef]
23. Xu, D.-P.; Li, Y.; Meng, X.; Zhou, T.; Zhou, Y.; Zheng, J.; Zhang, J.-J.; Li, H.-B. Natural antioxidants in foods and medicinal plants: Extraction, assessment and resources. *Int. J. Mol. Sci.* **2017**, *18*, 96. [CrossRef]
24. De Biaggi, M.; Rapalino, S.; Donno, D.; Mellano, M.G.; Beccaro, G. Genotype influence on chemical composition and sensory traits of chestnut in 18 cultivars grown on the same rootstock and at the same agronomic conditions. *Acta Hortic.* **2018**, 215–220. [CrossRef]
25. Cordell, G.A. Phytochemistry and traditional medicine; A revolution in process. *Phytochem. Lett.* **2011**, *4*, 391–398. [CrossRef]
26. Liang, Y.-Z.; Xie, P.; Chan, K. Quality control of herbal medicines. *J. Chromatogr. B* **2004**, *812*, 53–70. [CrossRef]
27. Steinmann, D.; Ganzera, M. Recent advances on HPLC/MS in medicinal plant analysis. *J. Pharm. Biomed. Anal.* **2011**, *55*, 744–757. [CrossRef]
28. Donno, D.; Cerutti, A.; Prgomet, I.; Mellano, M.; Beccaro, G. Foodomics for mulberry fruit (*Morus* spp.): Analytical fingerprint as antioxidants' and health properties' determination tool. *Food Res. Int.* **2015**, *69*, 179–188. [CrossRef]
29. De Biaggi, M.; Donno, D.; Mellano, M.G.; Riondato, I.; Rakotoniaina, E.N.; Beccaro, G.L. *Cornus mas* (L.) fruit as a potential source of natural health-promoting compounds: Physico-chemical characterisation of bioactive components. *Plant Foods Hum. Nutr.* **2018**, *73*, 89–94. [CrossRef]
30. Mok, D.K.; Chau, F.-T. Chemical information of Chinese medicines: A challenge to chemist. *Chemom. Intell. Lab. Syst.* **2006**, *82*, 210–217. [CrossRef]

31. Friedman, M. Nutritional value of proteins from different food sources. A review. *J. Agric. Food Chem.* **1996**, *44*, 6–29. [CrossRef]
32. Lynch, J.M.; Barbano, D.M. Kjeldahl nitrogen analysis as a reference method for protein determination in dairy products. *J. AOAC Int.* **1999**, *82*, 1389–1398. [CrossRef]
33. Schubert, O.T.; Röst, H.L.; Collins, B.C.; Rosenberger, G.; Aebersold, R. Quantitative proteomics: Challenges and opportunities in basic and applied research. *Nat. Protoc.* **2017**, *12*, 1289–1294. [CrossRef]
34. Tyanova, S.; Temu, T.; Carlson, A.; Sinitcyn, P.; Mann, M.; Cox, J. Visualization of LC-MS/MS proteomics data in MaxQuant. *Proteomics* **2015**, *15*, 1453–1456. [CrossRef] [PubMed]
35. Sikorski, Z.Z.; Kolakowska, A. *Chemical and Functional Properties of Food Lipids*; CRC Press: Boca Raton, FL, USA, 2010.
36. Zhang, C.; Wang, K.; Yang, L.; Liu, R.; Chu, Y.; Qin, X.; Yang, P.; Yu, H. Lipid metabolism in inflammation-related diseases. *Analyst* **2018**, *143*, 4526–4536. [CrossRef] [PubMed]
37. Beccaria, M.; Inferrera, V.; Rigano, F.; Gorynski, K.; Purcaro, G.; Pawliszyn, J.; Dugo, P.; Mondello, L. Highly informative multiclass profiling of lipids by ultra-high performance liquid chromatography; Low resolution (quadrupole) mass spectrometry by using electrospray ionization and atmospheric pressure chemical ionization interfaces. *J. Chromatogr. A* **2017**, *1509*, 69–82. [CrossRef]
38. Byrdwell, W.C. Multiple parallel mass spectrometry for liquid chromatography. In *Handbook of Advanced Chromatography/Mass Spectrometry Techniques*; Elsevier: Amsterdam, The Netherlands, 2017; pp. 365–405.
39. Bertozzi, C.R.; Rabuka, D.; Varki, A. *Essentials of Glycobiology*; Cold Spring Harbor Laboratory Press: Cold Spring Harbor, NY, USA, 2009; p. 784.
40. Barile, D.; Marotta, M.; Chu, C.; Mehra, R.; Grimm, R.; Lebrilla, C.B.; German, J.B. Neutral and acidic oligosaccharides in Holstein-Friesian colostrum during the first 3 days of lactation measured by high performance liquid chromatography on a microfluidic chip and time-of-flight mass spectrometry. *J. Dairy Sci.* **2010**, *93*, 3940–3949. [CrossRef] [PubMed]
41. Mamone, G.; Picariello, G.; Caira, S.; Addeo, F.; Ferranti, P. Analysis of food proteins and peptides by mass spectrometry-based techniques. *J. Chromatogr. A* **2009**, *1216*, 7130–7142. [CrossRef] [PubMed]
42. Soifoini, T.; Donno, D.; Jeannoda, V.; Rakotoniaina, E.; Hamidou, S.; Achmet, S.M.; Solo, N.R.; Afraitane, K.; Giacoma, C.; Beccaro, G.L. Bioactive compounds, nutritional traits, and antioxidant properties of Artocarpus altilis (Parkinson) fruits: Exploiting a potential functional food for food security on the Comoros Islands. *J. Food Qual.* **2018**, *2018*, 1–11. [CrossRef]
43. Hotchkiss, A.T., Jr.; Nuñez, A.; Strahan, G.D.; Chau, H.K.; White, A.K.; Marais, J.P.; Hom, K.; Vakkalanka, M.S.; Di, R.; Yam, K.L. Cranberry xyloglucan structure and inhibition of Escherichia coli adhesion to epithelial cells. *J. Agric. Food Chem.* **2015**, *63*, 5622–5633. [CrossRef] [PubMed]
44. Fusch, G.; Choi, A.; Rochow, N.; Fusch, C. Quantification of lactose content in human and cow's milk using UPLC mass spectrometry. *J. Chromatogr. B* **2011**, *879*, 3759–3762. [CrossRef]
45. Donno, D.; Mellano, M.G.; Prgomet, Z.; Cerutti, A.K.; Beccaro, G.L. Phytochemical characterization and antioxidant activity evaluation of mediterranean medlar fruit (*Crataegus azarolus* L.): Preliminary study of underutilized genetic resources as a potential source of health-promoting compounds for food supplements. *J. Food Nutr. Res.* **2017**, *56*, 18–31.
46. Canterino, S.; Donno, D.; Mellano, M.G.; Beccaro, G.L.; Bounous, G. Nutritional and sensory survey of *Citrus sinensis* (L.) cultivars grown at the most northern limit of the Mediterranean latitude. *J. Food Qual.* **2012**, *35*, 108–118. [CrossRef]
47. Spínola, V.; Llorent-Martínez, E.J.; Castilho, P.C. Determination of vitamin C in foods: Current state of method validation. *J. Chromatogr. A* **2014**, *1369*, 2–17. [CrossRef] [PubMed]
48. Spínola, V.; Mendes, B.; Câmara, J.S.; Castilho, P.C. Effect of time and temperature on vitamin C stability in horticultural extracts. UHPLC-PDA vs iodometric titration as analytical methods. *LWT* **2013**, *50*, 489–495. [CrossRef]
49. Chandra-Hioe, M.V.; Bucknall, M.; Arcot, J. Folate analysis in foods by UPLC-MS/MS: Development and validation of a novel, high throughput quantitative assay; folate levels determined in Australian fortified breads. *Anal. Bioanal. Chem.* **2011**, *401*, 1035–1042. [CrossRef] [PubMed]
50. Huang, M.; Winters, D. Application of ultra-performance liquid chromatography/tandem mass spectrometry for the measurement of vitamin D in foods and nutritional supplements. *J. AOAC Int.* **2011**, *94*, 211–223. [CrossRef] [PubMed]

51. De Biaggi, M.; Donno, D.; Mellano, M.G.; Gamba, G.; Riondato, I.; Rakotoniaina, E.N.; Beccaro, G.L. Emerging species with nutraceutical properties: Bioactive compounds from *Hovenia dulcis pseudofruits*. *Food Chem.* **2020**, *310*, 125816. [CrossRef]
52. Donno, D.; Beccaro, G.; Mellano, M.; Cerutti, A.; Bounous, G. Goji berry fruit (*Lycium* spp.): Antioxidant compound fingerprint and bioactivity evaluation. *J. Funct. Foods* **2015**, *18*, 1070–1085. [CrossRef]
53. Azevedo-Meleiro, C.H.; Rodriguez-Amaya, D.B. Confirmation of the identity of the carotenoids of tropical fruits by HPLC-DAD and HPLC-MS. *J. Food Compos. Anal.* **2004**, *17*, 385–396. [CrossRef]
54. Jeffery, E.; Brown, A.; Kurilich, A.; Keck, A.; Matusheski, N.; Klein, B.; Juvik, J. Variation in content of bioactive components in broccoli. *J. Food Compos. Anal.* **2003**, *16*, 323–330. [CrossRef]
55. Shahidi, F.; Ambigaipalan, P. Phenolics and polyphenolics in foods, beverages and spices: Antioxidant activity and health effects-A review. *J. Funct. Foods* **2015**, *18*, 820–897. [CrossRef]
56. Donno, D.; Turrini, F. Plant foods and underutilized fruits as source of functional food ingredients: Chemical composition, quality traits, and biological properties. *Foods* **2020**, *9*, 1474. [CrossRef]
57. Rice-Evans, C.; Miller, N.; Paganga, G. Antioxidant properties of phenolic compounds. *Trends Plant Sci.* **1997**, *2*, 152–159. [CrossRef]
58. Razafindrakoto, Z.R.; Donno, D.; Tombozara, N.; Andriamaniraka, H.; Andrianjara, C.; Ramanitrahasimbola, D.; Beccaro, G.L. Antioxidant, anti-inflammatory, and antidiabetic activities of leaves and stems of Uapaca bojeri Bail. (Euphorbiaceae), an endemic plant of Madagascar. *Pharmaceuticals* **2020**, *13*, 71. [CrossRef]
59. Tabart, J.; Franck, T.; Kevers, C.; Pincemail, J.; Serteyn, D.; Defraigne, J.-O.; Dommes, J. Antioxidant and anti-inflammatory activities of Ribes nigrum extracts. *Food Chem.* **2012**, *131*, 1116–1122. [CrossRef]
60. Tabart, J.; Kevers, C.; Evers, D.; Dommes, J. Ascorbic acid, phenolic acid, flavonoid, and carotenoid profiles of selected extracts from Ribes nigrum. *J. Agric. Food Chem.* **2011**, *59*, 4763–4770. [CrossRef] [PubMed]
61. Pavlović, A.V.; Papetti, A.; Zagorac, D.Č.D.; Gašić, U.M.; Mišić, D.; Tešić, Ž.L.; Natić, M. Phenolics composition of leaf extracts of raspberry and blackberry cultivars grown in Serbia. *Ind. Crop. Prod.* **2016**, *87*, 304–314. [CrossRef]
62. Natić, M.M.; Dabić, D.Č.; Papetti, A.; Akšić, M.M.F.; Ognjanov, V.; Ljubojević, M.; Tešić, Ž.L. Analysis and characterisation of phytochemicals in mulberry (*Morus alba* L.) fruits grown in Vojvodina, North Serbia. *Food Chem.* **2015**, *171*, 128–136. [CrossRef] [PubMed]
63. Gamba, G.; Donno, D.; Mellano, M.G.; Riondato, I.; De Biaggi, M.; Randriamampionona, D.; Beccaro, G.L. Phytochemical characterization and bioactivity evaluation of autumn olive (*Elaeagnus umbellata Thunb.*) pseudodrupes as potential sources of health-promoting compounds. *Appl. Sci.* **2020**, *10*, 4354. [CrossRef]
64. Donno, D.; Mellano, M.G.; De Biaggi, M.; Riondato, I.; Rakotoniaina, E.N.; Beccaro, G.L. New findings in Prunus padus L. fruits as a source of natural compounds: Characterization of metabolite profiles and preliminary evaluation of antioxidant activity. *Molecules* **2018**, *23*, 725. [CrossRef]
65. Cuadros-Rodríguez, L.; Ruiz-Samblás, C.; Valverde-Som, L.; Pérez-Castaño, E.; González-Casado, A. Chromatographic fingerprinting: An innovative approach for food 'identitation' and food authentication; A tutorial. *Anal. Chim. Acta* **2016**, *909*, 9–23. [CrossRef]
66. March, R.E.; Li, H.; Belgacem, O.; Papanastasiou, D. High-energy and low-energy collision-induced dissociation of protonated flavonoids generated by MALDI and by electrospray ionization. *Int. J. Mass Spectrom.* **2007**, *262*, 51–66. [CrossRef]
67. De Villiers, A.; Venter, P.; Pasch, H. Recent advances and trends in the liquid-chromatography; mass spectrometry analysis of flavonoids. *J. Chromatogr. A* **2016**, *1430*, 16–78. [CrossRef] [PubMed]
68. De Rijke, E.; Out, P.; Niessen, W.M.; Ariese, F.; Gooijer, C.; Brinkman, U.A. Analytical separation and detection methods for flavonoids. *J. Chromatogr. A* **2006**, *1112*, 31–63. [CrossRef] [PubMed]
69. Rak, G.; Fodor, P.; Abrankó, L. Three-step HPLC—ESI-MS/MS procedure for screening and identifying non-target flavonoid derivatives. *Int. J. Mass Spectrom.* **2010**, *290*, 32–38. [CrossRef]
70. Pilolli, R.; Monaci, L.; Visconti, A. Advances in biosensor development based on integrating nanotechnology and applied to food-allergen management. *TrAC Trends Anal. Chem.* **2013**, *47*, 12–26. [CrossRef]
71. Shefcheck, K.J.; Callahan, J.H.; Musser, S.M. Confirmation of peanut protein using peptide markers in dark chocolate using liquid chromatography-tandem mass spectrometry (LC-MS/MS). *J. Agric. Food Chem.* **2006**, *54*, 7953–7959. [CrossRef]

72. Shefcheck, K.J.; Musser, S.M. Confirmation of the allergenic peanut protein, Ara h 1, in a model food matrix using liquid chromatography/tandem mass spectrometry (LC/MS/MS). *J. Agric. Food Chem.* **2004**, *52*, 2785–2790. [CrossRef]
73. Ackerman, L.K.; Noonan, G.; Begley, T. Assessing direct analysis in real-time-mass spectrometry (DART-MS) for the rapid identification of additives in food packaging. *Food Addit. Contam. Part A* **2009**, *26*, 1611–1618. [CrossRef]
74. Dasenaki, M.E.; Thomaidis, N.S. Quality and authenticity control of fruit juices—A review. *Molecules* **2019**, *24*, 1014. [CrossRef]
75. Kyriacou, M.C.; Rouphael, Y. Towards a new definition of quality for fresh fruits and vegetables. *Sci. Hortic.* **2018**, *234*, 463–469. [CrossRef]
76. Dick, G.P. ISO 9000 certification benefits, reality or myth? *TQM Mag.* **2000**, *12*, 365–371. [CrossRef]
77. Siriamornpun, S.; Kaewseejan, N. Quality, bioactive compounds and antioxidant capacity of selected climacteric fruits with relation to their maturity. *Sci. Hortic.* **2017**, *221*, 33–42. [CrossRef]
78. Donno, D.; Cerutti, A.; Mellano, M.; Prgomet, Z.; Beccaro, G. Serviceberry, a berry fruit with growing interest of industry: Physicochemical and quali-quantitative health-related compound characterisation. *J. Funct. Foods* **2016**, *26*, 157–166. [CrossRef]
79. Gao, H.; Song, Z.; Wang, Z.; Qian, Z.; Zhang, Q. Overview on quantitative analysis of multi-components by single-marker. *China J. Chin. Mater. Med.* **2012**, *37*, 405–416.
80. Künsch, U.; Schärer, H.; Patrian, B.; Hurter, J.; Conedera, M.; Sassella, A.; Jermini, M.; Jelmini, G. Quality assessment of chestnut fruits. In Proceedings of the 2nd International Symposium on Chestnut, Bordeaux, France, 19–23 October 1998; pp. 119–128.
81. Beccaro, G.L.; Donno, D.; Lione, G.G.; De Biaggi, M.; Gamba, G.; Rapalino, S.; Riondato, I.; Gonthier, P.; Mellano, M.G. *Castanea* spp. Agrobiodiversity conservation: Genotype influence on chemical and sensorial traits of cultivars grown on the same clonal rootstock. *Foods* **2020**, *9*, 1062. [CrossRef] [PubMed]
82. Zhang, J.; Zhang, X.; Dediu, L.; Victor, C. Review of the current application of fingerprinting allowing detection of food adulteration and fraud in China. *Food Control.* **2011**, *22*, 1126–1135. [CrossRef]
83. Sun, F.; Ma, W.; Xu, L.; Zhu, Y.; Liu, L.; Peng, C.; Wang, L.; Kuang, H.; Xu, C. Analytical methods and recent developments in the detection of melamine. *TrAC Trends Anal. Chem.* **2010**, *29*, 1239–1249. [CrossRef]
84. Mazzotti, F.; Di Donna, L.; Maiuolo, L.; Napoli, A.; Salerno, R.; Sajjad, A.; Sindona, G. Assay of the set of all sudan azodye (i, ii, iii, iv, and para-red) contaminating agents by liquid chromatography-tandem mass spectrometry and isotope dilution methodology. *J. Agric. Food Chem.* **2007**, *56*, 63–67. [CrossRef]
85. Murty, M.; Chary, N.S.; Prabhakar, S.; Raju, N.P.; Vairamani, M. Simultaneous quantitative determination of Sudan dyes using liquid chromatography–atmospheric pressure photoionization mass spectrometry. *Food Chem.* **2009**, *115*, 1556–1562. [CrossRef]
86. Rebane, R.; Leito, I.; Yurchenko, S.; Herodes, K. A review of analytical techniques for determination of Sudan I–IV dyes in food matrixes. *J. Chromatogr. A* **2010**, *1217*, 2747–2757. [CrossRef]
87. Guo, T.; Fang, P.; Jiang, J.; Zhang, F.; Yong, W.; Liu, J.; Dong, Y. Rapid screening and quantification of residual pesticides and illegal adulterants in red wine by direct analysis in real time mass spectrometry. *J. Chromatogr. A* **2016**, *1471*, 27–33. [CrossRef]
88. Ziółkowska, A.; Wąsowicz, E.; Jeleń, H.H. Differentiation of wines according to grape variety and geographical origin based on volatiles profiling using SPME-MS and SPME-GC/MS methods. *Food Chem.* **2016**, *213*, 714–720. [CrossRef] [PubMed]
89. Flamini, R. Mass spectrometry in grape and wine chemistry. Part I: Polyphenols. *Mass Spectrom. Rev.* **2003**, *22*, 218–250. [CrossRef] [PubMed]
90. Wang, J.; Bryant-Genevier, J.; Nuñovero, N.; Zhang, C.; Kraay, B.; Zhan, C.; Scholten, K.; Nidetz, R.; Buggaveeti, S.; Zellers, E.T. Compact prototype microfabricated gas chromatographic analyzer for autonomous determinations of VOC mixtures at typical workplace concentrations. *Microsyst. Nanoeng.* **2018**, *4*, 17101. [CrossRef]
91. Wang, J.; Nuñovero, N.; Nidetz, R.; Peterson, S.J.; Brookover, B.M.; Steinecker, W.H.; Zellers, E.T. Belt-mounted micro-gas-chromatograph prototype for determining personal exposures to volatile-organic-compound mixture components. *Anal. Chem.* **2019**, *91*, 4747–4754. [CrossRef]

92. Wang, J.; Nuñovero, N.; Zhan, C.; Nidetz, R.; Steinecker, W.H.; Peterson, S.J.; Brookover, B.M.; Zellers, E.T. Microscale gas chromatography with microsensor array detection: Challenges and prospects. *Proceedings* **2017**, *1*, 633. [CrossRef]
93. Zhou, X.; Liu, J.; Cooks, R.G.; Ouyang, Z. Development of miniature mass spectrometry systems for bioanalysis outside the conventional laboratories. *Bioanalysis* **2014**, *6*, 1497–1508. [CrossRef]
94. Hibbert, D.B. Experimental design in chromatography: A tutorial review. *J. Chromatogr. B* **2012**, *910*, 2–13. [CrossRef]
95. Gong, F.; Liang, Y.-Z.; Xie, P.-S.; Chau, F.-T. Information theory applied to chromatographic fingerprint of herbal medicine for quality control. *J. Chromatogr. A* **2003**, *1002*, 25–40. [CrossRef]
96. Gong, F.; Wang, B.-T.; Liang, Y.; Chau, F.-T.; Fung, Y.-S. Variable selection for discriminating herbal medicines with chromatographic fingerprints. *Anal. Chim. Acta* **2006**, *572*, 265–271. [CrossRef]
97. Donno, D.; Beccaro, G.; Cerutti, A.; Mellano, M.G.; Bounous, M.M.A.G. Bud extracts as new phytochemical source for herbal preparations-quality control and standardization by analytical fingerprint. In *Phytochemicals-Isolation, Characterisation and Role in Human Health*; IntechOpen: Rijeka, Croatia, 2015; Volume 1, pp. 187–218.
98. Wang, X.; Wang, S.; Cai, Z. The latest developments and applications of mass spectrometry in food-safety and quality analysis. *TrAC Trends Anal. Chem.* **2013**, *52*, 170–185. [CrossRef]
99. Strupat, K.; Scheibner, O.; Bromirski, M. *High-Resolution, Accurate-Mass Orbitrap Mass Spectrometry-Definitions, Opportunities, and Advantages*; Thermo Fisher Scientific (Bremen) GmbH: Bremen, Germany, 2013.
100. Makarov, A.; Scigelova, M. Coupling liquid chromatography to Orbitrap mass spectrometry. *J. Chromatogr. A* **2010**, *1217*, 3938–3945. [CrossRef] [PubMed]
101. Hagberg, J. Analysis of brominated dioxins and furans by high resolution gas chromatography/high resolution mass spectrometry. *J. Chromatogr. A* **2009**, *1216*, 376–384. [CrossRef] [PubMed]
102. Kaufmann, A. The current role of high-resolution mass spectrometry in food analysis. *Anal. Bioanal. Chem.* **2012**, *403*, 1233–1249. [CrossRef] [PubMed]
103. LeDoux, M. Analytical methods applied to the determination of pesticide residues in foods of animal origin. A review of the past two decades. *J. Chromatogr. A* **2011**, *1218*, 1021–1036. [CrossRef]
104. Liu, J.; Liu, Y.; Wang, Y.; Abozeid, A.; Zu, Y.-G.; Tang, Z.-H. The integration of GC–MS and LC–MS to assay the metabolomics profiling in Panax ginseng and Panax quinquefolius reveals a tissue- and species-specific connectivity of primary metabolites and ginsenosides accumulation. *J. Pharm. Biomed. Anal.* **2017**, *135*, 176–185. [CrossRef]
105. Blay, P.; Hui, J.P.M.; Chang, J.; Melanson, J.E. Screening for multiple classes of marine biotoxins by liquid chromatography–high-resolution mass spectrometry. *Anal. Bioanal. Chem.* **2011**, *400*, 577–585. [CrossRef]
106. Gerssen, A.; Mulder, P.P.; De Boer, J. Screening of lipophilic marine toxins in shellfish and algae: Development of a library using liquid chromatography coupled to orbitrap mass spectrometry. *Anal. Chim. Acta* **2011**, *685*, 176–185. [CrossRef]
107. Herebian, D.; Zühlke, S.; Lamshöft, M.; Spiteller, M. Multi-mycotoxin analysis in complex biological matrices using LC-ESI/MS: Experimental study using triple stage quadrupole and LTQ-Orbitrap. *J. Sep. Sci.* **2009**, *32*, 939–948. [CrossRef]
108. Zachariasova, M.; Lacina, O.; Malachova, A.; Kostelanska, M.; Poustka, J.; Godula, M.; Hajslova, J. Novel approaches in analysis of Fusarium mycotoxins in cereals employing ultra performance liquid chromatography coupled with high resolution mass spectrometry. *Anal. Chim. Acta* **2010**, *662*, 51–61. [CrossRef]
109. Zachariasova, M.; Cajka, T.; Godula, M.; Malachova, A.; Veprikova, Z.; Hajslova, J. Analysis of multiple mycotoxins in beer employing (ultra)-high-resolution mass spectrometry. *Rapid Commun. Mass Spectrom.* **2010**, *24*, 3357–3367. [CrossRef]
110. De Brabander, H.F.; Noppe, H.; Verheyden, K.; Bussche, J.V.; Wille, K.; Okerman, L.; Vanhaecke, L.; Reybroeck, W.; Ooghe, S.; Croubels, S. Residue analysis: Future trends from a historical perspective. *J. Chromatogr. A* **2009**, *1216*, 7964–7976. [CrossRef] [PubMed]
111. Robert, C.; Gillard, N.; Brasseur, P.-Y.; Pierret, G.; Ralet, N.; Dubois, M.; Delahaut, P. Rapid multi-residue and multi-class qualitative screening for veterinary drugs in foods of animal origin by UHPLC-MS/MS. *Food Addit. Contam. Part A* **2013**, *30*, 443–457. [CrossRef] [PubMed]

112. Krauss, M.; Singer, H.P.; Hollender, J. LC-high resolution MS in environmental analysis: From target screening to the identification of unknowns. *Anal. Bioanal. Chem.* **2010**, *397*, 943–951. [CrossRef] [PubMed]
113. Díaz, R.; Ibáñez, M.; Sancho, J.V.; Hernández, F.H. Target and non-target screening strategies for organic contaminants, residues and illicit substances in food, environmental and human biological samples by UHPLC-QTOF-MS. *Anal. Methods* **2012**, *4*, 196–209. [CrossRef]
114. Fæste, C.K.; Rønning, H.T.; Christians, U.; Granum, P.E. Liquid chromatography and mass spectrometry in food allergen detection. *J. Food Prot.* **2011**, *74*, 316–345. [CrossRef]
115. Wittenberg, J.B.; Simon, K.A.; Wong, J.W. Targeted multiresidue analysis of veterinary drugs in milk-based powders using liquid chromatography Mass Spectrometry (LC-MS/MS). *J. Agric. Food Chem.* **2017**, *65*, 7288–7293. [CrossRef]
116. Ronsein, G.E.; Pamir, N.; Von Haller, P.D.; Kim, D.S.; Oda, M.N.; Jarvik, G.P.; Vaisar, T.; Heinecke, J.W. Parallel reaction monitoring (PRM) and selected reaction monitoring (SRM) exhibit comparable linearity, dynamic range and precision for targeted quantitative HDL proteomics. *J. Proteom.* **2015**, *113*, 388–399. [CrossRef]
117. Zhao, X.; Ma, F.; Li, P.; Li, G.; Zhang, L.; Zhang, Q.; Zhang, W.; Wang, X. Simultaneous determination of isoflavones and resveratrols for adulteration detection of soybean and peanut oils by mixed-mode SPE LC-MS/MS. *Food Chem.* **2015**, *176*, 465–471. [CrossRef]
118. Danezis, G.P.; Tsagkaris, A.S.; Brusic, V.; Georgiou, C.A. Food authentication: State of the art and prospects. *Curr. Opin. Food Sci.* **2016**, *10*, 22–31. [CrossRef]
119. Gallo, M.; Ferranti, P. The evolution of analytical chemistry methods in foodomics. *J. Chromatogr. A* **2016**, *1428*, 3–15. [CrossRef]
120. Everstine, K.; Spink, J.; Kennedy, S. Economically Motivated Adulteration (EMA) of food: Common characteristics of EMA incidents. *J. Food Prot.* **2013**, *76*, 723–735. [CrossRef] [PubMed]
121. Turrini, F.; Boggia, R.; Donno, D.; Parodi, B.; Beccaro, G.; Baldassari, S.; Signorello, M.G.; Catena, S.; Alfei, S.; Zunin, P. From pomegranate marcs to a potential bioactive ingredient: A recycling proposal for pomegranate-squeezed marcs. *Eur. Food Res. Technol.* **2020**, *246*, 273–285. [CrossRef]
122. Donno, D.; Mellano, M.G.; Riondato, I.; De Biaggi, M.; Andriamaniraka, H.; Gamba, G.; Beccaro, G.L. Traditional and unconventional dried fruit snacks as a source of health-promoting compounds. *Antioxidants* **2019**, *8*, 396. [CrossRef]
123. Donno, D.; Mellano, M.G.; Hassani, S.; De Biaggi, M.; Riondato, I.; Gamba, G.; Giacoma, C.; Beccaro, G.L. Assessing nutritional traits and phytochemical composition of artisan jams produced in comoros islands: Using indigenous fruits with high health-impact as an example of biodiversity integration and food security in rural development. *Molecules* **2018**, *23*, 2707. [CrossRef] [PubMed]
124. Canterino, S.; Donno, D.; Beccaro, G.; Bounous, G. Fruit quality and nutraceutical composition *in Citrus sinensis* (L.) Osbeck from northern Italy (Piedmont). *Acta Hortic.* **2011**, 383–388. [CrossRef]
125. Boggia, R.; Casolino, M.C.; Hysenaj, V.; Oliveri, P.; Zunin, P. A screening method based on UV-Visible spectroscopy and multivariate analysis to assess addition of filler juices and water to pomegranate juices. *Food Chem.* **2013**, *140*, 735–741. [CrossRef]
126. Donno, D.; Boggia, R.; Zunin, P.; Cerutti, A.K.; Guido, M.; Mellano, M.G.; Prgomet, Z.; Beccaro, G.L. Phytochemical fingerprint and chemometrics for natural food preparation pattern recognition: An innovative technique in food supplement quality control. *J. Food Sci. Technol.* **2016**, *53*, 1071–1083. [CrossRef]
127. Junot, C.; Fenaille, F.; Colsch, B.; Bécher, F. High resolution mass spectrometry based techniques at the crossroads of metabolic pathways. *Mass Spectrom. Rev.* **2014**, *33*, 471–500. [CrossRef]

MDPI
St. Alban-Anlage 66
4052 Basel
Switzerland
Tel. +41 61 683 77 34
Fax +41 61 302 89 18
www.mdpi.com

Foods Editorial Office
E-mail: foods@mdpi.com
www.mdpi.com/journal/foods